METHODS IN MOLECULAR BIOLOGY™

Series Editor
John M. Walker
School of Life Sciences
University of Hertfordshire
Hatfield, Hertfordshire, AL10 9AB, UK

For further volumes:
http://www.springer.com/series/7651

Microbial Carotenoids from Bacteria and Microalgae

Methods and Protocols

Edited by

José-Luis Barredo

Gadea BioPharma, Parque Tecnológico de León, León, Spain

Humana Press

Editor
José-Luis Barredo
Gadea BioPharma
Parque Tecnológico de León
León, Spain

ISSN 1064-3745 e-ISSN 1940-6029
ISBN 978-1-4939-5887-0 ISBN 978-1-61779-879-5 (eBook)
DOI 10.1007/978-1-61779-879-5
Springer New York Dordrecht Heidelberg London

Softcover re-print of the Hardcover 1st edition 2012

Printed on acid-free paper

Humana Press is part of Springer Science+Business Media (www.springer.com)

Preface

Carotenoids are a family of yellow to orange-red terpenoid pigments synthesized by photosynthetic organisms and by many bacteria and fungi. They have beneficial health effects, protecting against oxidative damage, and may be responsible for the colors associated with plants and animals. Carotenoids are also desirable commercial products used as colorants, feed supplements, and nutraceuticals in the food, medical, and cosmetic industries. Only a few of the more than 600 identified carotenoids are produced industrially, with β-carotene (a popular additive for butter, ice cream, orange juice, candies, etc.) the most prominent.

Commercial production of natural carotenoids from microorganisms is a new approach more eco-friendly than synthetic manufacture by chemical procedures. Despite the availability of a variety of natural and synthetic carotenoids, there is currently renewed interest in microbial sources. Due to its increasing importance, industrial biotechnological methods of carotenoids production have been developed with the algae *Dunaliella salina* and *Haematococcus pluvialis*, the fungus *Blakeslea trispora*, and the heterobasidiomycetous yeast *Xanthophyllomyces dendrorhous*.

This book is intended to provide practical experimental laboratory procedures for a wide range of carotenoids producing microorganisms. Although not an exhaustive treatise, it provides a detailed "step-by-step" description of the most recent developments in applied biotechnological processes useful for screening and selection of carotenoids producing microorganisms, construction of new carotenoids biosynthetic pathways, and extraction and analysis of carotenoids. The detailed protocols are cross-referenced in the Notes section, providing special details, little problems, troubleshooting, and safety comments that may not normally appear in journal articles and can be particularly useful for those not familiar with specific techniques.

The lead chapter of this volume represents an overview on the pathways of carotenoids biosynthesis in bacteria and microalgae. Next chapters show comprehensive experimental methods for the selection of new carotenoids producing bacteria and the engineering of several bacteria (including *Escherichia coli*, *Bradyrhizobium* sp., *Methylomonas* sp., *Bacillus subtilis*, and halophilic bacteria) for carotenoids production. Furthermore, methods for canthaxanthin production, construction of carotenoid reporter systems, directed evolution of carotenoid synthases, and improvement of β-carotene hydroxylase catalytic activity are described. Additionally, the book includes methods of DNA fingerprinting for the identification of carotenogenic *Dunaliella* species, ketocarotenoid biosynthesis in microalgae expressing the β-C-4-carotene oxygenase gene, characterization of carotenogenesis genes in *Anabaena* sp., obtaining lutein from microalgal biomass, NMR-based isotopologue profiling of microbial carotenoids, and analysis of diapocarotenoids.

This book has been written by outstanding experts in their fields and provides a reference source for laboratory and industrial professionals, as well as for graduate students in a number of biological disciplines (biotechnology, microbiology, genetics, molecular biology, etc.).

I am indebted to the authors that, in spite of their professional activities, agreed to participate in this book, to Dr. J. Walker, Series Editor, for his encouragement and advice in reviewing the manuscripts, and to the rest of the staff of The Humana Press, Inc., for their assistance in assembling this volume and their efforts in keeping this project on schedule. Last but not least, I warmly acknowledge my wife Natalia and our children Diego, José-Luis, Álvaro, and Gonzalo for their patience and support.

León, Spain *José-Luis Barredo*

Contents

Preface v
Contributors . . . . *ix*

1 Pathways of Carotenoid Biosynthesis in Bacteria and Microalgae 1
J. Paniagua-Michel, Jorge Olmos-Soto, and Manuel Acosta Ruiz

2 Selection and Taxonomic Identification of Carotenoid–Producing Marine Actinomycetes 13
Francisco Romero, Rosa Isabel Fernández-Chimeno, Juan Luis de la Fuente, and José-Luis Barredo

3 Isolation, Characterization, and Diversity of Novel Radiotolerant Carotenoid-Producing Bacteria 21
Dalal Asker, Tarek S. Awad, Teruhiko Beppu, and Kenji Ueda

4 A Novel Radio-Tolerant Astaxanthin-Producing Bacterium Reveals a New Astaxanthin Derivative: Astaxanthin Dirhamnoside 61
Dalal Asker, Tarek S. Awad, Teruhiko Beppu, and Kenji Ueda

5 Novel Zeaxanthin-Producing Bacteria Isolated from a Radioactive Hot Spring Water 99
Dalal Asker, Tarek S. Awad, Teruhiko Beppu, and Kenji Ueda

6 Novel Approach in the Biosynthesis of Functional Carotenoids in *Escherichia coli* 133
Hisashi Harada and Norihiko Misawa

7 Engineering *Escherichia coli* for Canthaxanthin and Astaxanthin Biosynthesis 143
Qiong Cheng and Luan Tao

8 Analysis of Canthaxanthin Production by *Gordonia jacobaea* 159
Patricia Veiga-Crespo, Teresa Vinuesa, Miquel Viñas, and Tomás G. Villa

9 Isolation and Light-Stimulated Expression of Canthaxanthin and Spirilloxanthin Biosynthesis Genes from the Photosynthetic Bacterium *Bradyrhizobium* sp. Strain ORS278 173
Eric Giraud and André Verméglio

10 Construction of Carotenoid Biosynthetic Pathways Through Chromosomal Integration in Methane-Utilizing Bacterium *Methylomonas* sp. Strain 16a 185
Rick W. Ye and Kristen Kelly

11 Genetic Modification in *Bacillus subtilis* for Production of C30 Carotenoids 197
Isamu Maeda

12 Carotenoids' Production from Halophilic Bacteria 207
María de Lourdes Moreno, Cristina Sánchez-Porro, María Teresa García, and Encarnación Mellado

13 Construction and Utilization of Carotenoid Reporter Systems: Identification of Chromosomal Integration Sites That Support Suitable Expression of Biosynthetic Genes and Pathways 219
Pamela L. Sharpe and Deana J. DiCosimo

14 Directed Evolution of Carotenoid Synthases for the Production of Unnatural Carotenoids 245
Maiko Furubayashi and Daisuke Umeno

15 A High-Throughput Screen for the Identification of Improved Catalytic Activity: β-Carotene Hydroxylase 255
Mark A. Scaife, Cynthia A. Ma, Phillip C. Wright, and Roberto E. Armenta

16 DNA Fingerprinting Intron-Sizing Method to Accomplish a Specific, Rapid, and Sensitive Identification of Carotenogenic *Dunaliella* Species. 269
Jorge Olmos-Soto, J. Paniagua-Michel, Rosalía Contreras, and Leonel Ochoa

17 Ketocarotenoid Biosynthesis in Transgenic Microalgae Expressing a Foreign β-C-4-carotene Oxygenase Gene 283
Marta Vila, Aurora Galván, Emilio Fernández, and Rosa León

18 Characterization of Carotenogenesis Genes in the Cyanobacterium *Anabaena* sp. PCC 7120. 297
Hajime Masukawa, Mari Mochimaru, and Shinichi Takaichi

19 Obtaining Lutein-Rich Extract from Microalgal Biomass at Preparative Scale 307
José M. Fernández-Sevilla, F. Gabriel Acién Fernández, and Emilio Molina Grima

20 NMR-Based Isotopologue Profiling of Microbial Carotenoids. 315
Eva Eylert, Adelbert Bacher, and Wolfgang Eisenreich

21 Analysis of Diapocarotenoids Found in Pigmented *Bacillus* Species 335
Laura Pérez-Fons and Paul D. Fraser

Index. *347*

Contributors

F. Gabriel Acién Fernández • *Department of Chemical Engineering, Universidad de Almería, Almería, Spain*
Manuel Acosta Ruiz • *Department of Marine Biotechnology, Centre for Scientific Research and Higher Education of Ensenada, Ensenada, BC, Mexico*
Roberto E. Armenta • *Fermentation and Metabolic Engineering Group, Ocean Nutrition Canada Ltd., Dartmouth, NS, Canada*
Dalal Asker • *Faculty of Agriculture, Food Science and Technology Department, Alexandria University, Alexandria, Egypt; Department of Molecular Biology & Biochemistry, Nelson Biological Labs, Rutgers University, Piscataway, NJ, USA*
Tarek S. Awad • *Department of Chemistry and Chemical Biology, Wright-Rieman Laboratories, Rutgers, The State University of New Jersey, New Brunswick, NJ, USA*
Adelbert Bacher • *Lehrstuhl für Biochemie, Technische Universität München, Munich, Germany*
José-Luis Barredo • *Gadea Biopharma, Parque Tecnológico de León, León, Spain*
Teruhiko Beppu • *Life Science Research Center, College of Bioresource Sciences, Nihon University, Fujisawa, Japan*
Qiong Cheng • *E. I. DuPont. de Nemours Inc., Wilmington, DE, USA*
Rosalía Contreras • *Department of Marine Biotechnology, Center for Scientific Research and Higher Education of Ensenada (CICESE), Ensenada, BC, Mexico*
Deana J. Dicosimo • *E.I. DuPont de Nemours, Inc., Wilmington, DE, USA*
Wolfgang Eisenreich • *Lehrstuhl für Biochemie, Technische Universität München, München, Germany*
Eva Eylert • *Lehrstuhl für Biochemie, Technische Universität München, München, Germany*
Rosa Isabel Fernández-Chimeno • *Department of Microbiology, Instituto Biomar, S.A., Parque Tecnológico de León, León, Spain*
Emilio Fernández • *Departamento de Bioquímica y Biología Molecular, Universidad de Córdoba, Córdoba, Spain*
José M. Fernández-Sevilla • *Department of Chemical Engineering, Universidad de Almería, Almeria, Spain*
Paul D. Fraser • *Centre for Systems and Synthetic Biology, School Biological Sciences, Royal Holloway, University London, Egham, Surrey, UK*
Juan Luis De La Fuente • *Gadea Biopharma, Parque Tecnológico de León, León, Spain*
Maiko Furubayashi • *Department of Applied Chemistry and Biotechnology, Chiba University, Chiba, Japan*
Aurora Galván • *Departamento de Bioquímica y Biología Molecular, Universidad de Córdoba, Córdoba, Spain*
María Teresa García • *Departamento de Microbiología y Parasitología, Facultad de Farmacia, Universidad de Sevilla, Seville, Spain*

Eric Giraud • *Laboratoire des Symbioses Tropicales et Méditerranéennes, IRD, UMR-IRD/SupAgro/INRA/UM2/CIRAD, Montpellier, France*
Hisashi Harada • *KNC Bio-Research Center, KNC Laboratories Co. Ltd., Kobe, Hyogo, Japan*
Kristen Kelly • *DuPont Experimental Station, Wilmington, DE, USA*
Rosa León • *Laboratorio de Bioquímica, Departamento de Química y Ciencia de Materiales, Universidad de Huelva, Huelva, Spain*
Cynthia A. Ma • *Fermentation and Metabolic Engineering Group, Ocean Nutrition Canada Ltd., Dartmouth, NS, Canada*
Isamu Maeda • *Faculty of Agriculture, Department of Bioproductive Science, Utsunomiya University, Utsunomiya, Tochigi, Japan*
Hajime Masukawa • *DOE Great Lakes Bioenergy Research Center, Michigan State University, East Lansing, MI, USA; Department of Microbiology and Molecular Genetics, Michigan State University, East Lansing, MI, USA; Research Institute for Photobiological Hydrogen Production, Kanagawa University, Hiratsuka, Japan; PRESTO, Japan Science and Technology Agency (JST), Kawaguchi, Japan*
Encarnación Mellado • *Departamento de Microbiología y Parasitología, Facultad de Farmacia, Universidad de Sevilla, Sevilla, Spain*
Norihiko Misawa • *Central Laboratories for Frontier Technology, Kirin Holdings Co. Ltd., i-BIRD, Ishikawa, Japan; Research Institute for Bioresources and Biotechnology, Ishikawa Prefectural University, Ishikawa, Japan*
Mari Mochimaru • *Department of Natural Sciences, Komazawa University, Tokyo, Japan*
Emilio Molina Grima • *Department of Chemical Engineering, Universidad de Almería, Almería, Spain*
María De Lourdes Moreno • *Departamento de Microbiología y Parasitología, Facultad de Farmacia, Universidad de Sevilla, Sevilla, Spain*
Leonel Ochoa • *Department of Marine Biotechnology, Center for Scientific Research and Higher Education of Ensenada (CICESE), Ensenada, BC, Mexico*
Jorge Olmos-Soto • *Department of Marine Biotechnology, Center for Scientific Research and Higher Education of Ensenada (CICESE), Ensenada, BC, Mexico*
J. Paniagua-Michel • *Department of Marine Biotechnology, Center for Scientific Research and Higher Education of Ensenada (CICESE), Ensenada, BC, Mexico*
Laura Pérez-Fons • *Centre for Systems and Synthetic Biology, School Biological Sciences, Royal Holloway, University London, Egham, Surrey, UK*
Francisco Romero • *Department of Microbiology, Instituto Biomar, S.A. Parque Tecnológico de León, León, Spain*
Cristina Sánchez-Porro • *Departamento de Microbiología y Parasitología, Facultad de Farmacia, Universidad de Sevilla, Sevilla, Spain*
Mark A. Scaife • *Fermentation and Metabolic Engineering Group, Ocean Nutrition Canada Ltd., Dartmouth, NS, Canada*
Pamela L. Sharpe • *E.I. DuPont de Nemours, Inc., Experimental Station, Wilmington, DE, USA*
Shinichi Takaichi • *Department of Biology, Nippon Medical School, Kawasaki, Japan*

LUAN TAO • *E.I. DuPont de Nemours, Inc., Experimental Station, Wilmington, DE, USA*
KENJI UEDA • *Life Science Research Center, College of Bioresource Sciences, Nihon University, Fujisawa, Japan*
DAISUKE UMENO • *Department of Applied Chemistry and Biotechnology, Chiba University, Chiba, Japan*
PATRICIA VEIGA-CRESPO • *Department of Microbiology and Parasitology, University of Santiago de Compostela, Santiago de Compostela, Spain*
ANDRÉ VERMÉGLIO • *CEA, DSV, IBEB, Lab Bioenerget Cellulaire, Saint-Paul-lez-Durance 13108, France*
MARTA VILA • *Laboratorio de Bioquímica, Departamento de Química y Ciencia de Materiales, Universidad de Huelva, Huelva, Spain*
TOMÁS G. VILLA • *Faculty of Pharmacy, Department of Microbiology and Parasitology, University of Santiago de Compostela, Santiago de Compostela, Spain*
TERESA VINUESA • *Department of Pathology and Experimental Therapeutics, University of Barcelona, Barcelona, Spain*
MIQUEL VIÑAS • *Department of Pathology and Experimental Therapeutics, University of Barcelona, Barcelona, Spain*
PHILLIP C. WRIGHT • *Department of Chemical and Biological Engineering, The ChELSI Institute, The University of Sheffield, Sheffield, UK*
RICK W. YE • *DuPont Experimental Station, Wilmington, DE, USA*

Chapter 1

Pathways of Carotenoid Biosynthesis in Bacteria and Microalgae

J. Paniagua-Michel, Jorge Olmos-Soto, and Manuel Acosta Ruiz

Abstract

The carotenoids, a subfamily of the isoprenoids, are among the most widespread, ancient, diverse, and rich class of all natural products and biomolecules. Microorganisms, as well as microalgae and bacteria synthesize isoprenoids from isopentenyl diphosphate (IPP) and its isomer dimethylallyl diphosphate (DMAPP). For long time, mevalonic acid was assumed to be the only natural precursor for IPP and DMAPP in the cytosolic acetate/mevalonate pathway for the biosynthesis of sterols, sesquiterpenes, triterpenoids, and carotenoids. At present, it is accepted that the relatively new route, the methylerythritol 4-phosphate (MEP), or 1-deoxy-D-xylulose-5-phosphate (DOXP) is the main pathway for the biosynthesis of plastidic isoprenoids, such as carotenoids, phytol (a side chain of chlorophylls), plastoquinone-9, isoprene, mono-, and diterpenes. Cytosolic isoprenoids (sterols) biosynthesized by MEP have been reported in eubacteria and algae (*Chlorella*, *Chlamydomonas*, *Scenedesmus*, and *Dunaliella*). This review summarizes current knowledge of the biosynthetic pathways leading to the formation of different isoprenoids and carotenoids in bacteria and microalgae. Particular attention was given to the last early steps of the biosynthesis of the key C_5-precursor and the final steps of the biosynthesis of carotenoids including selected examples in microalgae and bacteria as well as the recent advances in genomics and metabolic engineering.

Key words: Isoprenoids, Carotenoids, Mevalonate, 1-Deoxy-D-xylulose-5-phosphate, Bacteria, Microalgae, Metabolic engineering, Overview

1. The MVA Pathway of Isoprenoid and Carotenoid Biosynthesis

All isoprenoids and carotenoids derive from common five-carbon (C_5) building units and share common metabolic precursors in two independent pathways for the synthesis of isopentenyl diphosphate and dimethylallyl diphosphate (DMAPP) (Fig. 1): the mevalonate (MVA) pathway, and the relatively new 2-*C*-methyl-D-erythritol 4-phosphate (MEP) pathway (1). The mevalonate biosynthetic

José-Luis Barredo (ed.), *Microbial Carotenoids from Bacteria and Microalgae: Methods and Protocols*, Methods in Molecular Biology, vol. 892, DOI 10.1007/978-1-61779-879-5_1, © Springer Science+Business Media, LLC 2012

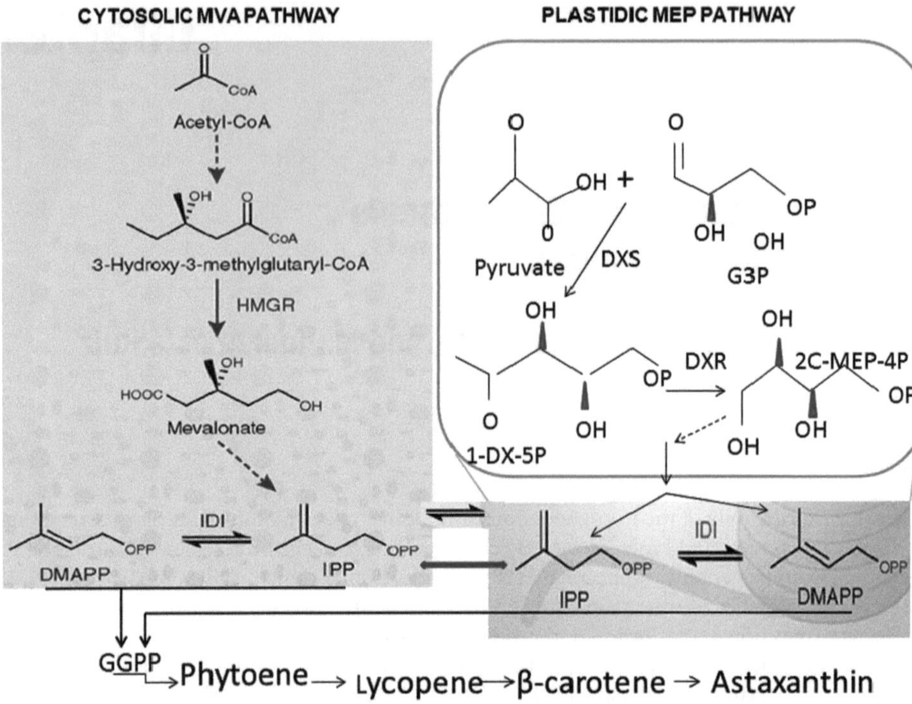

Fig. 1. Compartmentalized biosynthetic pathways of isoprenoids: the cytosolic acid mevalonic (MVA; *left*), and the plastidic (MEP; *right*); modified from Roberts CS (44). Both pathways produce isopentenyl diphosphate (IPP) and its isomer DMAPP, the universal building block for the synthesis of all carotenoids, exemplified here for *E. coli*.

pathway is present in the cytosol of Archaebacteria, fungi, and animals. This pathway begins from acetyl-CoA via hydroxymethylglutaryl-CoA. Mevalonic acid is produced in subsequent steps by two kinase reactions following decarboxylation. This pathway is associated to the formation of sterols, sesqui-terpenes, and triterpenes. The biosynthesis starts from three acetyl-CoAs, requires six enzymes, two NADPH, and three ATPs to finally yield isopentenyl-diphosphate (1), being regulated by the HMG-CoA reductase. The enzyme isopentenyl pyrophosphate isomerase (idi) promotes the isomerization of IPP to its isomer DMAPP. The biosynthesis of isoprenoids continues adding several IPP molecules (synthesized by either of two pathways) in a "head-to-tail" arrangement depending of the final product to be formed. The prenyl transferases and their different chain-length specificities develop chain elongation, and finally the short-chain prenyl transferases synthesize geranyl diphosphate (GPP, C10), farnesyl diphosphate (FPP, C15), or geranylgeranyl diphosphate (GGPP, C20), which are the direct precursors of mono-, di-, and tri-terpenes and carotenoids (2).

2. The DOXP/MEP Pathway of Isoprenoid–Carotenoid Biosynthesis

The MVA-independent pathway was discovered in the 1990s by labeling experiments in bacteria (3, 4). After new studies, its name was changed to pyruvate/glyceraldehyde 3-phosphate pathway or the first intermediate, deoxyxylulose 5-phosphate (DXP) pathway (5). At present, the pathway is named considering the presence of its first precursor, methylerythritol 4-phosphate (MEP), which is present in most eubacteria and the malaria parasite *Plasmodium falciparum*, but it is absent in Archaebacteria, fungi, and animals (5). In cyanobacteria, microalgae, and photosynthetic bacteria, the biosynthesis of isoprenoids follows the non-mevalonate, plastidic DOXP/MEP pathway. IPP biosynthesis starts from pyruvate and glycerinaldehyde-3-phosphate (seven enzymes, three ATP equivalents (ATP or CTP), and three NADPH (6, 7)). Finally, the HMBPP reductase achieves effective catalysis of substrates, IPP and its isomer DMAPP. DOXP is produced by the first enzyme, DOXP synthase, and is reduced by DOXP reductoisomerase to MEP. The exchange of intermediates and IPP appear to work between the plastids and the cytoplasm (3, 4). The C_{40} carotenoid pathway emerges (2) after phytoene formation by the synthase-catalyzed condensation of two molecules of geranylgeranyl diphosphate ($C_{20}PP$). Finally, different types and levels of modification of this C_{40} backbone may generate other carotenoids.

3. Biosynthesis of Carotenoids in Microorganisms

Carotenoids, C_{40} isoprenoids, biosynthesis initiates with the synthesis of the 5-carbon building block isopentenyl pyrophosphate (IPP). After IPP has been synthesized by either of the two routes (MVA or MEP), chain elongation proceeds by successive head-to-tail condensation of dimethylallyl pyrophosphate (DMAPP) to the growing polyprenyl pyrophosphate chain (2). Isomerization of IPP to DMAPP is catalyzed by the isopentenyl pyrophosphate isomerase. Chain elongation is catalyzed by prenyl transferases that synthesize geranyl PP (GPP, C_{10}), farnesyl PP (FPP, C_{15}) or geranylgeranyl PP (GGPP, C_{20}), which are the precursors of mono-, di-, and tri-terpenes and carotenoids (8). FPP (C_{15}) and GGPP (C_{20}) are the immediate precursors of C_{30} and C_{40} carotenoids (2, 9).

4. The Carotenoid Backbone

The first colorless C_{40} carotenoid, phytoene, is the product of two molecules of GGPP formed by tail-to-tail condensation rejecting two diphosphate groups, which are catalyzed by phytoene synthase (*crtB*, *al-2*). Colored carotenoids are synthesized by desaturation reactions of phytoene or dehydrosqualene, which creates conjugated double bonds (2). Different phytoene desaturase genes have been cloned that introduce two, three, four, or five double bonds into phytoene to produce ζ-carotene (plants, cyanobacteria, algae), neurosporene (*Rhodobacter*), o lycopene (most eubacteria and fungi) (10). After two cyclization reactions, the biosynthesized lycopene serves as a substrate for the formation of both α- and β-carotene. The former is converted in two steps to lutein, while the latter can then be converted to zeaxanthin by two successive hydroxylation steps. Zeaxanthin is epoxidized in two steps to give antheraxanthin and violaxanthin. In bacteria, only a single desaturase catalyzes the entire sequence of carotenoid desaturation steps, but in algae and cyanobacteria, carotenoid desaturation is accomplished by two distinct desaturases and a carotenoid isomerase (7). The common polyphene structures may allow carotenoids to isomerize (*cis–trans* isomers) (11). Further enzyme-catalyzed transformations (ketolation, hydroxylation, glycosylation, and oxidative cleavage) can influence on substrates derived from the C_{30} or C_{40} backbones to produce the already registered more than 700 known carotenoids.

5. MEP Pathway of Biosynthesis of Carotenoids in Microalgae

Studies on the biosynthetic pathways of isoprenoids and carotenoids on microalgae are still scarce and related to specific groups. In Chlorophyte algal groups, the synthesis of sterols and carotenoids follows the DOXP pathway (1). The marine diatoms *Nitzschia* sp. and *Phaeodactylum* sp., follow the MVA pathway for the biosynthesis of the cytosolic sterols, while the biosyntheses of carotenoids, phytol, and isoprenes are achieved by the DOXP/MEP pathway (1). In the case of *Chlorella vulgaris*, the non-mevalonate pathway is involved in the biosynthesis of isopentenyl diphosphate, after inhibiting brassinosteroid formation (12). The non-mevalonate pathway for isopentenyl diphosphate biosynthesis was evidenced in this Chlorophyte by Mevinolin, a specific inhibitor of HMG-CoA reductase, and clomazone. The null action of mevinolin in the accumulation of photosynthetic pigments (chlorophylls and carotenoids) in S*cenedesmus* cultures corroborated that the biosynthesis of the plastidic isoprenoids as well as that of the sterols proceeds via the nonmevalonate route (13). In diatoms, isoprenoid biosynthesis is species dependent and cannot simply be grouped according to structural type (14). The MVA pathway is located in the cytoplasm and leads to the biosynthesis of sesquiterpenes and triterpenes, whereas the MEP pathway is located in the plastids and accounts for the biosynthesis of mono-, di-, sester-, tri-, and tetraterpenes. Thus, although phytol in the chloroplast is made exclusively by the MEP pathway (13), the presence of the non-mevalonate 2-*C*-methyl-D-erythritol 2-phosphate (MEP) pathway for synthesis of isopentenyl diphosphate (IPP) and DMAPP occurs in the photosynthetic cyanobacterium *Synechocystis* sp. (15). The presented data support alternative routes of entry of pentose phosphate cycle substrates derived from photosynthesis. The labeling experiments performed with diatoms pointed out the key role of CO_2 and photosynthesis in the MEP pathway for isoprenoid biosynthesis (16), while mixotrophic growth conditions pointed out the competition for the carbon sources (acetate or CO_2) for different biosynthetic routes.

6. The Phytoene Synthase, a Key Enzyme of Carotenoid Biosynthesis

The phytoene synthase (PSY) is considered as the enzyme performing the rate-limiting entry reaction into the carotenoid biosynthesis pathway in photosynthetic organisms (17, 18). Some green algae (*Chlamydomonas reinhardtii*, *C. vulgaris*, *Volvox carteri*), as well as diatoms (*Thalassiosira pseudonana* and *Phaeodactylum tricornutum*), have been associated with only one class of PSY gene (19). It seems that other gene copies were lost in these species. Recent gene duplication of PSY has been reported in the green alga *Dunaliella salina/bardawil.* The discovery of two classes of PSY gene families in some algae suggests that carotenoid biosynthesis in these algae is differentially regulated in response to development and environmental stress as well (20). PSY was previously investigated, for example, in *C. reinhardtii* (21, 22) and *Haematococcus pluvialis/lacustris.* The genome of *C. reinhardtii* contains only one functional *psy* gene (22). Under stress conditions, carotenogenic genes of *H. pluvialis*, viz, phytoene synthase (PSY), phytoene desaturase (PDS), lycopene cyclase (LCY), β-carotene ketolase (BKT), and β-carotene hydroxylase (CHY) can be up-regulated (23). In the biosynthesis of carotenoids, the first committed step, the head-to-head condensation of GGPP to phytoene, is mediated by PSY. The subsequent steps of the pathway leading to the synthesis of colored carotenoids are carried out by membrane-localized enzymes such as phytoene desaturase (PDS) and lycopene β-cyclase

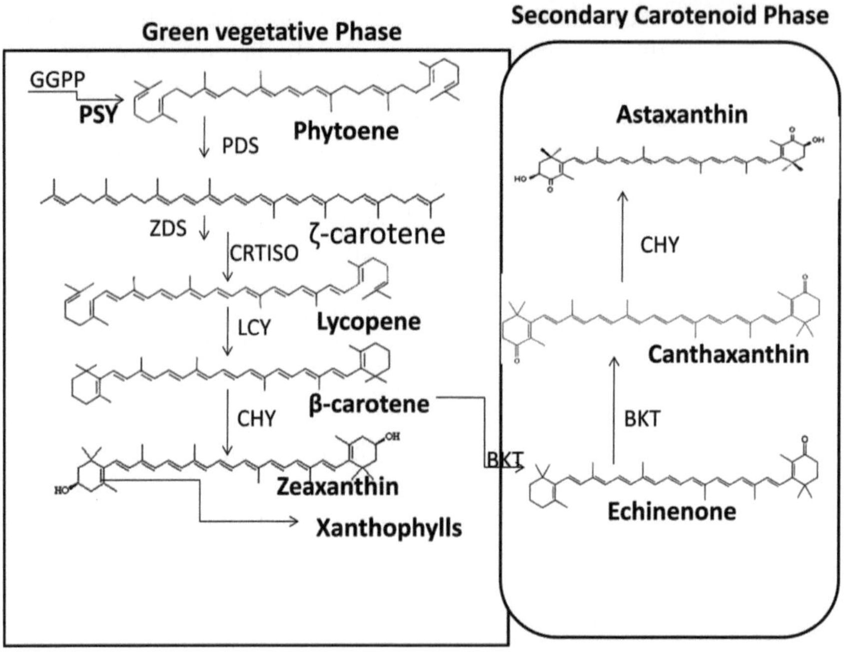

Fig. 2. Pathway of secondary carotenoid synthesis in *Haematococcus pluvialis* modified from Vidhyavathi et al. (23). *PSY* phytoene synthase, *PDS* phytoene desaturase, *ZDS* ζ-carotene desaturase, *CRTISO* carotenoid isomerase, *LCY* lycopene cyclase, *BKT* β-carotene ketolase, *CHY* β-carotene hydroxylase. Current intermediates were omitted for simplification.

(LCY) (24). The biosynthesis of astaxanthin in *Haematococcus* sp. follows the pathway up to β-carotene formation. Some specific steps of astaxanthin biosynthesis are located in the cytoplasm, but enzymatic machinery of the general carotenoid pathway appears to be localized in the chloroplast (25, 26) (Fig. 2).

7. MEP Pathway of Phytoene and β-Carotene in *D. salina*

Among carotenogenic strains, the green alga *D. salina* represents an interesting model of natural source of industrially important β-carotene (10% of algal dry weight) in numerous market applications (27, 28). Carotenoid biosynthesis in *D. salina* depend on the supply of the MEP-derived precursors (29, 30). Recent advances in carotenogenesis of *Dunaliella* species based on the PSY and PDS sequences have been reported (31). Sequences for the key enzyme leading to carotenogenesis, PSY of *D. salina* and *D. bardawil*, can be obtained from data of the National Center for Biotechnology Information (NCBI; http://www.ncbi.nlm.nih.gov) (20). It is well known that PSY catalyzes the head-to-head combination of two molecules of GGPP leading to generation of the first colorless carotenoid phytoene. Once the phytoene is

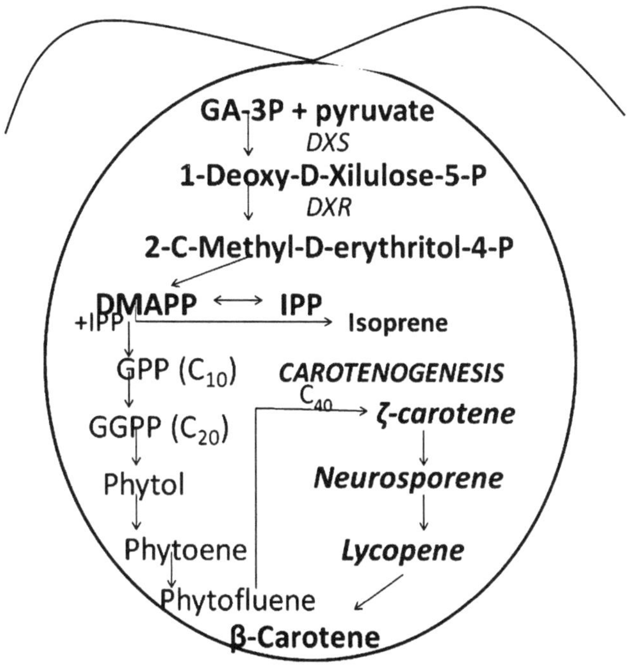

Fig. 3. MEP pathway of isoprenoids and carotenoid biosynthesis in *Dunaliella salina* (20, 29, 30). Enzyme and intermediate abbreviations are explained in the text.

biosynthesized, the next steps in carotenogenesis are the stepwise desaturation reactions (or dehydrogenation reactions) resulting in the conversion from phytoene (nine double bonds molecule) to lycopene (has 13 double bonds and a chromophore of 11 conjugated double bonds) via phytofluene, ζ-carotene, and neurosporene as intermediates. The major component of *Dunaliella* is β-carotene (32), which contains a β cyclohexene ring at each end. Cyclohexene rings added to either or both ends of lycopene and catalyzed by lycopene cyclase (LYC) leads to the formation of δ-, α-, γ-, and β-carotene (Fig. 3).

8. Isoprenoid–Carotenoid Biosynthesis in Bacteria

Although most of the members of the bacteria use exclusively the DOXP/MEP pathway of IPP formation and DMAPP synthesis, a minority uses the mevalonate pathway and only a few have both pathways (33). Moreover, the use of alternative enzymes for the same biosynthetic step can take place for isoprenoid and carotenoid biosynthesis. *E. coli* is probably the best-studied bacterium for the elucidation of the MEP pathway as in the case of the photosynthetic cyanobacteria, as well as green sulfur bacteria (*Chlorobium*), the purple bacteria (e.g., *Rhodobacter*, proteobacteria), and many non-photosynthetic bacteria (Aquificales) (4). In certain bacteria (*Borrelia burgdorferi, Myxococcus fulvus, Paracoccus zeaxanthinifaciens*), exception from this general rule can be associated to the lost the DOXP/MEP pathway and replaced it by the MVA pathway acquired by lateral gene transfer of the full or part of the genes of this pathway (4). The second group is represented by members of the Gram-positive bacteria with a high G+C content, i.e., several *Streptomyces* species possessing both the MVA and the DOXP/MEP route. The initial precursors utilized in the pathway are pyruvate (PYR) and glyceraldehyde 3-phosphate (GA3P) generating DXP (15). The initial reaction is catalyzed by DXP synthase, MEP is finally formed by a reduction and rearrangement catalyzed by 1-deoxy-D-xylulose-5-phosphate reductoisomerase. Subsequently, MEP is converted to 4-diphosphocytidyl-2-*C*-methyl-D-erythritol (34, 35). The chain elongation proceeds by successive head-to-tail condensation of IPP to DMAPP and next to the growing polyprenyl diphosphate chain (Fig. 1) involved in the synthesis of GPP (C_{10}), FPP (C_{15}), and GGPP (C_{20}). GPP and FPP are synthesized by FPP synthase encoded by *ispA* in *E. coli*, whereas GGPP formation is catalyzed by GGPP synthase encoded by *crtE*. Colorless C_{40} phytoene is formed by condensation of two molecules of GGPP, which is catalyzed by phytoene synthase encoded by *crtB*. Subsequent desaturation of phytoene by phytoene desaturase (*crtI*) introduces four double bonds into phytoene to produce red-colored lycopene. Synthesis of cyclic carotenoids involves cyclization of one or both end groups of lycopene. A lycopene β-cyclase (*crtY*) catalyzes the cyclization of both ends of lycopene to make β-carotene, which in turn is converted into zeaxanthin and canthaxanthin by β-hydroxylase (*crtZ*) and ketolase (*crtW*) enzymes, respectively, and finally to astaxanthin (36).

9. Isoprenoid and Carotenoid Biosynthesis in Archaea and Eucarya

Generally, Archaea and eukaryotes follow the MVA pathway of IPP and DMAPP biosynthesis (7), although this has been commonly explained as a horizontal gene transfer acquisition from archaeal or eukaryotic donors. It is known that most archaeal species lack the three last enzymes of the classic (eukaryotic-like) MVA pathway (37). These enzymes are the phosphomevalonate kinase (PMK), the mevalonate-5-decarboxylase (MDC), and the isopentenyl diphosphate isomerase (IDI). Recent phylogenomic analyses have already confirmed the MEP pathway to be restricted to bacteria and plastid-bearing eukaryotes. In the three domains of life, the preference of the MEP pathway in bacteria, the classical MVA pathway in eukaryotes, and the alternative MVA pathway in archaea has been reported (38).

10. Genomics and Metabolic Engineering of Carotenoids Microbial Production

The increasing scientific and industrial importance of carotenoids has led to find new ways of producing additional carotenoid structures throughout metabolic engineering of microbial production. Metabolomics, genomics, transcriptomics, and proteomics contribute with metabolic engineering efforts to determine the presence of genes for specific biosynthetic steps for obtaining specific or novel carotenoids and higher yields. Some needed premises, to obtain high-yield production of carotenoids in engineered microbial hosts have been pointed out (2), viz, a robust and optimized pool of the available isoprenoid precursor, expression of carotenogenic genes for efficient transformation of precursors to desired carotenoid compounds, and sufficient storage capabilities for the lipophilic carotenoids, either by engineering or by selecting a suitable heterologous host with known high storage capabilities for hydrophobic compounds.

Several studies showed that overexpression of 1-deoxy-D-xylose 5-phosphate synthase (*dxs*), which catalyzes the first step in the non-mevalonate pathway, improved lycopene production by two- to threefold (2, 10, 39). Overexpression of *dxs* and 1-deoxy-D-xylose 5-phosphate reducto-isomerase (*dxr*), the enzyme next to DXS in the pathway, further improved lycopene yields by 1.4–2-fold (40). An even larger effect on carotenoid production was observed when *idi* was overexpressed, indicating that the isomerization of IPP to DMAPP is indeed a rate-limiting step in isoprenoid biosynthesis. Controlling the expression of conditions limiting isoprenoid and carotenoid biosynthetic enzymes increases the flux through the carotenoid pathway. Actually, genome information is being generated for sequenced cyanobacteria (http://img.jgi.doe.gov/pub/main.cgi?Page=restricte; http://www.ncbi.nlm.nih.gov/) which allowed (41) to obtain data set of genes encoding enzymes in the carotenoid biosynthetic pathway. The expression of diatoms *T. pseudonana and P. tricornutum* carotenoid biosynthesis-related genes were assessed experimentally and in silico by means of quantitative real-time RT-PCR and by analysis of the *Phaeodactylum* digital gene expression database (http://www.biologie.ens.fr/diatomics/EST3) (42). Non-carotenogenic *E. coli* cells transformed with heterologous carotenoid genes produced carotenoids concentrations considerably low (~1 mg/g dry cell weight) when compared to up to the 100-fold higher levels produced by carotenogenic algae and microbial strains such as *Dunalliella*, *Haematococcus*, *Flavobacterium*, and *Xanthophyllomyces dendrorhous* (*Phaffia rhodozyma*) (2). By combination of genes from different organisms with different carotenoid biosynthetic branches, novel carotenoids not found in any other pathway can be also synthesized (7). Applications of genetic engineering allowed the construction of transgenic microalgae able to synthesize ketocarotenoids, such as the β-carotene ketolase (*bkt1*) cDNA from *H. pluvialis*, involved in the biosynthesis of astaxanthin (43). It has been reported in green algae that the final step of the plastidial methylerythritol phosphate (MEP) pathway is catalyzed by 4-hydroxy-3-methylbut-2-enyl diphosphate reductase, an enzyme proposed to play a key role in the regulation of isoprenoid biosynthesis (38). DNA transformation methods (44) and the current genome sequencing efforts applied in *D. salina* enhance the biotechnological potential of β-carotene production. The genome sequences for a number of different microalgae became available from the DOE joint genome institute (http://www.jgi.doe.gov). The expression of carotenogenic genes: phytoene synthase (PSY), phytoene desaturase (PDS), lycopene cyclase (LCY), β-carotene ketolase (BKT), and β-carotene hydroxylase (CHY) in *H. pluvialis* are regulated under stress conditions (23). Actually, sequences of main enzymes of the metabolic pathways associated to isoprenoid biosynthesis from *Dunaliella* species can be obtained from the National Centre for Biotechnology Information (NCBI, http://www.ncbi.nlm.nih.gov).

Metabolomics considers the following major approaches for obtaining recombinant valuable products such as novel carotenoids: target analysis, metabolite profiling, and metabolic fingerprinting (11). The available pool of carotenoid biosynthetic genes from different organisms combined into recombinant pathways opens numerous possibilities for the synthesis of structurally diverse carotenoids in recombinant microorganisms to produce a rich and diverse platform of new carotenoids challenging those obtained by chemical synthesis.

References

1. Lichtenthaler HK (2004) Evolution of carotenoid and isoprenoid biosynthesis in photosynthetic and non-photosynthetic organisms. 16th international plant lipid symposium, Budapest, Hungary
2. Lee PC, Schmidt-Dannert C (2002) Metabolic engineering towards biotechnological production of carotenoids in microorganisms. Appl Microbiol Biotechnol 60:1–11
3. Rohmer M, Knani M, Simonin P (1993) Isoprenoid biosynthesis in bacteria: a novel pathway for the early steps leading to isopentenyl diphosphate. Biochem J 295:517–524
4. Lichtenthaler HK (1999) The 1-deoxy-D-xylulose-5-phosphate pathway of isoprenoid biosynthesis in plants. Annu Rev Plant Physiol Plant Mol Biol 50:47–65
5. Rodríguez-Concepción M, Boronat A (2002) Elucidation of the methylerythritol phosphate pathway for isoprenoid biosynthesis in bacteria and plastids. A metabolic milestone achieved through genomics. Plant Physiol 130:1079–1089
6. Wanke M, Skorupinska K, Sweizewska E (2001) Isoprenoid biosynthesis via 1-deoxy-D-xylulose 5-phosphate/2C-methyl-D-erithritol4-phosphate (DOXP/MEP) pathway. Acta Biochim Pol 48:663–672
7. Umeno DM, Tobias A, Arnold FH (2005) Diversifying carotenoid biosynthesis pathways by direct evolution. Microbiol Mol Biol Rev 69:51–78
8. Hunter WN (2007) The non-mevalonate pathway of isoprenoid precursor biosynthesis. J Biol Chem 282:21573–21557
9. Lu S, Li L (2008) Carotenoid metabolism: biosynthesis, regulation, and beyond. J integr Plant Biol 50:778–785

10. Verdoes J, Van Ooyen A (1999) Isolation of the isopentenyl diphosphate isomerase encoding gene of *Paffia rhodozyma* improved carotenoid production in *Escherichia coli*. Acta Bot Gallica 146:43–53
11. Liu GN, Zhu YH, Jiang JG (2009) The metabolomics of carotenoids in engineered cell factory. Appl Microbiol Biotechnol 83:989–999
12. Bajguz A, Asami T (2004) Effects of brassinazole, an inhibitor of brassinosteroid biosynthesis, on light and dark grown *Chlorella vulgaris*. Planta 218:869–877
13. Schwender J, Gemunden C, Lichtenthaler H (2001) Chlorophyta exclusively use the 1-deoxyxylulose 5-phosphate/2-C-methylerythritol 4-phosphate pathway for the biosynthesis of isoprenoids. Planta 212:416–423
14. Massé G, Belt S, Rowland S, Rohmer M (2004) Isoprenoid biosynthesis in the diatoms *Rhizosolenia setigera* (Brightwell) and *Haslea ostrearia* (Simonsen). Proc Natl Acad Sci U S A 101:4413–4418
15. Ershov Y, Gantt R, Cunningham F, Gant E (2002) Isoprenoid biosynthesis in *Synechocystis* sp. strain PCC6803 is stimulated by compounds of the pentose phosphate cycle but not by pyruvate or deoxyxylulose-5-phosphate. J Bacteriol 184:5045–5051
16. Cvejic J, Rohmer M (2000) CO_2 as main carbon source for isoprenoid biosynthesis via the mevalonate-independent methylerythritol 4-phosphate route in the marine diatoms *Phaeodactylum tricornutum* and *Nitzschia ovalis*. Phytochemistry 53:21–28
17. Salvini M, Salvinia M, Bernini A, Fambrini M, Pugliesi M (2005) cDNA cloning and expression of the phytoene synthase gene in sunflower. Plant Physiol 162:479–484
18. Li S, Li L (2008) Carotenoid metabolism: biosynthesis, regulation and beyond. J Integr Plant Physiol 50:778–785
19. Tran D, Haven JH, Qiu WG (2009) An update on carotenoid biosynthesis in algae: phylogenetic synthase evidence for the existence of two classes of phytoene synthase. Planta 229:723–729
20. Ye ZW, Jiang JG, Wu H (2008) Biosynthesis and regulation of carotenoids in Dunaliella: progresses and prospects. Biotechnol Adv 25:352–360
21. Bohne F, Linden H (2002) Regulation of carotenoid biosynthesis genes in response to light in *Chlamydomonas reinhardtii*. Biochim Biophys Acta 1579:26–34
22. Lohr M, Grossman AR (2005) Genome-based examination of chlorophyll and carotenoid biosynthesis in *Chlamydomonas reinhardtii*. Plant Physiol 138:490–515
23. Vidhyavathi R, Venkatachalam L, Sarada R, Ravishankar GA (2009) Regulation of carotenoid biosynthetic gene expression and carotenoid accumulation in the green alga *Haematococcus pluvialis* under nutrient stress conditions. J Exp Bot 59:1409–1418
24. Cunningham FX, Gantt E (2000) Genes and enzymes of carotenoid biosynthesis in plants. Annu Rev Plant Physiol Plant Mol Biol 49:557–583
25. Grunewald K, Eckert M, Hirschberg J, Hagen C (2000) Phytoene desaturase is located exclusively in the chloroplast and up-regulated at the mRNA level during accumulation of secondary carotenoids in *Haematococcus pluvialis* (Volvocales, Chlorophyceae). Plant Physiol 122:1261–1268
26. Jin E, Lee CG, Polle JE (2006) Secondary carotenoid accumulation in Haematococcus biosynthesis, regulation and biotechnology. J Microbiol Biotechnol 16:821–831
27. Ben-Amotz A, Polle JEW, Rao DV (2009) The Alga Dunaliella: biodiversity, physiology, genomics, and biotechnology. Enfield Science, Enfield, NH
28. Borowitzka MA, Borowitzka LJ (1988) *Dunaliella*. In: Borowitzka MA, Borowitzka LJ (eds) Micro-algal Biotechnology. Cambridge University Press, Cambridge, p 27
29. Paniagua-Michel J, Capa-Robles W, Olmos-Soto J, Gutierrez-Millan LE (2009) The carotenogenesis pathway via the isoprenoid-β-carotene interference approach in a new strain of *Dunaliella salina* isolated from Baja California Mexico. Mar Drugs 7:45–56
30. Capa-Robles W, Paniagua-Michel J, Olmos Soto J (2009) The biosynthesis and accumulation of beta-carotene in *Dunaliella salina* proceed via the glyceraldehyde 3-phosphate/pyruvate pathway. Nat Prod Res 23:1021–1028
31. Ramos A, Marquesa R, Rodrigues M, Heriques N, Baumgartner A, Castilho R, Brening B, Varela JC (2009) Molecular and functional characterization of a cDNA encoding 4-hydroxy-3-methylbut-2-enyl diphosphate reductase from *Dunaliella salina*. J Plant physiol 166:968–977
32. Paniagua-Michel J, Dujardin E, Sironval C (1995) Growth of *Dunaliella bardawil* under carotenogenic conditions. J Mar Biotech 2:101–104
33. Sangari FJ, Perez-Gil J, Carretero-Paulet L, Garcia-Lobo JM, Rodriguez-Concepcion M (2010) A new family of enzymes catalyzing the first committed step of the methylerythritol 4-phosphate (MEP) pathway for isoprenoid biosynthesis in bacteria. Proc Natl Acad Sci U S A 107:14081–14086

34. Rohmer M (1999) The discovery of a mevalonate-independent pathway for isoprenoid biosynthesis in bacteria, algae and higher plants. Nat Prod Rep 16:565–574
35. Lichtenthaler HK (2009) Biosynthesis and accumulation of isoprenoid carotenoids and chlorophylls and emission of isoprene by leaf chloroplasts. Bull Georgian Acad Sci 3:82–94
36. Das A, Yoon SH, Lee SH (2007) An undate on microbial carotenoid production: application of recent metabolic engineering tools. Appl Microbiol Biotechnol 77:505–512
37. Lange BM, Ketchumre M, Croteau RB (2001) Isoprenoid biosynthesis. Plant Physiol 127:305–314
38. Lombard J, Moreira D (2011) Origins and early evolution of the mevalonate pathway of isoprenoid biosynthesis in the three domains of life. Mol Biol Evol 28:87–99
39. Mattthews PD, Wurtzel ET (2000) Metabolic engineering of carotenoid accumulation in *Escherichia coli* by modulation of the isoprenoid precursor pool with expression of deoxyxylulose phosphate synthase. Appl Microbiol Biotechnol 53:396–400
40. Kim SW, Keasling JD (2001) Metabolic engineering of nonmevalonate isopentenyl diphosphate synthesis pathway in *Escherichia coli* enhances lycopene production. Biotechnol Bioeng 72:408–415
41. Liang C, Zhao F, Wei W, Wen Z, Qin S (2006) Carotenoid biosynthesis in cyanobacteria: structural and evolutionary scenarios based on comparative genomics. Int J Biol Sci 2:197–207
42. Coesel S, Oborn M, Varela J, Falciatore A, Bowler C (2008) Evolutionary origins and functions of the carotenoid biosynthetic pathway in marine diatoms. Plos One 3:1–15
43. León R, Couso I, Fernández E (2007) Metabolic engineering of ketocarotenoid biosynthesis in the unicelullar microalga *Chlamydomonas reinhardtii*. J Biotechnol 15:143–152
44. Roberts CS (2007) Production and engineering of terpenoids in plant cell culture. Nat Chem Biol 3:87–395

Chapter 2

Selection and Taxonomic Identification of Carotenoid-Producing Marine Actinomycetes

Francisco Romero, Rosa Isabel Fernández-Chimeno, Juan Luis de la Fuente, and José-Luis Barredo

Abstract

Carotenoids are important pigments produced by plants and many microorganisms, including fungi, microalgae, cyanobacteria, and bacteria. Marine actinomycetes are a group of bacteria that produce a variety of metabolites with economic potential. Here, we describe a general method of selecting marine actinomycetes as carotenoids' producers. The screening is carried out at two levels: the first one involves a quick selection of strains by visual color inspection, and the second consists in the analysis of the extracts by HPLC. The taxonomic analysis of the producing strains gives us an overview of the groups of actinomycetes in which carotenoids can be found.

Key words: Carotenoids, Actinomycetes, Marine bacteria, Color, Taxonomic analysis

1. Introduction

Carotenoids are a family of yellow to orange-red terpenoid pigments synthesized by photosynthetic organisms and by some microorganisms, such as fungi, microalgae, cyanobacteria, and heterotrophic and prototrophic bacteria (1). These compounds are used as colorants, feed supplements, and nutraceuticals in the food, medical, and cosmetic industries (2). Carotenoids are naturally found in the chloroplasts and chromoplasts of plants and some other photosynthetic organisms, such as algae. Carotenoids are potent antioxidants that may have protective effects on the microorganisms against oxidative damage (3, 4). Animals do not produce carotenoids, so they have to get them through their diets.

Carotenoids conform a numerous group of compounds with over 600 known molecules. Their main function in phototrophic organisms is to absorb light energy for its use in photosynthesis, as well as to protect chlorophyll from photodamage. In humans, they act

José-Luis Barredo (ed.), *Microbial Carotenoids from Bacteria and Microalgae: Methods and Protocols*, Methods in Molecular Biology, vol. 892, DOI 10.1007/978-1-61779-879-5_2, © Springer Science+Business Media, LLC 2012

as antioxidants, some of them as precursors in the biosynthesis of vitamin A (β-carotene, α-carotene, γ-carotene, and β-cryptoxanthin), and to protect the macula lutea of the eye. Carotenoids are used too as natural pigments in the aquaculture and poultry industry.

Actinomycetes in general and marine actinomycetes in particular are the most important producers of secondary metabolites as well as other products of industrial and pharmaceutical interest (5). While the terrestrial actinomycetes have been thoroughly searched since the discovery of the streptomycin by Waskman in 1943, the marine counterpart has focused little attention until recently. Nowadays, this group of organisms is being intensively studied, and new taxons have been isolated in sediments (6) or associated to invertebrates (7). Besides the production of secondary metabolites and enzymes, actinomycetes can produce non-soluble pigments that make them a target for the screening in search of new carotenoids or as overproducers of known ones.

The methods described here are of general applicability with any collection of marine bacteria. The growth media are suitable for the growth of the majority of marine actinomycetes. We describe first a quick identification of cultures by visual inspection with the potential for producing carotenoids. This procedure greatly diminishes the number of cultures that have to be screened by HPLC. A method for the identification of carotenoids is detailed. Due to the great number of different carotenoids, the complete identification of the compound will usually require methodologies as mass spectrometry and NMR, which are out of the scope of this chapter. Finally, a procedure for identifying the microorganisms through 16S rDNA sequence is given. This procedure works well with most marine actinomycetes.

2. Materials

2.1. Growth of Marine Actinomycetes

1. Artificial seawater salts (Hobby-Marin, Dohse Aquaritik AG, Gelsdorf, Germany).
2. 172b liquid medium: Add about 800 mL distilled water to a glass beaker. Weigh in 10 g of glucose, 20 g of soluble starch, 5 g of yeast extract, 5 g of tryptone, and 10 g of artificial seawater salts. Adjust to pH 7, and add 1 g of calcium carbonate (see Note 1). Make up to 1 L with distilled water. Autoclave at 121°C for 20 min.
3. 172b solid medium: 172b liquid medium and 15 g of agar. Autoclave at 121°C for 20 min.
4. ISP1 with marine salts: 5 g of tryptone, 3 g of yeast extract, and 10 g of artificial seawater salts. Make up to 1 L with distilled water. pH should be 7.0–7.2. Autoclave for 15 min at 121°C.

2.2. Liquid Fermentation and Extraction

1. Orbital shaker with temperature control. Additional humidity control is desirable.
2. Freezing and lyophilizing equipment.
3. Microcentrifuge.

2.3. Carotenoids' Analysis

1. Sample extraction cocktail: Dichloromethane: methanol: acetone (1:1:2; v/v).
2. Nylon filters 0.45 μm.
3. Mobile phase: Ethyl acetate:*n*-hexane (1:1; v/v).
4. Reverse-phase HPLC column: Nucleosil 5 μm NH_2 100A (250×4.6 mm) (Phenomenex, Torrance, CA, USA).
5. HPLC equipment with PDA detector.

2.4. DNA Amplification

1. 27f primer: 5′-AGAGTTTGATCMTGGCTCAG-3′, where M=C:A (1:1).
2. 1492r primer: 5′-TACGGYTACCTTGTTACGACTT-3′, where Y=C:T (1:1).
3. DNeasy kit (Qiagen, Valencia, CA, USA).
4. Plastic microtubes (Eppendorf, Hamburg, Germany).
5. 10× Tuning buffer containing 25 mM Mg^{2+}.
6. dNTP mix: 0.5 mM each dATP, dTTP, dGTP, and dCTP.
7. Taq DNA polymerase.
8. Thermocycler.

2.5. Agarose Gel Electrophoresis

1. TBE: 10.8 g of Tris base, 5.5 g of boric acid, 20 mL of 0.5 M EDTA, and distilled water up to 1 L. pH 8.0 (see Note 2).
2. Ethidium bromide 10 mg/mL (see Note 3).
3. Agarose low electroendosmotic flow, electrophoresis grade (8).
4. Electrophoresis equipment.

2.6. 16S rDNA Sequencing

1. QIAquick gel extraction kit (Qiagen, Valencia, CA, USA).
2. DNA ladder 100 pb (BioLabs, Beverly, MA, USA).

3. Methods

3.1. Quick Identification of Producing Actinomycetes

The actinomycetes can be screened in a fast way by the color of their colonies. The producing genera are spread in different suborders and classes of the order *Actinomycetales*. A non-exhaustive list of genera and colors (yellow, pink, orange, and red) that can be searched is shown in Table 1.

1. Grow the bacteria in solid medium 172b (see Note 4). Incubate the plates at 28°C for 7 days.

Table 1
List of genera and colors (yellow, pink, orange, and red) that can be found in carotenoid-producing marine actinomycetes

Basic color	Shade	Genus
Yellow	Pale-cream	*Actinoplanes, Agrococcus, Arsenicicoccus, Cellulomonas, Kocuria, Microbacterium, Rhodococcus, Serinicoccus, Streptomyces*
	Intense	*Rhodococcus*
	Yellow-orange	*Microcella, Rhodococcus*
	Yellow-greenish	*Kocuria, Leifsonia, Microbacterium, Rhodococcus*
Pink		*Arthrobacter, Gordonia*
Orange-red		*Blastococcus, Dietzia, Microcella, Micromonospora, Nocardia, Rhodococcus, Salinibacterium, Streptomyces, Verrucosispora*

2. Check the color of the colonies visually and make a photographic record (see Note 5).
3. Select the colonies that show the colors described in Table 1.
4. Preserve these cultures frozen to be used as inocula for liquid fermentation (see Note 6) (9).

3.2. Liquid Fermentation

Liquid fermentation for carotenoids' detection can be made in one step in flasks. 5 mL broths are enough for HPLC analysis. The broths are subjected to lyophilization for eliminating the water content. The desiccated broths are then extracted with a solvent mixture of adequate polarity in order to extract the carotenoids.

1. To prepare a fresh culture of the bacteria, inoculate the bacteria onto Petri dishes (9-cm diameter) containing 172b agar medium.
2. Incubate the Petri dishes at 28°C for 7 days.
3. Inoculate a 250-mL Erlenmeyer flask containing 30 mL of 172b liquid medium with the fresh culture (see Note 7). Incubate the flask at 28°C on a rotary shaker at 250 rpm for 96 h.
4. After 96 h of incubation, transfer a 5 mL sample of the fermentation broth to a 25-mL amber vial. Cap the vial with a lyophilization butyl rubber stopper and freeze it at –20°C. Store the vial at –20°C until needed for lyophilization (see Note 8).
5. Lyophilize the vial (see Note 9).

3.3. Carotenoids' Analysis

1. Add 15 mL of extraction cocktail to a lyophilized vial and shake for 20 min in an orbital shaker.
2. Filter through 0.45-μm nylon filters in HPLC vials.

3. Adjust sample temperature to 10°C and column temperature to 30°C.
4. Adjust mobile-phase flow to 1 mL/min in an isocratic elution mode.
5. Inject samples of 10 μL and run for 60 min.
6. Monitor the OD in the wavelength range from 200 to 700 nm.
7. Compare the HPLC profile obtained (RF and maximum of absorption) with the HPLC profile of the standards (Fig. 1). A typical HPLC profile of carotenoids is shown in Fig. 1.

3.4. DNA Amplification

The carotenoid-producing strains can be taxonomically identified by their 16S rDNA sequence. For that, the breakage of the cells is needed. The isolation of DNA from actinomycetes sometimes has problems since they are Gram-positive bacteria and posses a thick peptidoglycan layer in their cell walls. This thick layer can render some actinomycetes insensitive to lysozyme treatment. When this happens, several procedures can be used either for weakening the peptidoglycan layer so that it becomes sensitive to the lysozyme treatment or using another protocol for breaking the cells (10). Here, we describe a method for obtaining the DNA that works well with most actinomycetes. It usually produces DNA of enough quality to get good PCR amplifications of the 16S rDNA.

1. Grow the bacteria in ISP1 liquid medium with marine salts (50-mL-capacity flasks with 10 mL of medium) for 72 h at 28°C and 200 rpm.
2. Harvest the cells by centrifuging 1 mL of culture in a plastic microtube for 10 min at 5,000 × *g* (see Note 10).
3. Extract the DNA using the DNeasy kit.
4. Monitor the quality and amount of DNA by 0.8% half-strength agarose gel electrophoresis in TBE buffer and 5 μL of ethidium bromide (from a stock solution of 10 mg/mL) per 100 mL of gel solution.
5. Amplify by PCR the 16S rDNA genes using the eubacterial primers 27f and 1492r (11). Reaction mixture: full-strength tuning buffer, 200 nM primer 27f, 200 nM primer 1492r, 100 nM of each dNTP, 0.05 U of Taq polymerase, and 5 μL of DNA sample. PCR thermal cycle: 1 min at 94°C, and then 30 cycles of 10 s at 98°C, 30 s at 48°C, and 3 min at 72°C. Complete the reaction by incubating during 7.5 min at 72°C.

3.5. rDNA Sequencing

1. Purify the amplified fragments of DNA from agarose gels using the QIAquick gel extraction kit.
2. Analyze by electrophoresis in 0.8% agarose gel using a standard preparation (100 pb DNA ladder) to quantify the DNA by comparison. For that, a sample of the standard 100 pb DNA ladder is run in parallel to some dilutions of the DNA sample.

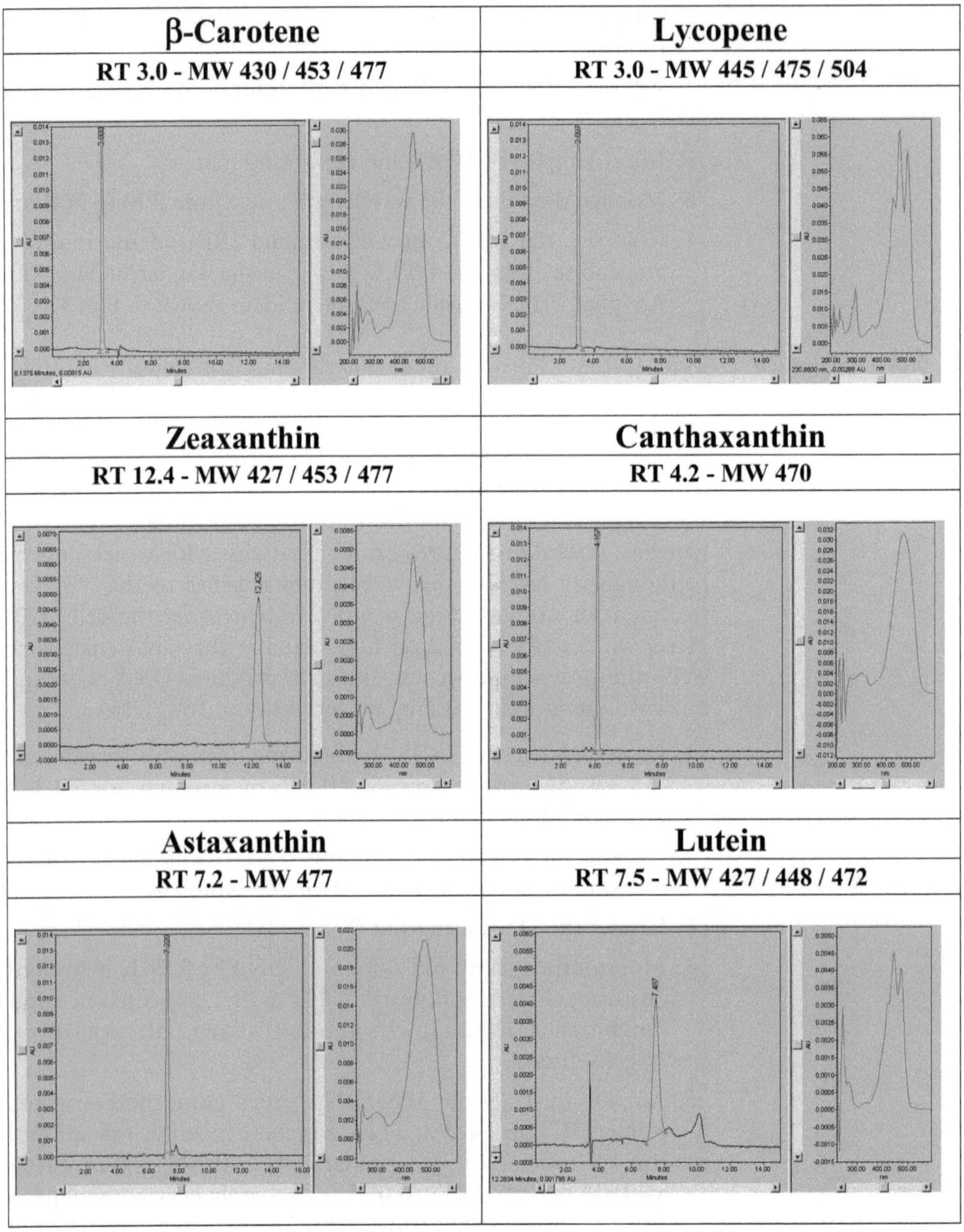

Fig. 1. Typical HPLC profile of some carotenoids produced by marine actinomycetes. *RT* retention time (min). *MW* maximum wavelength.

The amount of the DNA is estimated for the intensity of the band compared to the standard.

3. Prepare the purified DNA fragments for sequencing according to the advice of the sequencing service (see Note 11).
4. Analyze the sequences for the identification of the producing strains.

4. Notes

1. Calcium carbonate acts as a buffer counteracting the production of acids with carbonic anhydride. It must be always added after pH adjustment.
2. Prepare the solution five times more concentrated (5× TBE) and store it at room temperature to avoid precipitation.
3. Ethidium bromide is highly mutagenic. Wear double gloves. Decontaminate the solutions following Lunn and Sansone (12) or by incineration since ethidium bromide decomposes at 262°C.
4. 172b is a rich, non-sporulating medium. Aerial mycelium development masks the color of the colony, making more difficult to identify the cultures producing the carotenoids.
5. The photographs are very useful for comparing the colors of the colonies with the carotenoids' production and taxonomy of the cultures. In a continuous screen, it helps improving the selection of the colonies.
6. Actinobacteria, as most Gram-positive bacteria, survive well to freezing. For not very long periods of time, they can be frozen in the absence of protecting agents. Freezing the cultures on agar plugs is then recommended since the agar acts as a cryopreservative.
7. The aeration of the cultures when fermenting in 250-mL flasks with 30 mL of medium is good enough for most applications that demand high aeration rate. 30 mL of broth exceeds the amount needed simply for carotenoids' detection, but produces enough volume for other analysis. If only carotenoids' detection is needed, then 50-mL-capacity flasks containing 10 mL of medium would provide with a well-aerated culture.
8. Different lyophilizers will produce vacuum at different rates depending on the pump used and the volume of lyophilizing chamber. At the same time, the amount of filling of the vials and the amount of heat received by the vials to maintain the evaporation of the water are important factors in the lyophilization. Tests will have to be performed to ensure that the frozen broths do not melt during the process. Different volumes of broth could be tested.
9. The production of foaming substances during the fermentation can cause bubbling in some vials at the beginning of the lyophilization.
10. Young cultures of actinomycetes are more susceptible to lysozyme than older ones since the cells of the younger cultures

are actively reproducing and the level of cross-linkage in the peptidoglycan is smaller.

11. The sequencing service used in this work is Secugen S.L. (Madrid, Spain). It requests for 450 ng of amplified DNA and 1.5 μL of 5 μM primer 27f.

Acknowledgment

This work was in part supported by the grant IAP-580000-2008-52 of the Spanish Ministerio de Industria, Turismo y Comercio.

References

1. Britton G (1998) Overview of carotenoid biosynthesis. In: Britton G, Liaaen-Jensen S, Pfander H (eds) Carotenoids: biosynthesis and metabolism. Birkhäuser, Basel, pp 13–147
2. Bauernfeind JC (1981) Carotenoids as colorants and vitamin A precursors: technical and nutritional applications. Academic, New York, NY
3. Schroeder WA, Johnson EA (1993) Antioxidant role of carotenoids in *Phaffia rhodozyma*. J Gen Microbiol 139:907–912
4. Kobayashi M, Kakizono T, Nishio N, Nagai S, Kurimura Y, Tsuji Y (1997) Antioxidant role of astaxanthin in the green algae *Haematococcus pluvialis*. Appl Microbiol Biotechnol 48:351–356
5. Bull AT, Stach JEM, Ward AC, Goodfellow M (2005) Marine actinobacteria: perspectives, challenges, future directions. Antonie Van Leeuwenhoek 87:65–79
6. Mincer TJ, Jensen PR, Kauffman CA, Fenical W (2002) Widspread and persistent populations of a major new marine actinomycete taxon in ocean sediments. Appl Environ Microbiol 68:5005–5011
7. Montalvo NF, Mohamed NM, Enticknap JJ, Hill RT (2005) Novel actinobacteria from marine sponges. Antonie Van Leeuwenhoek 87:29–36
8. Sambrook J, Fritsch EF, Maniatis T (1989) Molecular cloning, 2nd edn. Cold Spring Harbor Laboratory, Cold Spring Harbor, NY
9. Dietz A, Currie SA (1996) Actinomycetes. In: Hunter-Cevera JC, Belt A (eds) Maintaining cultures for biotechnology and industry. Academic, San Diego, pp 85–99
10. Johnson JL (1994) Similarity analysis of DNAs. In: Gerhardt P, Murray RGE, Wood WA, Krieg NR (eds) Methods for general and molecular bacteriology. American Society for Microbiology, Washington, DC, pp 655–682
11. Lane DJ (1991) 16S/23S rRNA. In: Stackebrandt E, Goodfellow M (eds) Nucleic acid techniques in bacterial systematics. Wiley, Chichester, pp 115–175
12. Lunn G, Sansone EB (1987) Ethidium bromide: destruction and decontamination of solutions. Anal Biochem 162:453–458

Chapter 3

Isolation, Characterization, and Diversity of Novel Radiotolerant Carotenoid-Producing Bacteria

Dalal Asker, Tarek S. Awad, Teruhiko Beppu, and Kenji Ueda

Abstract

Carotenoids are natural pigments that exhibit many biological functions, such as antioxidants (i.e., promote oxidative stress resistance), membrane stabilizers, and precursors for vitamin A. The link between these biological activities and many health benefits (e.g., anticarcinogenic activity, prevention of chronic diseases, etc.) has raised the interest of several industrial sectors, especially in the cosmetics and pharmaceutical industries. The use of microorganisms in biotechnology to produce carotenoids is favorable by consumer and can help meet the growing demand for these bioactive compounds in the food, feed, and pharmaceutical industries. This methodological chapter details the development of a rapid and selective screening method for isolation and identification of carotenoid-producing microorganisms based on UV treatment, sequencing analysis of 16S rRNA genes, and carotenoids' analysis using rapid and effective High-Performance Liquid Chromatography–Diodearray–MS methods. The results of a comprehensive 16S rRNA gene-based phylogenetic analysis revealed a diversity of carotenoid-producing microorganisms (104 isolates) that were isolated at a high frequency from water samples collected at Misasa (Tottori, Japan), a region known for its high natural radioactivity content. These carotenoid-producing isolates were classified into 38 different species belonging to 7 bacterial classes (*Flavobacteria*, *Sphingobacteria*, α-*Proteobacteria*, γ-*Proteobacteria*, *Deinococci*, *Actinobacteria*, and *Bacilli*). The carotenoids produced by the isolates were zeaxanthin (6 strains), dihydroxyastaxanthin (24 strains), astaxanthin (27 strains), canthaxanthin (10 strains), and unidentified molecular species that were produced by the isolates related to *Deinococcus*, *Exiguobacterium*, and *Flectobacillus*. Here, we describe the methods used to isolate and classify these microorganisms.

Key words: Carotenoids, Irradiation, Misasa, Bacteria, Phylogeny, Diversity, High-performance liquid chromatography, HPLC, Photodiode array detection

1. Introduction

1.1. Structure, Function, and Importance of Carotenoids

Carotenoids are yellow-, orange-, or red-colored pigments, formed by the condensation of isoprenyl units, which provide the pleasing color of plants and animals. Carotenoids are generally classified into two groups: the carotenes (non-oxygen-substituted hydrocarbon carotenoids, such as α-carotene, β-carotene, etc.) and the xanthophylls

José-Luis Barredo (ed.), *Microbial Carotenoids from Bacteria and Microalgae: Methods and Protocols*, Methods in Molecular Biology, vol. 892, DOI 10.1007/978-1-61779-879-5_3, © Springer Science+Business Media, LLC 2012

(oxygen-substituted carotenoids, such as zeaxanthin, canthaxanthin, astaxanthin, etc.). All photosynthetic (e.g., plants and algae) and some non-photosynthetic organisms (e.g., yeast, fungi, and bacteria) synthesize carotenoids (1). In photosynthetic organisms, these substances serve as accessory pigments in light-capturing complexes by absorbing light and transferring the energy to chlorophyll. In both phototrophic and non-phototrophic organisms, carotenoids protect cells from harmful oxygen radicals (2). Members of the animal kingdom including humans, birds, fish, and invertebrates cannot synthesize carotenoids and must thus obtain from the diet. Carotenoids, especially β-carotene, canthaxanthin, and astaxanthin, have become commercially important as color-enhancing additives to animal feed in the aquaculture, poultry, and swine feed industries. Carotenoids are also used as food colorants (3, 4).

Many health benefits are linked to the biological activities of carotenoids, such as vitamin A precursors and antioxidants (5–7). The consumption of diets rich in carotenoids is associated with enhancement of the immune system, and reduction of the risk of degenerative diseases such as cancer, cardiovascular diseases, macular degeneration, and cataract (8–14). These benefits diversified the applications of carotenoids in foods, cosmetics, and pharmaceuticals.

Among more than 700 carotenoids reported so far (1), β-carotene, astaxanthin, canthaxanthin, lutein, lycopene, and zeaxanthin (Fig. 1) are the most important and high-value carotenoids, which are commonly used in the supplements, foods, and feeds industries. The global market of these carotenoids has increased dramatically from 2004 to 2009, and is expected to increase further in the coming years (15, 16).

1.2. Microbial Production of Carotenoids

Currently, the commercial production of carotenoids is mostly carried out by chemical synthesis (17) or extraction from plant tissue and food industrial wastes (18). Although chemical synthesis produces carotenoids with high purity and at low cost, they often contain some reaction precursors or by-products of nonbiological nature that may have undesired side effects (19, 20) and are not eco-friendly. In addition, consumer awareness of food safety and the health benefits of natural products have increased the demand for biological rather than synthetic production. The major advantages of the biological production of carotenoids are the wide range of their biosynthetic capability and ability to produce only the naturally occurring stereoisomers. Food (i.e., plants or food-processing wastes) and microorganisms are examples of carotenoids' biological sources. Unlike microorganisms, the production of carotenoids from food has many disadvantages, such as season fluctuation, limited resources, competition with the food industry, and requirement of land beside complicated extraction and purification process. For example, the carotenoids produced from plants contain mixtures of carotenoids, fats, oils, waxes, and unsaponifiable

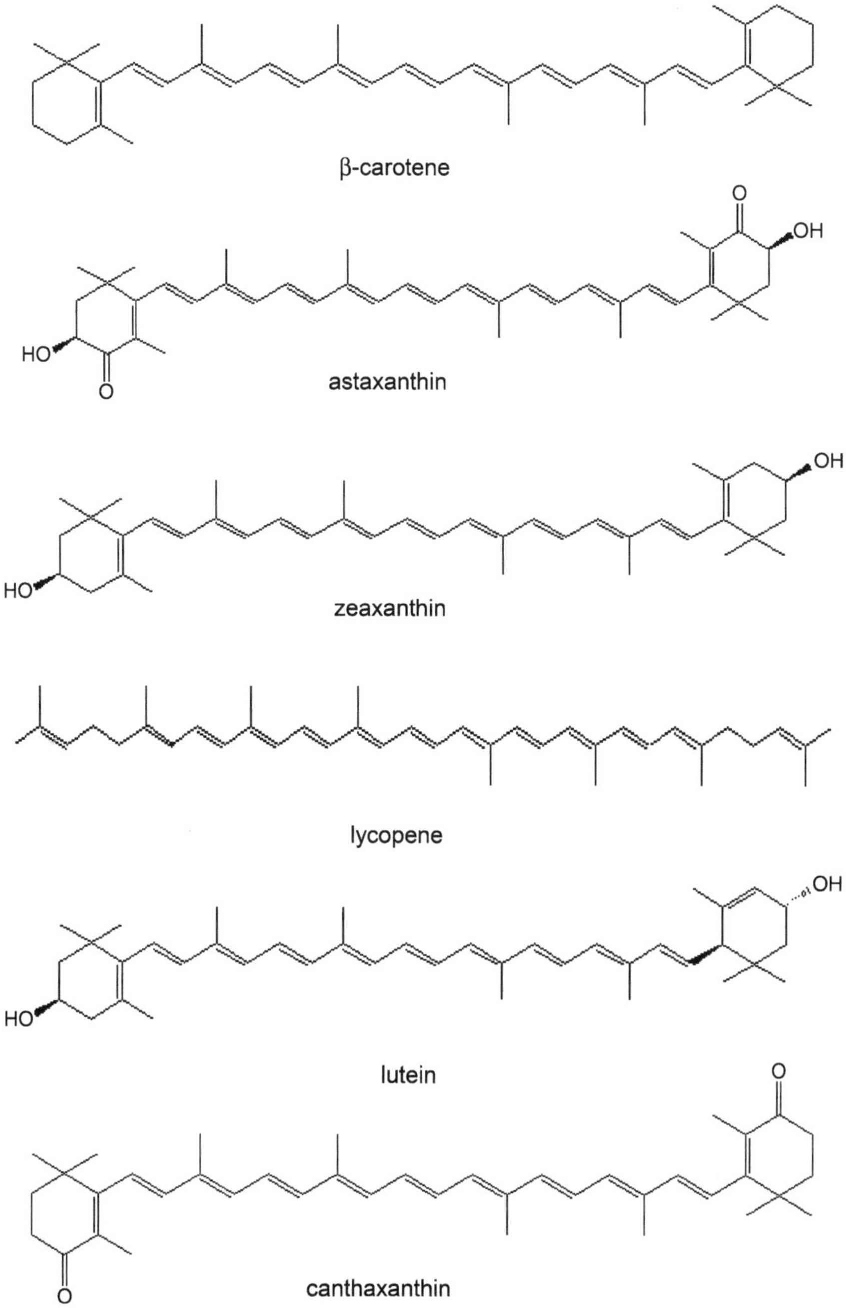

Fig. 1. Chemical structure of high-value carotenoids.

compounds (20, 21). Nevertheless, the microbial production of carotenoids is still awaiting lot of challenges to reduce cost and simultaneously increase productivity possibly by increasing biomass production and/or carotenoid synthesis, for example by selecting a strain that grows fast, accumulates high amount of carotenoids, facilitates the extraction and purification processes of carotenoids, and has a lower cost production. In addition, the application of modern tools of bioprocessing and recombinant DNA technology could potentially increase the production of carotenoids.

Table 1 lists the microbial sources of carotenoids. As can be noticed, the commercial production of β-carotene has been carried out using *Dunaliella salina* (22) and *Blakeslea trispora* (23), while

Table 1
Microorganisms reported for production of high-value carotenoids

Carotenoid	Type of microbe	Microorganism	Yield	Commercial producer	References
β-Carotene	Green algae	*Dunaliella Salina*[a]	3–5%	Australia, China, India, Israel, Japan, the USA	(22)
	Green algae	*Dunaliella bardawil*[a]	10–14%		(103)
	Blue-green algae	*Spirulina*	0.8–1%		
	Mold	*Blakeslea trispora*[a]	30 mg/g	Russia, Ukraine, Spain	(23)
	Mold	*Phycomyces blakesleeanus*	10–35 mg/g		(104, 105)
	Mold	*Mucor circinelloides*			(23)
Lycopene	Mold	*B. trispora*[a]	24 mg/g		(106)
	Mold	*Fusarium sporotrichioides*	0.5 mg/g		(107)
Astaxanthin	Microalgae	*Haematococcus pluvialis*[a]	1–3%	India, Japan, Sweden, the USA	(108, 109)
	Yeast	*Xanthophyllomyces dendrorhous*[a,b]	4–8.1 mg/L	China, Japan, the USA	(110, 111)
	Archaea	*Halobacterium salinarium*	265 μg/g		(112)
	Bacteria	*Paracoccus* sp. N81106[c]	0.3 mg/6 L		(75)
	Bacteria	*Paracoccus carotinifaciens* E-396	–		(113)
	Bacteria	*Brevundimonas* sp. SD212[d]	1.5 mg/27 L		(114)
	Bacteria	*Paracoccus* sp. PC1[e]	–		(115)
	Bacteria	*Paracoccus bogoriensis*	0.4 mg/g of wet cells		(116)

(continued)

Table 1 (continued)

Carotenoid	Type of microbe	Microorganism	Yield	Commercial producer	References
	Bacteria	*Brevundimonas* spp. (wild type)	27.6–365 μg/g		(63)
	Bacteria	*Brevundimonas* sp. M7 (mutant)	1.3 mg/g		(63)
Zeaxanthin[f]	Bacteria	*Flavobacterium* sp.	16 mg/g		(117)
	Bacteria	*Sphingobacterium multivorum*[g]	10.6 μg/mL		(118)
	Bacteria	*Synechocystis* sp. PCC 6803	0.98 μg/mL		(119)
	Bacteria	*Erwinia herbicola*			(120)
	Bacteria	*Paracoccus zeaxanthinifaciens*[h]			(76, 121)
Canthaxanthin	Bacteria	*Micrococcus roseus*	1.7 μg/mL		(122)
	Bacteria	*Gordonia jacobaea* MV-1	13,373 μg/mL		(123, 124)
	Archaea	*Haloferax alexandrinus* TM	2 mg		(27, 125)
	Bacteria	*Bradyrhizobium* sp.	1.34 mg/g		(126)
β-Cryptoxanthin	Bacteria	*Brevibacterium linens*	0.3 μg/mL		(127)
Lutein	*Chlorophyta*	*Chlorella zofingiensis*	21 μg/mL		(128)
		Chlorella protothecoides CS-41	225 μg/mL		(129)
		Muriellopsis sp.	35 μg/mL		(130)

[a]Effective commercial production
[b]Formerly *Phaffia rhodozyma*
[c]Formerly *Agrobacterium aurantiacum* N81106
[d]Formerly *Pseudomonas* sp.
[e]Formerly *Alcaligenes* sp. PC-1
[f]Please see Table 1, Chapter 5, for a complete list of reported zeaxanthin-producing microorganisms
[g]Formerly *Flavobacterium* spp.
[h]Formerly *Flavobacterium* sp. strain R-1 512

astaxanthin has been commercially produced from *Haematococcus pluvialis* (24) or *Phaffia rhodozyma* (25). However, there is no effective bacterial production of these carotenoids, with only a few bacteria that synthesize astaxanthin, zeaxanthin, canthaxanthin, and cryptoxanthin. To meet the increasing global and industrial demands of natural high-value carotenoids, there are needs to develop screening strategies to enhance the isolation of promising bacterial strains capable of maximizing the production of carotenoids to a commercial level.

1.3. Screening of Carotenoid-Producing Bacteria

In the following sections, we demonstrate a strategy for the screening of carotenoid-producing bacteria. As a result, a number of potential carotenoid-producing bacteria were isolated from hot spring waters with high radioactivity (26). On the basis of 16S rRNA sequencing, we will show how the isolates were identified and classified. We then describe three high-performance liquid chromatography (HPLC) methods for the analysis of the carotenoids produced by the isolates (26). Finally, we examine the relationship between radioactivity resistance and carotenoid production (carotenogenesis) (26).

1.4. Source of Isolation

The natural occurrence of colored microbial mats and dominancy of pigmented microbes in extreme habitats may serve as guides for potential sources of microbial carotenoids. For example, the red coloration of the surface water and soil of salt crystallizer ponds is associated with the growth of extremely halophilic red archaea (27) or bacteria (28). The yellow and orange colors at Yellowstone hot spring in the USA are carotenoids developed by thermophilic archaea (29–31). These microbes produce carotenoids to protect their cells against photooxidation induced by the intense sunlight in summer, and stabilize their cell membranes at extremely high temperatures (32–35). The exposure to solar UV radiations, such as type A (320–400 nm) and type B (280–320 nm), may damage biological cells. UV-A radiations are able to generate free radicals that damage DNA, and the exposure to UV-B can lead to direct damage of DNA and proteins. As antioxidants, carotenoids protect cells by quenching free radicals, and some carotenoids have been reported to have the ability of absorbing UV-A (36). The protective role of carotenoids against UV-B irradiation in some microbial systems was evidenced (30, 37–39). In the Gram-negative gliding bacterium *Myxococcus*, carotenoid production occurs in a light-dependent manner, probably to protect the cells from photooxidative damages (40). Another specific environmental stimulus that is possibly correlated to microbial carotenogenesis is ionizing radiations (e.g., X-ray), which damage living cells mainly by inducing oxidative stress, reactive oxygen species, and free radicals that attack biological macromolecules, such as nucleic acids, proteins, and lipids (41). Most members of *Deinococcus* spp. are radiotolerant, and produce carotenoids. Studies have shown that the colorless mutant of *D. radiodurans* is more sensitive to ionizing radiation, hydrogen peroxide, and desiccation than the wild type, suggesting an association between carotenoids and tolerance to environmental stresses (42, 43). However, pigmentation strategies are not obligatory as some microbes have highly efficient DNA-repair mechanisms as an

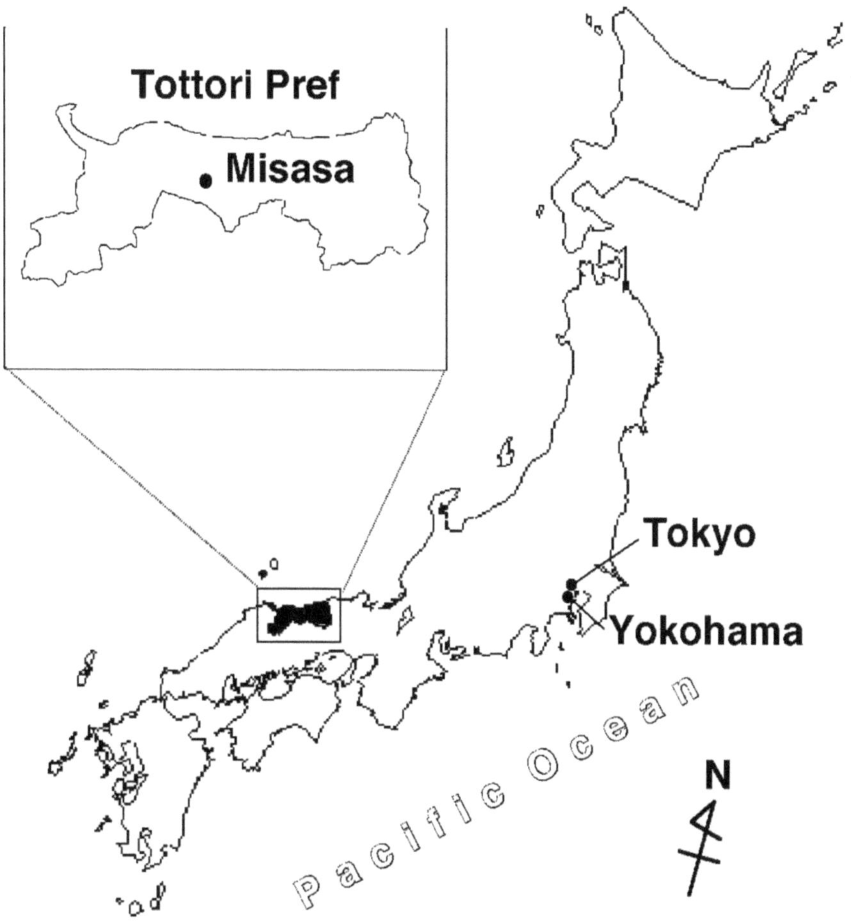

Fig. 2. Geographical location of Misasa (Tottori, Japan) (N35.41, E133.88). Adapted from Asker et al. 2007 (26).

alternative to damage prevention (44). Many carotenoid-producing bacteria have been isolated from various extreme environments, such as very low temperatures (33), high salinity (27, 45), strong light (27, 36), acidic and alkaline (46), and thermophilic conditions (29, 32, 35, 47). One may hypothesize, based on these evidences, that the oxidative stresses in extreme environments are selective factors associated with pigmented microorganisms, which are able to synthesize antioxidants (i.e., carotenoids) to protect their vital molecules (e.g., proteins and nucleic acids). Misasa (Tottori, Japan; Fig. 2) is a radioactive region that is characterized by high radon activity (^{226}Ra, 0.60 Bq/L; ^{228}Ra, 0.41 Bq/L) (48). This region was, thus, selected as a potential source for the isolation of carotenoid-producing bacteria (26). Our preliminary observations showed many bacterial colonies that exhibited the typical color of carotenoids. The result of an extensive screening study revealed the widespread distribution of carotenoid-producing bacteria in this region as well as their marked taxonomic diversity (26).

1.5. Screening Strategy

Microorganisms produce a wide variety of pigments, such as carotenoids, melanins, flavins, quinones, monascins, violacein, phycocyanin, or indigo (49). In general, there are two types of pigments: cell-bound and non-cell-bound pigments. Cell-bound pigments are nondiffusible pigments (e.g., carotenoids and chlorophyll) that can be recovered from the cells by disruption and solvent extraction. The second class of pigments (i.e., diffusible pigments) are secreted into the broth or agar media, and can be recovered by extraction from the media with large quantities of suitable solvents. Carotenoids are cell-bound pigments that are produced inside the cells and do not diffuse in the agar. Carotenoids give the microbial colonies their distinctive color (i.e., colonies on agar may be yellow, orange, pink, or red). Therefore, one can screen environmental water samples for carotenoid-producing microorganisms in two stages (Fig. 3). The first stage includes primary screening and microbial library construction. Primary screening is done visually on cell-bound pigmented colonies (i.e., colonies that produce nondiffusible pigments) that are grown on an appropriate agar media and incubated at various temperatures of 30–60°C (Fig. 3). After incubation, the yellow-, orange-, or red-colored microbial colonies that had formed are isolated, and each organism is

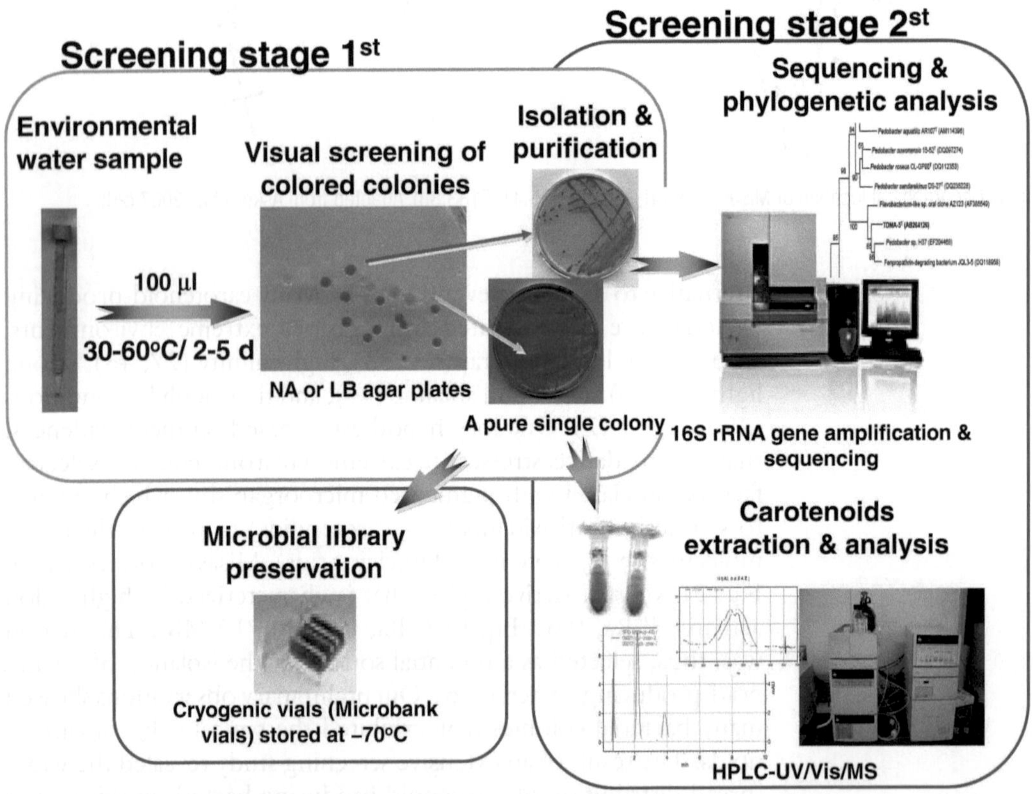

Fig. 3. Screening strategy of carotenoid-producing bacteria.

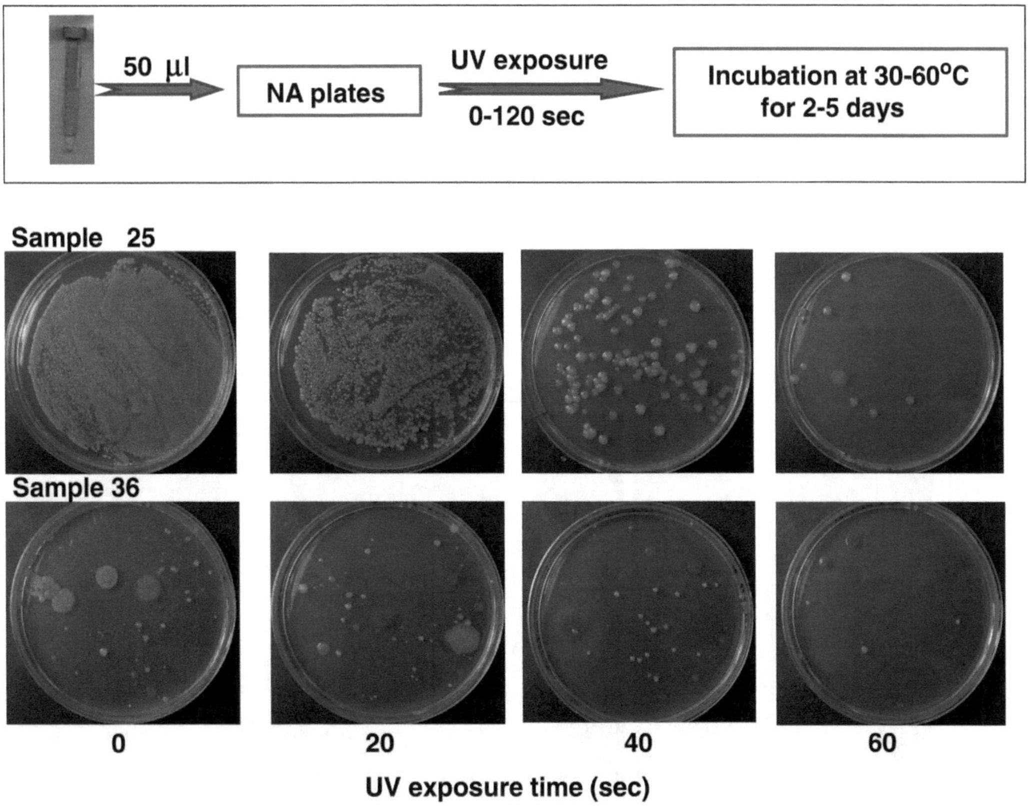

Fig. 4. UV treatment for the screening of carotenoid-producing bacteria from the water samples collected at Misasa. Each sample was diluted 1:10, and 50 μL of each dilution was inoculated onto NA agar plates and incubated at 30–60°C for 3 days. Prior to incubation, the plates were exposed to UV for the indicated periods.

purified using the standard microbial manipulation technique. As a selective screening technique, agar plates can be UV-irradiated for 1–3 min before incubation. The UV treatment helps to diminish the uninterested (uncolored) colonies, which have no UV resistance (Fig. 4). A microbial library that includes the pigmented isolates is then constructed. The second stage involves sequencing of 16S rRNA genes and carotenoids' analysis by using a rapid and reliable set of HPLC complementary methods (Fig. 3). By using this screening strategy, we attempted to screen each strain for the three following objectives: (1) discovery of novel microbial sources of carotenoids at levels of genus and species; (2) biodiversity of carotenoid-producing bacteria in a specific environment (e.g., radioactive hot spring water); and (3) discovery of novel carotenoids with novel properties.

Many yellow-, orange-, pink-, or red-colored colonies were obtained after inoculating and cultivating the water samples collected from Misasa (a representative result is shown in Fig. 5). The frequency at which the colored colonies were obtained from the samples appeared markedly higher than that from the samples collected from normal environments.

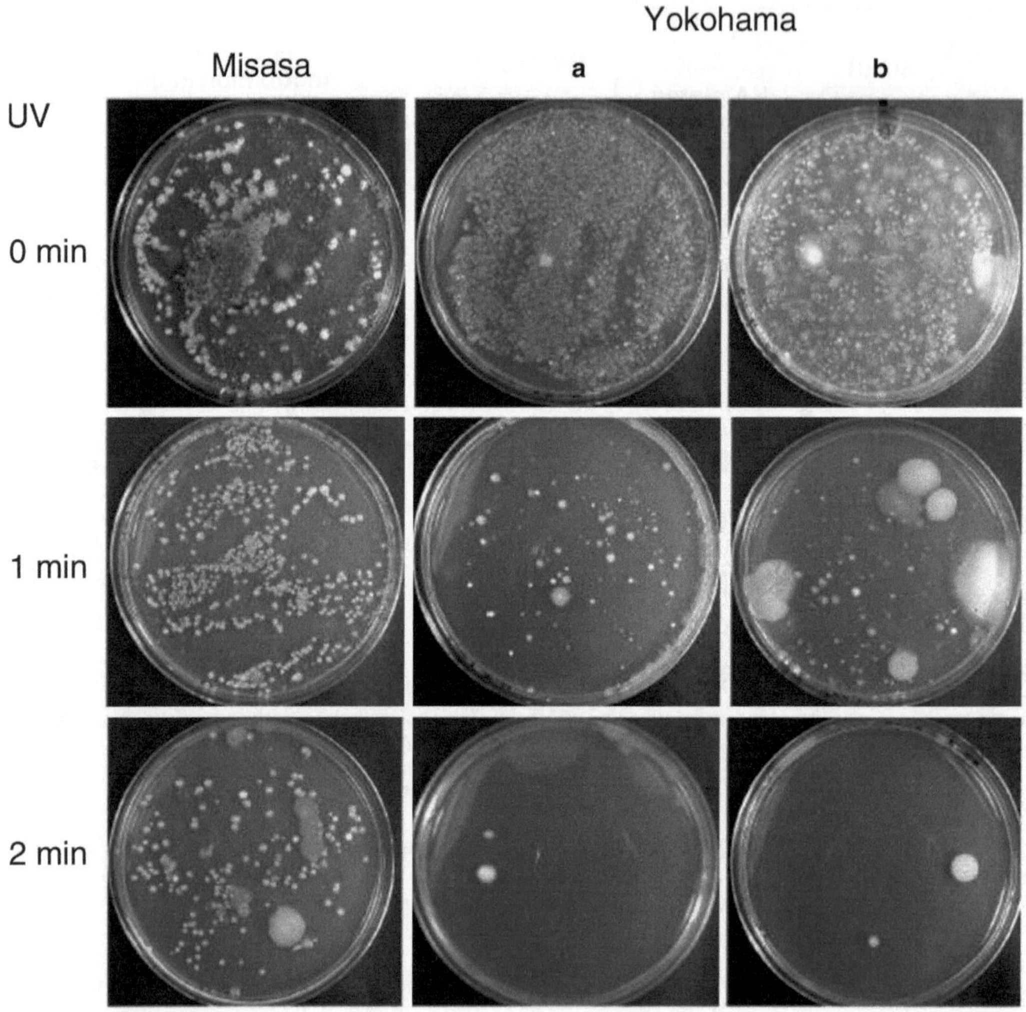

Fig. 5. A representative result of colony isolation from the water samples collected at Misasa. Each sample was diluted 1:10, and 50 μL of each dilution was inoculated onto LB agar plates and incubated at 30°C for 3 days. Prior to incubation, the plates were exposed to UV for the indicated periods. Typical results of the same treatment using samples collected from normal environments: (**a**) Miya River, Yokohama, Kanagawa. (**b**) O-oka River, Yokohama, Kanagawa. Adapted from Asker et al. 2007 (26).

1.6. Identification of Microbial Carotenoids

Carotenoids are lipophilic compounds (i.e., insoluble in water while soluble in nonpolar solvents) that are located within the cell attached to cellular membranes. The extraction of carotenoids from microbial cells involves two major steps: cells' disruption (i.e., break down or dissolve cell walls) followed by extraction by organic solvents. Disruption of the microbial cells is traditionally done by mechanical, biological, and chemical methods. Mechanical disruption can be achieved by high-pressure homogenization or grinding with glass beads (or silica gel). These methods are, however, unfavorable because they are slow, require large volumes of cells, have low productivity of carotenoids, and require high energy (50, 51). Enzymatic method has been widely used for microbial disruption (51–54); however, it is also time consuming to break the cells, and may simultaneously result in serious degradation of carotenoids. On the other hand, chemical extraction is a simple, fast, and low-cost method (55–57). Nevertheless, it is important to avoid the use of alkaline chemicals as they induce irreversible conversion of astaxanthin to astacene (58). Other alternative methods for microbial cells' disruption including freezing and thawing, ultrasonication, and freeze-drying are also slow and may result in serious degradation of carotenoids. The second step is to extract carotenoids from the microbial cell debris by successive repeated extraction in organic solvents, such as acetone, petroleum ether, hexane, chloroform, ethanol, methanol, and dimethyl sulfoxide (DMSO). While a single organic solvent can be used, the use of mixtures of solvents is advantageous as it yields high carotenoids recovery (59, 60). The simultaneous use of mixtures of some solvents demonstrated synergistic interactions, which resulted in enhanced carotenoids yield (61). A relatively rapid method for extracting astaxanthin from *P. rhodozyma* was developed by Sedmak et al. (62) who used hot DMSO to extract carotenoids and then transferred the dissolved carotenoids into chloroform. However, the later step complicates the process, especially when a large number of samples are extracted. A modification of this method was used by Asker and Isaka (63) to extract carotenoids from the Gram-negative bacteria *Brevundimonas* spp. (63) (Fig. 6). In the modified method, the

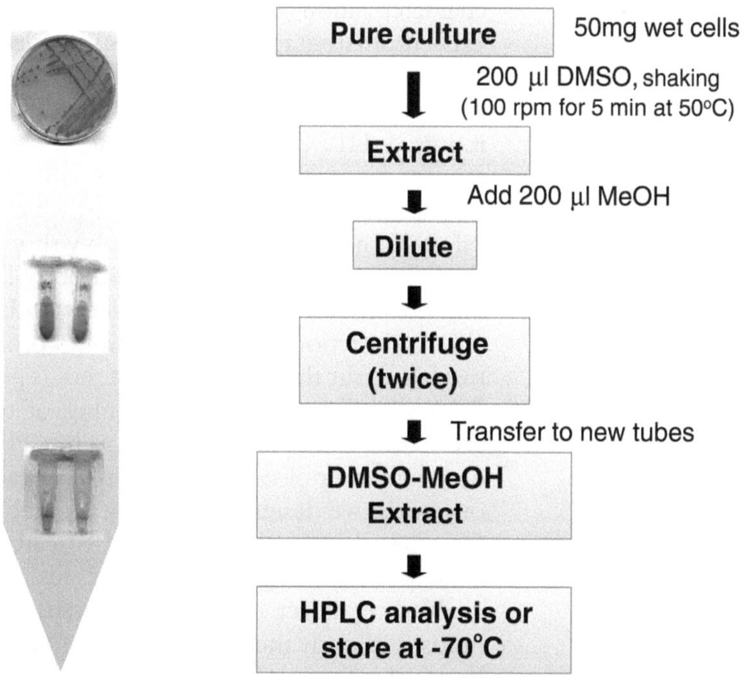

Fig. 6. Outline of a simple and rapid method for bacterial carotenoid extraction.

cells (approximately 50 mg) are suspended in 200 μL DMSO and agitated at 50°C by rotary shaking in the dark (5–10 min) to initiate lysis of the cells. An equal volume of methanol is then mixed with the lysed cells and centrifuged to obtain the carotenoid extract. This method is simple, rapid, and straightforward as it does not involve any cell disruption or carotenoids' concentration after extraction, and excludes the step of carotenoids' transfer into chloroform. In addition, it requires a very small amount of the microbial cells for the identification of carotenoids. Importantly, this method enables the extraction of carotenoids from both Gram-negative and Gram-positive bacteria, as well as a wide variety of microorganisms including yeast, fungi, and microalgae (unpublished data). Accordingly, the modified method is ideal for the high-throughput screening of carotenoid-producing microorganisms and their diversity.

Screening for interested carotenoid(s) from microbial sources can be a difficult task possibly due to (1) the presence of complex mixture of carotenoids whose structures are slightly different, (2) low content of the targeted carotenoids, (3) large number of samples, and (4) instability of carotenoids to light, oxygen, and heat. Conventional chromatographic methods (e.g., thin-layer chromatography (TLC) and column chromatography) can be used for separation and identification of carotenoids as color permits visual monitoring of the separation of carotenoids. However, these methods have many disadvantages, such as requirement of large sample volumes, low resolution, time consuming, carotenoids' degradation, and low recovery of carotenoids that are needed for further characterization (e.g., spectra analysis and molecular weight) (64). Gas chromatography (GC) is unsuitable because of low volatility and thermolability of carotenoids (64). Therefore, a suitable high-performance separation followed by a sensitive detection is required. HPLC has been utilized for the routine analysis of carotenoids in complex mixtures (64–68). It is characterized by its reproducibility and high sensitivity, which provide reliable analytical data. In addition, the reasonably short analysis time minimizes the isomerization and decomposition of labile compounds such as carotenoids. HPLC can be coupled with UV/Vis photodiode array (PDA) detection. The HPLC elution profile provides useful information about the classes of carotenoids present in the sample. The PDA allows for online determining of the UV/Vis spectrum of each component. In addition, using HPLC/MS system allows determination of molecular weight of each component. To meet our goals, we designed a rapid and reliable set of HPLC methods for separation and identification of carotenoids that may have a wide polarity range. These methods (see below) have greatly enabled to detect high-value carotenoids, discover novel carotenoids, and study the diversity of carotenoids produced by unknown bacterial isolates.

In our screening approach, we use three HPLC methods (A, B, and C) (see Subheading 3) (Figs. 7–9) for the separation and identification of carotenoids from our DMSO/MeOH library extracts. HPLC method A is primarily used to examine the presence of any kind of carotenoid (i.e., polar and apolar) in a short time (Fig. 7a). In method A, a short reverse-phase (RP-C18) HPLC column and moderate polarity of mobile phase (acetonitrile/methanol/tetrahydrofurane) is used to separate high polar carotenoids (e.g., astaxanthin and zeaxanthin), as well as apolar carotenoids (e.g., β-carotene) within less than 10 min (Fig. 7a). This system is coupled with UV/Vis PDA detector. The HPLC elution profile provides useful information about the classes of carotenoids present in the sample (Fig. 7a). The PDA allows for online determining of the UV/Vis spectrum of each component (Fig. 7b1–5). Although method A is very useful in a rapid screening approach, its drawback is the overlapping of slightly similar polar carotenoids such as astaxanthin and zeaxanthin (Fig. 7a). To solve this problem, we use a second method (method B) to separate and identify the overlapping polar carotenoids (Fig. 7b). In method B, using the same column and a high polar solvent such as methanol (90%),

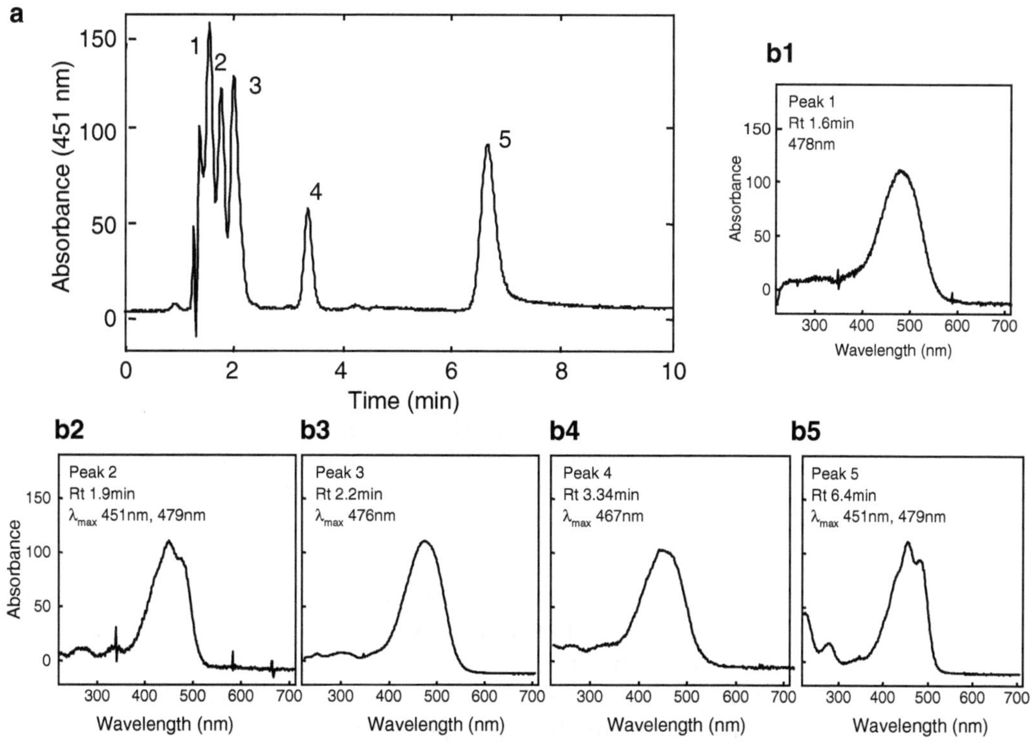

Fig. 7. HPLC method A for carotenoids' analysis. Elution profile (λ_{max} 451 nm) of carotenoids standards, including astaxanthin, zeaxanthin, canthaxanthin, echinenone, and β-carotene (**a**), and absorption spectrum of each peak (**b**) are shown. The carotenoid fraction was analyzed by reverse-phase liquid chromatography using method A.

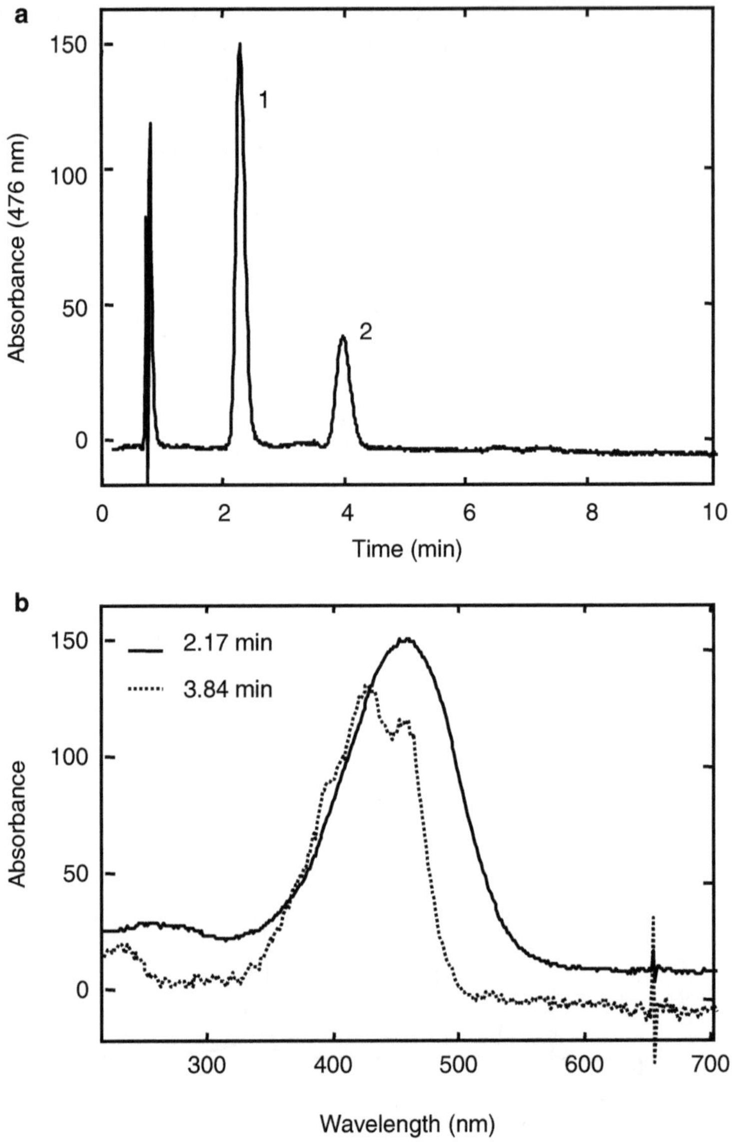

Fig. 8. HPLC method B for carotenoids' analysis. Elution profile (λ_{max} 476 nm) of carotenoids standards astaxanthin and zeaxanthin (**a**) and absorption spectrum of each peak (**b**) are shown. The carotenoid fraction was analyzed by reverse-phase liquid chromatography using method B.

it is possible to separate between overlapping polar carotenoids such as zeaxanthin and astaxanthin in a short time (1–3 min) (Fig. 8a). The same UV/Vis PDA detection system is also used in method B (Fig. 8b). To determine the molecular weight of the carotenoids, method C is performed on an LC/MS system (see Subheadings 3 and 4) (Fig. 9a, b). In method C, a long column provides more separation between the carotenoids species, while MS allows the determination of molecular weight of individual

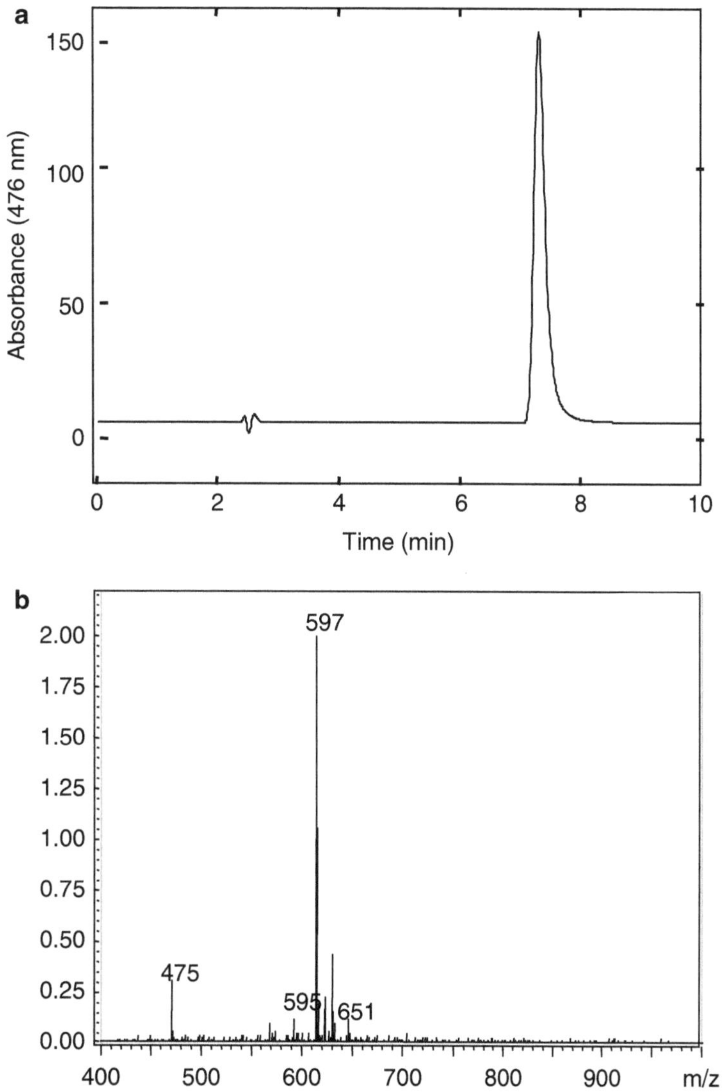

Fig. 9. HPLC method C for carotenoids' analysis. Elution profile (λ_{max} 476 nm) of carotenoids standards astaxanthin (**a**) and its mass fragmentation pattern (**b**) are shown. The carotenoid fraction was analyzed by reverse-phase liquid chromatography using method C.

carotenoids. Therefore, carotenoids can be identified based on retention time (R_t), absorption spectrum, and molecular weight. Commercial β-carotene, astaxanthin, zeaxanthin, canthaxanthin, and echinenone (see Subheading 3) are used as standards.

We studied the spectral characteristics of the pigments isolated from the cells of the isolates. Based on their solubility in organic solvent and spectroscopic properties, it was clear that all the pigments were carotenoids. Table 2 shows the spectroscopic properties of the carotenoids isolated from the representative strains of each species. Based on the absorption and mass spectral data, we successfully identified the major carotenoids produced by the nine

Table 2
Spectroscopic and HPLC characteristics of the analyzed carotenoids

Carotenoids	Absorption (λ_{max}, nm)	HPLC, retention time (R_t, min)[a]			Mol. wt (M^++1)
		Method A	Method B	Method C	
Zeaxanthin	451, 476	0.98	3.85	10.3	569
Nostoxanthin	449, 475	0.89	2.29	6.5	601
Astaxanthin	476	0.81	2.31	7.05	597
Hydroxyastaxanthin	476	0.79	1.79	5.4	613
Dihydroxyastaxanthin	476	0.65	1.34	4.21	629
Canthaxanthin	476	2.04	7.49	20.3	565

[a](See Note 6)

strains that belonged to α-*Proteobacteria*, i.e., astaxanthin and canthaxanthin (TDMA-8, TDMA-9, and TDMA-10); nostoxanthin and zeaxanthin (TDMA-12, TDMA-14, TDMA-15, and TDMA-16); and astaxanthin, hydroxyastaxanthin, and dihydroxyastaxanthin (TDMA-7 and TDMA-17). The major carotenoid produced by TDMA-2 (*Flavobacteria*) and TDMA-5 (*Sphingobacteria*) was zeaxanthin. The strains with noteworthy production efficiency were TDMA-5 (0.8 mg/g dry cells of zeaxanthin), TDMA-7 (1.4 mg/g dry cells of astaxanthin), TDMA-8 (1.1 mg/g dry cells of canthaxanthin), TDMA-16 (1.7 mg/g dry cells of a mixture of zeaxanthin, nostoxanthin, and an unknown carotenoid), and TDMA-17 (2.8 mg/g dry cells of a mixture of astaxanthin, dihydroxyastaxanthin, and an unknown carotenoid). On the other hand, we could not identify the molecular species produced by the other isolates, which showed spectroscopic characteristics different from those of the known carotenoids (1). It is noteworthy that the isolates belonging to *Deinococci* produced a common carotenoid species that was characterized by its absorption at approximately 476 and 510 nm.

1.7. Identification of Carotenoid-Producing Isolates

For interested carotenoid producers, bacterial identification is carried out to determine whether isolates belong to approved taxa (genus or species) or unknown. Traditionally, identification of unknown bacterial isolates entails numerous morphological, biochemical, cultural, antigenic, and other phenotypic characteristics, which are materials, time, and effort consuming. For rapid identification of a large number of samples, these methods are thus inappropriate. In contrast, the use of molecular methods based on PCR and DNA sequencing can shorten the identification time. Among these methods, 16S rRNA gene sequencing is used for identification at both the genus and species levels, and to study the relatedness of microbial taxa by phylogenetic relationships (69–72). Recently, high-throughput sequencing and screening methods have been developed to study microbial diversity and biotechnological potential of various microbial communities (73).

In our screening approach, a conventional 16S rRNA gene-based analysis is used to identify the isolates that produce carotenoids (Fig. 3). After obtaining the sequencing results, 16S rRNA gene sequences are correlated with those in the GenBank/EMBL/DDBJ nucleotide sequence databases by using the BLASTN program (http://www.ncbi.nlm.nih.gov/BLAST/). The strains that share 99% or higher 16S rRNA gene sequence similarity can be grouped into a single species. If the determined sequence of an unknown isolate exhibits high similarity (>97% and ≤100%) with the reference sequence of an approved species, then the unknown isolate may be assigned to this species. When the similarity is <97% and ≥95%, the unknown isolate is assigned to the corresponding genus with the possibility of claiming a novel species. When the similarity is <95%, the unknown isolate is assigned to a family with possibility of claiming a novel genus. The next step is to construct a phylogenetic tree based on the complete sequences of 16S rRNA gene for all pigmented isolates. The major physiological properties, enzymatic activity, and γ-ray resistance are also important characteristics when describing novel isolates.

We isolated 104 potential carotenoid producers and studied their phylogenetic affiliation by using a conventional 16S rRNA gene-based study. By grouping the strains that shared 99% or higher 16S rRNA gene sequence similarity to each other into a single species, the isolates were classified into 38 species, which belonged to 7 different bacterial classes (*Flavobacteria*, *Sphingobacteria*, α-*Proteobacteria*, γ-*Proteobacteria*, *Deinococci*, *Actinobacteria*, and *Bacilli*). Table 3 summarizes the taxonomic affiliations as well as the isolation conditions and colony appearances of potential carotenoid producers. Table 4 summarizes several physiological characteristics of the representative strains. Most isolates were obtained on Nutrient agar, but LB medium was effective for the isolation of some strains, possibly due to its relatively high salt concentration (0.5% NaCl). Strains TDMA-32 and TDMA-33 grew on both Nutrient agar and LB medium, but they produced yellow pigments only on LB medium.

The screening culture was carried out at various temperatures between 30 and 60°C. Most isolates were mesophilic, except for three moderately thermophilic strains (TDMA-21, TDMA-32, and TDMA-33). The most frequently isolated species were *Brevundimonas nasdae* (corresponding to TDMA-7) (20 isolates) and *Deinococcus grandis* (TDMA-20) (12 isolates). The group that belonged to or was related to the genera *Chryseobacterium*, *Paracoccus*, *Sphingomonas*, and *Deinococcus* exhibited diversity to some extent (Table 3).

Table 3
Phylogenetic characteristics of the isolates

Representative strain	Total no.	Isolation conditions: Media[a]	Isolation conditions: Temperature (°C)	Colony appearance[b]	Closest identified relative (accession number)	Similarity (%)	Closest GenBank relative (accession number)	Similarity (%)
Flavobacteria								
TDMA-1	1	N	30	Ys	*Chryseobacterium formosense* (AY315443)	97	*Chryseobacterium stagni* (DQ314741)	99
TDMA-2	1	N	30	Yg	*Cloacibacterium normanense* (AJ575430)	99	*Cloacibacterium normanense* (AJ575430)	99
TDMA-3	1	N	30	Oc/L	*Chryseobacterium halotolerans* (AY883416)	98	*Chryseobacterium taichungense* (DQ363429)	100
TDMA-39	1	N	30	Y/L	*Chryseobacterium taichungense* (AJ843132)	96	*Chryseobacterium indologenes* (AY050493)	98
TDMA-34	1	N	30	Y/S	*Terrimonas ferruginea* (AM230484)	93	Uncultured bacterium clone (AF502210)	97
Sphingobacteria								
TDMA-4	1	N	30	Oc	*Flectobacillus major* (M62787)	98	*Flectobacillus* sp. (DQ145723)	99
TDMA-5	2	L	30	Ysg/S	*Pedobacter piscium* (AJ438174)	92	Fenpropathrin-degrading bacterium (DQ118958)	97
α-Proteobacteria								
TDMA-6	3	N, L	30	O/S	*Brevundimonas aurantiaca* (AJ227787)	100	*Brevundimonas* sp. (AJ227796)	100
TDMA-7	20	L	30	O/S	*Brevundimonas nasdae* (AB071954)	99	*Brevundimonas* sp. (AJ244704)	100
TDMA-8	2	N, L	30	O/S	*Paracoccus solventivorans* (AY014175)	96	*Paracoccus* sp. (DQ108402)	98

TDMA-9	3	N	30	O/S	*Paracoccus zeaxanthinifaciens* (AF461158)	98	*Paracoccus zeaxanthinifaciens* (AF461158)	98
TDMA-10	3	L	30	O/S	*Paracoccus denitrificans* (Y16930)	96	*Paracoccus* sp. (AB025192)	99
TDMA-36	1	N	30	R/S	*Porphyrobacter tepidarius* (AF465839)	99	*Porphyrobacter tepidarius* (AY048657)	99
TDMA-37	1	N	30	R/S	*Porphyrobacter cryptus* (AF465834)	99	Alpha proteobacterium (AJ505839)	99
TDMA-14	3	N	30	Y/S	*Novosphingobium aromaticivorans* (AB025012)	98	Uncultured bacterium (AY328758)	98
TDMA-48	2	N	37	Y	*Novosphingobium capsulatum* (D16147)	98	*Sphingomonas* sp. (AJ001053)	98
TDMA-15	1	N	30	Yd/L	*Sphingomonas azotifigens* (AB217473)	99	*Sphingomonas azotifigens* (AB217473)	99
TDMA-16	3	N	40	Yd/S	*Sphingomonas mali* (Y09638)	95	Uncultured soil bacterium (AF423289)	97
TDMA-17	4	N	40	R,S	*Sphingomonas aerolata* (AJ429240)	94	*Kaistobacter koreensis* (AY769084)	99
TDMA-12	1	N	30	Yc/S	*Sphingomonas melonis* (AB055863)	99	*Sphingomonas* sp. (AF131295)	100
γ-Proteobacteria								
TDMA-47	1	N	30	Ys/S	*Pseudoxanthomonas suwonensis* (AY927994)	95	Gram-negative bacterium (AY864080)	95
TDMA-19	3	N	30	O	*Shewanella putrefaciens* (AJ000213)	99	*Shewanella* sp. (DQ307734)	99
TDMA-38	2	N	30	P-R/L	*Serratia marcescens* (AJ233431)	99	Uncultured γ-proteobacteria (AF529354)	100
Deinococci								
TDMA-20	12	N	30,40	P-Rc	*Deinococcus grandis* (Y11329)	98	*Deinococcus* sp. (DQ003317)	99
TDMA-21	2	N	55	Rc	*Deinococcus murrayi* (Y13041)	97	*Deinococcus murrayi* (Y13043)	98

(continued)

Table 3 (continued)

Representative strain	Total no.	Isolation conditions		Colony appearance[b]	Closest identified relative (accession number)	Similarity (%)	Closest GenBank relative (accession number)	Similarity (%)
		Media[a]	Temperature (°C)					
TDMA-24	4	N	30,37	R	*Deinococcus indicus* (AJ549111)	96	*Deinococcus* sp. (AB022911)	98
TDMA-25	1	N	45	Rc	*Deinococcus maricopensis* (AY743276)	91	*Deinococcus maricopensis* (AY743276.1|)	91
TDMA-uv51	1	N	40	LP/S	*Deinococcus murrayi* (Y13043)	90	Uncultured *Deinococcus* sp. (AY569298)	91
TDMA-uv53	1	N	40	LP/S	*Deinococcus deserti* (AY876379)	94	*Deinococcus* sp. (AB244334)	99
Actinobacteria								
TDMA-13	1	N	30	Yc/S	*Aureobacterium testaceum* (X77445)	99	Glacial ice bacterium (AF479366)	99
TDMA-26	1	N	30	Yc/S	*Curtobacterium citreum* (X77436)	99	*Curtobacterium flaccumfaciens* pv. Basellae (AY273210)	99
TDMA-41	1	N/L	30, 40	Pc,S	*Dietzia natronolimnaea* (X92157)	99	*Dietzia natronolimnaea* (AJ717373)	99
TDMA-59	1	N/L	30	O,c	*Dietzia maris* (X79291)	99	*Dietzia sp. CR1-3* (AY205297)	100
TDMA-42	4	N, L	30	Ow	*Kocuria erythromyxa* (Y11330)	99	*Kocuria* sp. (DQ060377)	100

Bacilli								
TDMA-27	5	N, L	30, 37, 40, 45	O/L	*Exiguobacterium aestuarii* (AY594264)	99	*Exiguobacterium* sp. (AY205564)	100
TDMA-33	1	N, L	55	LY/L	*Bacillus flavothermus* (AF004589)	98	*Bacillus* sp. YMY1010 (AY198416)	99
TDMA-14	6	L	30	Y-O/L	*Bacillus aquaemaris* (AF483625)	98	*Bacillus* sp. (DQ084469)	99
TDMA-32	2	N, L	60	Y/L	*Anoxybacillus kamchatkensis* (AF510985)	99	*Anoxybacillus kualawohkensis* (DQ401072)	99

[a] *N* Nutrient agar, *L* LB agar
[b] *Y* yellow, *O* orange, *P* pink, *R* red, *s* shiny, *g* gummy, *c* creamy, *w* wet, *d* dry, *L* large, *S* small
Adapted from Asker et al. 2007 (26)

Table 4
Physiological characteristics of the isolates

Strain name	Growth temperature (°C)	Anaerobic growth	β-Galacto-sidase	Amy lase	Lipase	Caseinase	Gelatinase	DNase	Carotenoid spectrum (λ_{max}, nm)
Flavobacteria									
TDMA-1	18–40	–	–	+	+	+	–	+	446, 469
TDMA-2	18–45	+	–	+	–	+	–	–	450, 475
TDMA-3	18–45	+	+	+	+	+	+	–	449
TDMA-39	18–40	–	–	+	–	+	+	–	450
TDMA-34	18–40	–	–	+	–	–	–	–	448, 466
Sphingobacteria									
TDMA-4	18–40	+	–	–	+	+	+	+	442
TDMA-5	18–40	–	+	+	+	+	+	+	451, 472
α-*Proteobacteria*									
TDMA-6	18–40	–	–	+	+	+	+	–	476
TDMA-7	18–40	–	–	+	+	+	+	–	476
TDMA-8	20–40	–	–	+	–	+	–	–	470
TDMA-9	18–40	–	–	+	–	–	+	–	472
TDMA-10	18–40	–	–	+	–	–	–	+	471
TDMA-14	18–45	–	–	–	+	–	–	–	450, 476
TDMA-48	18–40	–	+	+	+	–	–	–	451, 477
TDMA-15	18–45	+	+	+	+	+	+	–	451, 477
TDMA-16	22–40	–	–	–	+	+	–	–	451, 477
TDMA-17	22–45	–	–	+	+	+	+	–	476
TDMA-12	18–40	–	–	+	+	+	–	–	450, 476
γ-*Proteobacteria*									
TDMA-47	20–45	–	–	+	+	+	+	+	418, 443, 462
Deinococci									
TDMA-20	18–40	–	+	+	+	+	+	+	476, 510
TDMA-21	20–55	–	+	+	+	+	+	+	476, 510

TDMA-24	18–40	–	+	+	+	+	+	+	477, 505
TDMA-25	20–45	–	+	+	+	+	+	–	479, 510
TDMA-uv51	18–40	–	+	+	+	+	+	–	479, 510
TDMA-uv53	20–45	–	+	+	+	+	+	–	465, 490
Actinobacteria									
TDMA-13	18–40	–	–	+	–	+	+	+	450, 476
TDMA-26	18–40	–	+	+	+	+	+	–	451, 478
TDMA-41	16–40	–	–	–	+	+	+	–	476
TDMA-59	22–37	–	–	–	–	–	–	–	476
TDMA-42	20–40	–	–	+	+	+	+	–	468, 498
Bacilli									
TDMA-27	20–45	–	+	+	–	+	+	+	453, 480
TDMA-33	40–60	–	–	+	+	–	–	–	429, 452, 482
TDMA-14	20–40	–	–	+	+	+	–	+	468, 498
TDMA-32	40–60	–	–	+	+	–	–	–	429, 452, 482

Adapted from Asker et al., 2007 (26)

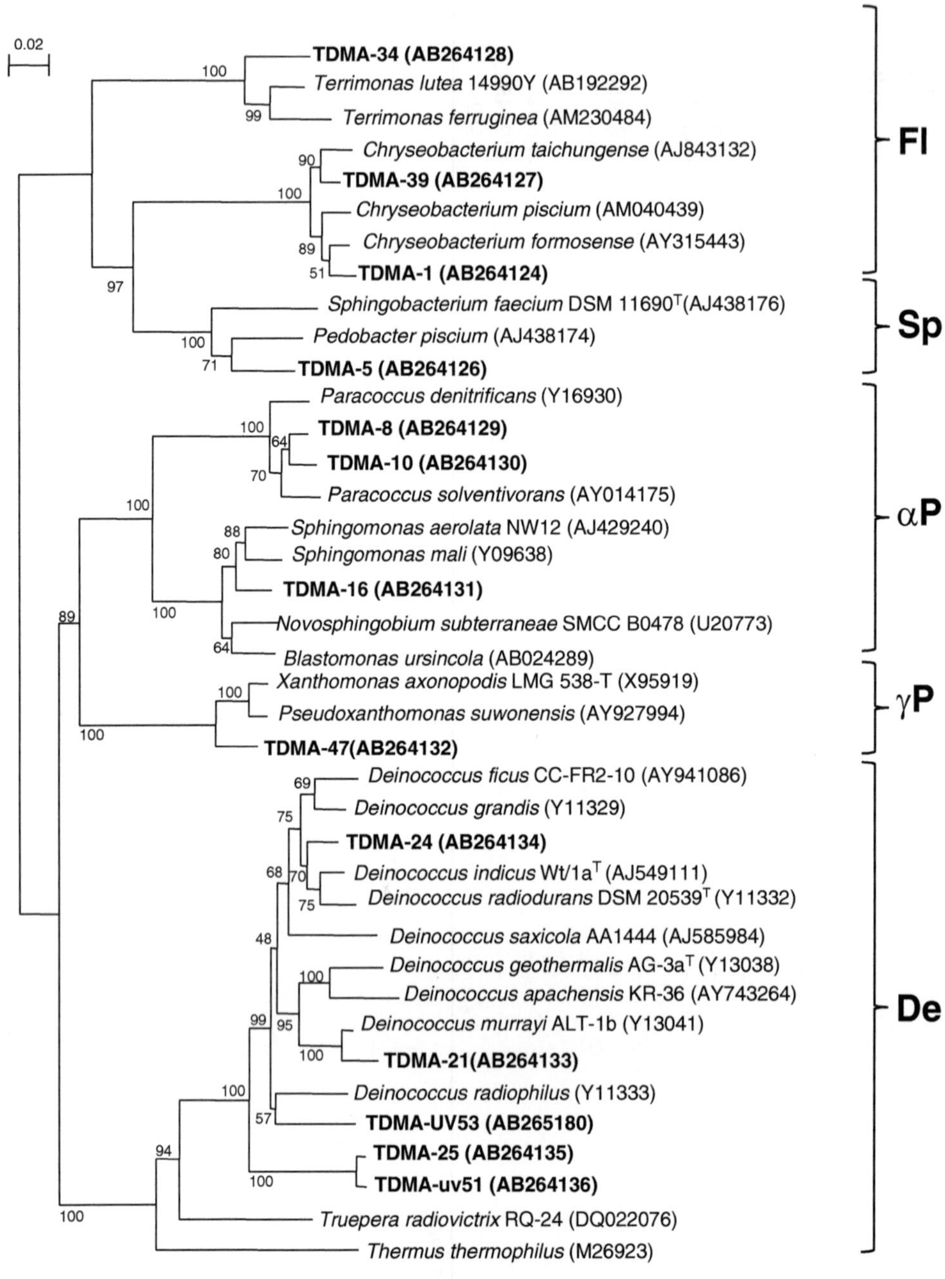

Fig. 10. The 16S rRNA gene-based phylogeny of the taxonomically new carotenoid producers isolated in this study. The tree was constructed using the maximum-likelihood method with ClustalW (131). The bootstrap value expressed as a percentage of 1,000 replicates is given at each node; values above 50% are shown. Nucleotide sequence accession numbers are indicated in parentheses. Fl, *Flavobacteria*; Sp, *Sphingobacteria*; αP, α-*Proteobacteria*; γP, γ-*Proteobacteria*; De, *Deinococci*. Adapted from Asker et al. 2007 (26). Recently, strains TDMA-16 and -17 have been described as new *Sphingomonas* spp. (85, 86), and strain TDMA-5 has been described as a new genus and new species of the family *Sphingobacteriaceae* (87). In addition, strains TDMA-21, -24, -25, -54, -uv53, and -uv51 have been described as novel *Deinococcus* spp. (82–84).

We found 14 species that exhibited low 16S rRNA gene similarities (<97%) with their closest identified relatives. As shown in the phylogenetic tree presented (Fig. 10), these isolates exhibited a distinct phylogenetic location from that of the closest identified relatives. These results, thus, indicated that the isolates are new in terms of both taxonomy and carotenoid production. In addition to the above 14 species, we also found 9 species that belong to new genera in terms of carotenoid production: *Chryseobacterium* (TDMA-1, TDMA-2, TDMA-3, and TDMA-39), *Flectobacillus* (TDMA-4), *Novosphingobium* (TDMA-14 and TDMA-48), *Curtobacterium* (TDMA-26), and *Kocuria* (TDMA-42). Other isolates belonged to genera or specific species that are known for their ability to produce carotenoids, including *Brevundimonas* (74), *Paracoccus* (formerly named *Agrobacterium aurantiacum* and *Flavobacterium* sp. strain R-1512) (75, 76), *Porphyrobacter* (29), *Sphingomonas* (77), *Aureobacterium* (78), and *Dietzia* (79). Currently, *D. radiodurans* (80) is the only species in the genus *Deinococcus* that has been described for its ability to produce carotenoids; however, our isolates comprised several different species from this genus. Similarly, in *Bacillus* and related genera, *Bacillus flavothermus* (81) is the only species known for carotenoid production, but our isolates comprised several other species. The detailed phylogenetic and physiological characteristics of each new taxon were described elsewhere (82–87).

1.8. Irradiation Resistance of Carotenoid Producers

Some of the above isolates were studied for their irradiation resistance to γ-rays and UV rays (Table 5). The five isolates belonging to *Deinococci* (TDMA-20, TDMA-24, TDMA-uv53, TDMA-25, and TDMA-uv51) showed marked resistance; the ratios of their survival after exposure to γ-rays and UV light were higher than those exhibited by the type strain of *D. radiodurans* (NBRC 15346;

Table 5
γ-Ray and UV resistance of the isolates

	Survival (%)	
Strain	**γ-Ray**	**UV**
D. radiodurans	76	50
TDMA-20	100	88
TDMA-24	100	83
TDMA-uv53	95	78
TDMA-25	85	64
TDMA-uv51	82	61
TDMA-17	2.1	1.6
TDMA-16	0.45	0.23
TDMA-8	0.02	0.02

Adapted from Asker et al., 2007 (26)

positive control). The other irradiation-resistant isolates were the three strains belonging to α-*Proteobacteria* (TDMA-17, TDMA-16, and TDMA-8) whose survival ratios were very low (0.02–2%), yet distinct from those (<0.01%) of other strains including TDMA-7, TDMA-41, TDMA-59, and *E. coli* C600 (negative control).

Further, to characterize the correlation between irradiation resistance and carotenogenesis, we treated the water samples with UV light prior to cultivation and observed its effect on the phenotypes of the microbial colonies that appeared after incubation. Results showed that the frequency of occurrence of UV-resistant isolates from the samples collected at Misasa was markedly high (a representative result is shown in Fig. 5), and that most of the UV-resistant colonies exhibited yellow, orange, pink, or red color. In contrast, UV treatment markedly reduced the microbial population that was cultured from river samples collected from normal environments (representative results are shown in Fig. 5). In these cases, however, the small number of colonies that were obtained after UV treatment contained high ratios of yellow or orange pigments.

We also screened for potential carotenoid producers among the UV-resistant isolates derived from the Misasa samples. Based on the 16S rRNA gene sequence similarity, the 44 resultant UV-resistant isolates were classified into 9 species, which included 7 species (corresponding to TDMA-7, TDMA-14, TDMA-16, TDMA-20, TDMA-24, TDMA-27, and TDMA-42) that were characterized in the above study and 2 species that occupy new phylogenetic positions in *Deinococci* (TDMA-uv51 and TDMA-uv53; see Table 3 and Fig. 10).

2. Discussion

In this study, we discovered that carotenoid-producing bacteria frequently occur in water samples collected at Misasa, a radioactive region in Japan. In addition, these microorganisms showed marked taxonomic diversity content. When we similarly screened for carotenoid producers from fresh (plain) water samples collected from normal environments, the phylogenetic diversity was relatively poor and a large proportion of the isolates comprised *Brevundimonas* and *Erythrobacter*, which are well-known genera in terms of carotenoid production (our unpublished observations). On the other hand, the strains isolated in this study belonged to various taxa in seven bacterial classes. Furthermore, they contained a high proportion of taxonomically new strains (14 strains/104 isolates) of which strains TDMA-16 and -17 have been described as new *Sphingomonas* spp. (85, 86) and strain TDMA-5 has been described as a new genus and new species of the family *Sphingobacteriaceae* (87), while strains TDMA-21, -24, -25, -54, -uv53, and -uv51 have been described as novel *Deinococcus* spp. (82–84). While there is currently no clear explanation for the occurrence of this unique diversity, it may be due to the following reasons: (1) the microbial population in this region has not been known just because it has been left uninvestigated and/or (2) the total number of microbial cells contained in this region is relatively low due to the high radioactivity, so we could access minor population without examining large number of dominating strains. Alternatively, we may think that the evolution under the highly mutative environment has generated the marked phylogenetic diversity in the microorganisms living there.

Irradiation resistance has been recognized as a feature of several bacteria, including strains of genera such as *Deinococcus* (44), *Rubrobacter* (88), *Acinetobacter* (89), *Chroococcidiopsis* (90), *Hymenobacter* (91), *Kineococcus* (92), *Methylobacterium* (93), and *Halobacterium* (94). The unusual resistance of *D. radiodurans* to ionizing radiation, UV light, hydrogen peroxide, and numerous other agents that damage DNA has been attributed to its effective DNA repair capacity (44, 95). However, certain lines of evidence have suggested that this resistance also involves the functions of its antioxidant defense apparatus including enzymes, such as superoxide dismutase (96), peroxidase, and catalase (97), and nonenzymatic systems, such as vitamins A and E and carotenoids (42). Saito et al. (98) suggested that the radioresistance of *Halobacterium salinarium* is partially due to carotenoids which act as effective in vitro antioxidants. It is known that *D. radiodurans* and *H. salinarium* mutants that are defective in carotenoid production are more sensitive to ionizing radiation and hydrogen peroxide than their parental strains (42, 94).

Some extremely radioresistant bacteria are known to contain a red pigment (99); for example, *D. radiodurans* and *Rubrobacter radiotolerans* exhibit the characteristic reddish color because they contain deinoxanthin and bacterioruberin, respectively (80, 100). *H. salinarium* also produces bacterioruberin (94, 101). We also observed that the radiotolerant isolates obtained in this study (Table 5) exhibited a very apparent red color. Among them, the slightly radiotolerant strain TDMA-17 was unique in terms of astaxanthin production; this is the first organism from the family *Sphingomonadaceae* that produces this kind of carotenoid (85), maybe useful in industrial applications due to its strong antioxidant activity.

Overall, the evidence indicates a clear correlation between irradiation resistance and carotenoid production. Accordingly, irradiation with γ-rays or UV rays can be an effective procedure in the isolation of carotenoid producers. We also suppose that the strains isolated from this radioactive region exhibit diversity not only in taxonomy but also in carotenoid production. Interestingly, all the isolates belonging to *B. nasdae* (corresponding to TDMA-7) were characterized by their orange color and high carotenoid production (1.1–1.3 mg/g dry cells) despite the fact that the type strain of *B. nasdae* does not produce carotenoids (102). It appears worthwhile to continue exploring new microbes and compounds derived from the radioactive region.

3. Materials

3.1. Water Samples' Collection

1. 15-mL sterilized (disposable) polystyrene tubes with cap.
2. Rack.
3. Markers for labeling.
4. Refrigerated storage (4°C).

3.2. Primary Screening and Microbial Library Construction

1. Analytical balance.
2. 15-mL culture tubes.
3. UV lamp (GL-15, 53 μW/cm^2) (Toshiba, Tokyo, Japan).
4. Micropipettes and sterile tips (0–20, 20–200, and 100–1,000μL).
5. Vortexer.
6. Refrigerated microcentrifuge (13,500 and 14,500×g max speed, rotor holds 1.5-mL tubes).
7. Incubators (28, 30, 37, 40, and 50°C).
8. Sterile loops, sterile toothpicks, and sterile cotton swabs.
9. Clean bench equipped with UV light.
10. Nutrient agar: 3 g/L beef extract, 5 g/L peptone, and 15 g/L agar (Difco Laboratories, Detroit, MI, USA).
11. Luria-Bertani (LB) solid medium: 10 g/L tryptone, 5 g/L yeast extract, 5 g/L NaCl, and 15 g/L agar.
12. Peptone water (Difco Laboratories, Detroit, MI, USA).
13. Dilution tubes with sterile peptone water (9 mL) in each tube.

3.3. Strains' Preservation and Library Construction

1. Microbank kit (Pro-Lab Diagnostics, Toronto, Canada).
2. Glycerol 20%.
3. Freezer −70°C.

3.4. Carotenoids' Analysis

1. HPLC-grade solvents.
2. β-Carotene (Sigma, St. Louis, MO, USA).
3. Astaxanthin (Sigma, St. Louis, MO, USA).
4. Zeaxanthin (DHI; Water & Environment, Hørsholm, Denmark).
5. Canthaxanthin (DHI; Water & Environment, Hørsholm, Denmark).
6. Echinenone (DHI; Water & Environment, Hørsholm, Denmark).
7. Butylhydroxytoluene (BHT) (Sigma, St. Louis, MO, USA).
8. Sterile spatula or toothpick.

9. 1.5-mL Eppendorf tubes.
10. Racks.
11. Balance.
12. Eppendorf table top vial shaker.
13. Spectrophotometer.
14. Syringe filter (0.45-μm hydrophobic Teflon membrane).
15. Syringe 1 mL.
16. 96-well plate.
17. HPLC (Class-VP; Shimadzu, Kyoto, Japan) with PDA detector.
18. HPLC/MS system (LCMS-2010EV, Shimadzu).
19. CAPCELLPAC C18 HPLC column (35 mm×4.6 mm; particle size, 5 μm) (Shiseido, Tokyo, Japan).
20. Shim-Pack FC-ODS HPLC column (150 mm×4.6 mm; particle size, 5 μm) (Shimadzu, Kyoto, Japan).
21. Acetonitrile:methanol:tetrahydrofurane (5.8:3.5:0.7).
22. Methanol:water (9:1).

3.5. Identification of Bacterial Isolates:16S rRNA Gene Sequences

1. Sterile material, including Eppendorf tubes, toothpicks, and deionized water.
2. DNA purification kit (Edge BioSystems, Gaithersburg, MD, USA).
3. Primer B8F: 5′-AGAGTTATCCTGGCTCAG.
4. Primer B1492R: 5′-GGTTACCTTGTTACGACTT.
5. Thermocycler.
6. Taq polymerase.
7. Agarose.
8. 10× TBE buffer: 108 g Tris base, 55 g boric acid, and 40 mL 0.5 M EDTA (pH 8.0).
9. 1× TBE buffer: 100 mL of 10× TBE buffer, and 900 mL of deionized water.
10. 6×blue loading dye: 30% glycerol in distilled water, and 0.25% bromophenol blue.
11. Ethidium bromide (see Note 1).
12. Agarose gel electrophoresis system.
13. UV safety goggles and UV box.
14. PCR tubes (0.2 mL thin-walled).
15. BigDye Terminator v3.1 Cycle Sequencing Kit (Applied Biosystems, Foster City, CA, USA).
16. Pipettes and tips (0–10, 20–200, 100–1,000 μL).
17. Gene Clean kit II (Funakoshi, Tokyo, Japan).

4. Methods

4.1. Water Samples' Collection

1. Collect water samples from the surface water layer of various places in selected environments.
2. Store samples at 4°C immediately until screening.

4.2. Primary Screening and Microbial Library Construction

Screening and isolation of carotenoid-producing microorganisms from natural water samples can be done using a direct agar series dilution method. Colored colonies grown on an appropriate media agar plate can be primarily isolated by visual screening.

1. Do a series of dilution of 10^{-1}, 10^{-2}, 10^{-3}, and 10^{-4} for each water sample. Spread 0.1 mL of each dilution on LB or Nutrient agar in duplicate (see Note 2).
2. In case of UV-irradiated plates, before incubation and under sterilization condition (inside the clean bench), open the plates and treat them using a UV lamp (GL-15, 53 μW/cm^2) for 0, 1, 2, and 3 min.
3. Incubate aerobically the agar plates at various temperatures (28, 30, 37, 40, and 50°C).
4. To isolate a single colony: Prepare LB or NA agar plates, and then classify each plate into 6–8 sections using the marker. Using a sterile inoculating loop (or a sterile toothpick), touch a single-colored colony (yellow, orange, pink, or red) and then streak it into a single section in the appropriate plate (LB or NA) to produce a single colony. Incubate the inoculated plates at an appropriate temperature for 2–3 days.
5. To obtain batches of isolates (biomass): Prepare and classify LB or NA agar plates as in step 4. Transfer a single colony from each isolate (using a sterile toothpick) into a fresh single section in the appropriate plate (LB or NA) and streak it as a thick batch (see Note 3).

4.3. Strains' Preservation and Library Construction

To preserve isolated strains, use the Microbank kit or the glycerol stock method, and then store at –70°C (see Note 4).

1. To preserve isolated strains using the Microbank kit, follow the following instructions, which are also provided with the kit manual.
   (a) Use a permanent marker to label the vial as desired, one organism per vial to be inoculated.
   (b) Under aseptic conditions, inoculate the cryopreservative fluid with pure young colonial growth (18–24 h).
   (c) Close vial tightly and invert 4–5 times to emulsify organism (do not vortex).

(d) At this point, the microorganisms will be bound to the porous beads. The excess cryopreservative should be well aspirated leaving the inoculated beads as free of liquid as possible.

(e) Record the inoculation coding on the printed Microbank™ Storage Box lid or on any other suitable recording system of choice.

(f) Store the inoculated cryovial at –70°C for best long-term results.

2. To preserve isolated strains in glycerol: Aliquot 1 mL of sterilized glycerol solution (20%) into sterilized Eppendorf tubes. Inoculate the glycerol with pure young bacterial colonial growth (18–24 h). Mix thoroughly and store at –20 or –70°C.

4.4. Extraction of Carotenoids

1. Using a sterilized spatula or toothpick, transfer about 50 mg of wet microbial cells into the 1.5-mL Eppendorf tube.
2. Add 200 μL of DMSO and agitate the suspensions at 50°C by rotary shaking at 100 rpm in the dark until the cells are bleached (5–10 min) (see Note 5).
3. Add 200 μL of methanol and vortex for 30 s at 15,000 rpm.
4. Transfer the supernatants to new Eppendorf tubes and centrifuge at 15,000 × *g* for 10 min.
5. Filter the supernatants (carotenoid extracts) using a 0.45-μm hydrophobic Teflon membrane syringe filter into a 96-well plate for automated HPLC analysis.
6. Proceed with HPLC analysis or flush carotenoid extracts using N_2 gas and store at –70°C until analysis.

4.5. HPLC Method A

1. Analyze carotenoid extracts using HPLC method A to screen for all carotenoid species, ranging from polar carotenoids such as astaxanthin to apolar carotenoids such as β-carotene. You can use an HPLC with automated injection samples for samples loaded into a 96-well plate.
2. Analyze standards samples of carotenoids, such as β-carotene, astaxanthin, zeaxanthin, and canthaxanthin, under the same conditions.
3. Compare HPLC profiles of screened unknown samples with those of the standards.
4. Using PDA, compare the spectral absorption of interested HPLC peaks with those of corresponding standards.
5. Solubilize carotenoids standards in DMSO and dilute with methanol (1:1, v/v).
6. Prepare a standard curve for each carotenoids using HPLC.

7. Place all samples into the 96-well plate. Add carotenoids standards solutions in the beginning wells.
8. Using the automated injection system, inject 10 μL of each DMSO–methanol extract onto the CAPCELLPAC C18 column.
9. Elute with a mixed solvent containing acetonitrile:methanol:te trahydrofurane (5.8:3.5:0.7), at a flow rate of 1 mL/min.
10. Analyze the HPLC profiles for both R_t and PDA Spectra of each peak.
11. Compare both R_t and PDA Spectra of each interested peak with those of the carotenoids standards.
12. Select interested carotenoid extract and store at −80°C for further analysis.
13. Regularly wash HPLC column (see Note 6).

4.6. HPLC Method B

Subject interested carotenoid extract from method A to method B for screening of interested polar carotenoids such as astaxanthin, zeaxanthin, and carotenoid glucosides such as astaxanthin glucoside.

1. Analyze the standard samples of carotenoids under the same condition.
2. Compare HPLC profiles of screened samples with those of the standards.
3. Use PDA to compare the absorption spectra of interested HPLC peaks with those of the corresponding standards.
4. Apply 10 μL of interested samples onto the same column as above (column temperature 35°C).
5. Elute with a mixed solvent containing methanol:water (9:1) at a flow rate of 1.0 mL/min.
6. Analyze the HPLC profiles for both R_t and PDA Spectra of each peak.
7. Compare both R_t and PDA Spectra of each interested peak with those of the carotenoids standards.
8. Select the interested carotenoid extract for more analysis using Method C.
9. Regularly wash HPLC column (see Note 6).

4.7. HPLC Method C

Subject interested carotenoid extracts from methods A and B to method C to determine the molecular weight of interested carotenoids as follows.

1. Inject 10 μL of interested DMSO–methanol extract onto the Shim-Pack FC-ODS.
2. Elute with 95% methanol at a flow rate of 1.0 mL/min.

3. Analyze the HPLC profiles for both R_t, molecular peak, and fragment pattern of interested peak.
4. Compare R_t, molecular peak, and fragment pattern of interested peak with those of the corresponding carotenoids standards.
5. Regularly wash HPLC column (see Note 6).

4.8. 16S rRNA Gene Sequences

1. Carefully transfer 50 mg wet cells from each bacterial isolate into Eppendorf tubes.
2. Extract genomic DNA from each bacterial isolate using the bacterial genomic DNA purification kit.
3. Check the quality of extracted genomic DNAs by agarose gel electrophoresis (steps 4–9).
4. Prepare 1% gel solution by melting 1 g of agarose in 100 mL of 0.5× TBE buffer in a microwave for approximately 2 min. Allow to cool for 5 min, then add 2.5 μL ethidium bromide, and stir to mix (see Note 7).
5. Cast the gel using the supplied tray and comb. Allow the gel to set for a minimum of 30–40 min at room temperature.
6. Load into separate wells: 10 μL of 1 kb ladder, diluted samples (5 μL sample + 5 μL water + 2 μL 6× loading dye).
7. Run the gel for 30 min at 100 V.
8. Expose the gel to UV light and photograph (wear UV goggles).
9. Confirm DNA quality.
10. Set up PCR reaction mixtures in a final volume of 50 μL: 10 μL 10× PCR buffer, 1 μL PCR Nucleotide Mix, 50 U Taq DNA polymerase, 1 μL (10 μM) primer B8F, 1 μL (10 μM) primer B1500R, 2 μL DNA extract, and 33 μL deionized and sterilized water.
11. Perform PCR using a protocol that includes an initial denaturation for a period of 5 min at 94°C; 30 cycles of 94°C for 1 min, 55°C for 1 min, and 72°C for 1 min; 72°C for 3 min; and incubation at 4°C until further processing.
12. Purify PCR amplicons using agarose gel electrophoresis (steps 13–19).
13. Prepare 2% agarose gel, and let it to solidify at room temperature for 30–40 min.
14. Add 10 μL 6× blue loading dye to each PCR product (50 μL).
15. Carefully load all 60 μL into wide gel well.
16. Run the gel for 30 min at 100 V.

17. Wear UV goggles. Remove gel and place on UV box to identify region with PCR products (see Note 7).
18. Cut DNA bands from agarose gel and transfer into sterilized Eppendorf tubes.
19. Purify PCR products from the gel using the Gene Clean kit.
20. Determine the 16S rRNA gene sequences using the BigDye Terminator v3.1 cycle sequencing kit.

4.9. Phylogenetic Analysis

1. Compare the nearly complete 16S rRNA gene (about 1,396–1,450 bp) sequences of isolates to those available in GenBank/EMBL/DDBJ nucleotide sequence database using the BLASTN program (http://www.ncbi.nlm.nih.gov/BLAST/).
2. Using ClustalW, align the obtained sequences with those of closed relatives.
3. Construct a neighbor-joining phylogenetic tree (see Fig. 10).
4. Use NJ plot to draw and edit the obtained tree.

5. Notes

1. Ethidium bromide is mutagenic and moderately toxic. You must wear gloves and handle this chemical with care.
2. We collected about 40 water samples from numerous locations throughout the spa, including the water close to the hot spring well, the downstream of the well, and the river. Dilution was dependant on the nature of samples. Samples collected close to the hot spring well (source of radioactivity) did not require dilution because they were pure, low in organic matter, and free of microbial cells. In contrast, samples collected from the water stream on the river were very rich in organic matter and microbial cells, which required serial dilution plating of 10^{-1}, 10^{-2}, 10^{-3}, and 10^{-4}.
3. You can transfer 6–8 colonies on a single plate; each will yield more than 200 mg wet cells. Each section will produce enough biomass for strain preservation, carotenoid extraction, and DNA extraction.
4. You can use 15–20% glycerol stock.
5. Carotenoids are highly unstable and sensitive to light, oxygen, and temperature, which induce their oxidation. To minimize the oxidation of carotenoids, the extraction must be quickly done without exposure to light, oxygen, and high temperatures. Therefore, all the experimental procedures should be carried out in a dark room. Alternatively, dark tubes or tubes foiled with aluminum foil can be used. Nitrogen gas should be

also used for flushing the pigment extracts before fridge storage at −20°C to avoid oxidation. In addition, the use of an antioxidant such as BHT (18) can significantly reduce the oxidation of carotenoids during extraction.

6. DMSO–MeOH carotenoid extracts usually contain impurities, such as salts, lipids, fatty compounds, proteins, and other cell compounds. These contaminants are usually adsorbed on the stationary phase of the reversed-phase HPLC column, which becomes contaminated by the repeated injection of samples. This can cause the retention times to shift and tailing to occur. This may also cause the column backpressure to build up to intolerably high levels, which overpressures the pump and causes the column to settle and create a void depending upon where the blockage occurs. To prevent this problem, we clean the columns on a regular basis throughout our work. Washing the HPLC columns with 100% acetonitrile and then with 100% methanol was sufficient to clean and regenerate the column. We strongly recommend using a pre-column.
7. The PCR product will show up as a single bright glowing band of about 1.5 kb. You can also take photographs if needed.

References

1. Britton G, Liaaen-Jensen S, Pfander H (2004) Carotenoids Handbook. Birkhäuser, Basel
2. Krinsky NI (1989) Antioxidant functions of carotenoids. Free Radic Biol Med 7:617–635
3. Klaui H, Bauernfeind JC (eds) (1981) Carotenoids as food colors. Academic, New York, pp 48–317
4. Francis FJ (2000) Carotenoids as food colorants. Cereal Food World 45:198–203
5. Johnson EJ (2002) The role of carotenoids in human health. Nutr Clin Care 5:56–65
6. Paiva SA, Russell RM (1999) Beta-carotene and other carotenoids as antioxidants. J Am Coll Nutr 18:426–433
7. Krinsky NI (2001) Carotenoids as antioxidants. Nutrition 17:815–817
8. Schalch W, Weber P (1994) Vitamins and carotenoids—a promising approach to reducing the risk of coronary heart disease, cancer and eye diseases. Adv Exp Med Biol 366: 335–350
9. Agarwal S, Rao AV (2000) Carotenoids and chronic diseases. Drug Metabol Drug Interact 17:189–210
10. Faure H, Fayol V, Galabert C, Grolier P, Moel GL, Stephens J et al (1999) Carotenoids: 2. Diseases and supplementation studies. Ann Biol Clin (Paris) 57:273–282
11. Peng YM, Peng YS, Childers JM, Hatch KD, Roe DJ, Lin Y et al (1998) Concentrations of carotenoids, tocopherols, and retinol in paired plasma and cervical tissue of patients with cervical cancer, precancer, and noncancerous diseases. Cancer Epidemiol Biomarkers Prev 7:347–350
12. Preobrazhenskii VN, Vasilenko VV, Matveev AA (1997) Beta-carotenoids and their potential use in diseases of the digestive tract. Klin Med (Mosk) 75:43–45
13. Jacques PF, Chylack LT Jr (1991) Epidemiologic evidence of a role for the antioxidant vitamins and carotenoids in cataract prevention. Am J Clin Nutr 53:352S–355S
14. Seddon JM, Ajani UA, Sperduto RD, Hiller R, Blair N, Burton TC et al (1994) Dietary carotenoids, vitamins A, C, and E, and advanced age-related macular degeneration. Eye disease case-control study group. JAMA 272:1413–1420
15. März U (2005) GA-110R - the global market for carotenoids. Business Communications Co., Norwalk
16. März, U. (2008) FOD025C - The Global Market for Carotenoids. BCC Research
17. Ernst H (2002) Recent advances in industrial carotenoid synthesis. Pure Appl Chem 74: 1369–1382

18. Meyers SP, Bligh D (1981) Characterization of astaxanthin pigments from heat-processed crawfish waste. J Agric Food Chem 29: 505–508
19. Bhosale P (2004) Environmental and cultural stimulants in the production of carotenoids from microorganisms. Appl Microbiol Biotechnol 63:351–361
20. Ausich RL (1997) Commercial opportunities for carotenoid production by biotechnology. Pure Appl Chem 69:2169–2173
21. Stankovic I (2004) Zeaxanthin chemical and technical assessment (CTA). In: FAO (ed) 63rd JECFA
22. Jahnke LS (1999) Massive carotenoid accumulation in *Dunaliella bardawil* induced by ultraviolet-A radiation. J Photochem Photobiol 48:68–74
23. Iturriaga EA, Papp T, Breum J, Arnau J, Eslava AP (2005) Strain and culture conditions improvement for β-carotene production with Mucor. Humana, Totowa
24. Bubrick P (1991) Production of astaxanthin from *Haematococcus*. Bioresour Technol 38:237–239
25. Miller MW, Yoneyama M, Soneda M (1976) *Phaffia*, a new yeast genus in the *Deuteromycotina* (*Blastomycetes*). Int J Syst Bacteriol 26:286–291
26. Asker D, Beppu T, Ueda K (2007) Unique diversity of carotenoid-producing bacteria isolated from Misasa, a radioactive site in Japan. Appl Microbiol Biotechnol 77:383–392
27. Asker D, Ohta Y (1999) Production of canthaxanthin by extremely halophilic bacteria. J Biosci Bioeng 88:617–621
28. Oren A, Rodriguez-Valera F (2001) The contribution of halophilic bacteria to the red coloration of saltern crystallizer ponds. FEMS Microbiol Ecol 36:123–130
29. Hanada S, Kawase Y, Hiraishi A, Takaichi S, Matsuura K, Shimada K et al (1997) *Porphyrobacter tepidarius* sp. nov., a moderately thermophilic aerobic photosynthetic bacterium isolated from a hot spring. Int J Syst Bacteriol 47:408–413
30. Kilian O, Steunou AS, Fazeli F, Bailey S, Bhaya D, Grossman AR (2007) Responses of a thermophilic *Synechococcus* isolate from the microbial mat of Octopus Spring to light. Appl Environ Microbiol 73:4268–4278
31. Allewalt JP, Bateson MM, Revsbech NP, Slack K, Ward DM (2006) Effect of temperature and light on growth of and photosynthesis by *Synechococcus* isolates typical of those predominating in the octopus spring microbial mat community of Yellowstone National Park. Appl Environ Microbiol 72:544–550
32. Lutnaes BF, Strand A, Petursdottir SK, Liaaen-Jensen S (2004) Carotenoids of thermophilic bacteria - *Rhodothermus marinus* from submarine Icelandic hot springs. Biochem Syst Ecol 32:455–468
33. Strand A, Shivaji S, LiaaenJensen S (1997) Bacterial carotenoids 55. C-50-carotenoids 25. Revised structures of carotenoids associated with membranes in psychrotrophic Micrococcus roseus. Biochem Syst Ecol 25:547–552
34. Lazrak T, Wolff G, Albrecht AM, Nakatani Y, Ourisson G, Kates M (1988) Bacterioruberins reinforce reconstituted *Halobacterium* lipid-membranes. Biochim Biophys Acta 939:1 60–162
35. Yamano Y, Sakai Y, Hara M, Ito M (2002) Carotenoids and related polyenes. Part 9. Total synthesis of thermozeaxanthin and thermocryptoxanthin and the stabilizing effect of thermozeaxanthin on liposomes. J Chem Soc Perk T 1:2006–2013
36. Stafsnes MH, Josefsen KD, Kildahl-Andersen G, Valla S, Ellingsen TE, Bruheim P (2010) Isolation and characterization of marine pigmented bacteria from Norwegian coastal waters and screening for carotenoids with UVA-blue light absorbing properties. J Microbiol 48:16–23
37. Hirabayashi H, Ishii T, Takaichi S, Inoue K, Uehara K (2004) The role of carotenoids in the photoadaptation of the brown-colored sulfur bacterium *Chlorobium phaeobacteroides*. Photochem Photobiol 79:280–285
38. Moline M, Libkind D, Dieguez Mdel C, van Broock M (2009) Photoprotective role of carotenoids in yeasts: Response to UV-B of pigmented and naturally-occurring albino strains. J Photochem Photobiol B 95:156–161
39. Libkind D, Moline M, Sampaio JP, van Broock M (2009) Yeasts from high-altitude lakes: influence of UV radiation. FEMS Microbiol Ecol 69:353–362
40. Martinez-Laborda A, Balsalobre JM, Fontes M, Murillo FJ (1990) Accumulation of carotenoids in structural and regulatory mutants of the bacterium *Myxococcus xanthus*. Mol Gen Genet 223:205–210
41. Halliwell B (1996) Free radicals, proteins and DNA: oxidative damage versus redox regulation. Biochem Soc Trans 24:1023–1027
42. Carbonneau MA, Melin AM, Perromat A, Clerc M (1989) The action of free radicals on *Deinococcus radiodurans* carotenoids. Arch Biochem Biophys 275:244–251

43. Tian B, Xu Z, Sun Z, Lin J, Hua Y (2007) Evaluation of the antioxidant effects of carotenoids from *Deinococcus radiodurans* through targeted mutagenesis, chemiluminescence, and DNA damage analyses. Biochim Biophys Acta 1770:902–911
44. Battista JR (1997) Against all odds: the survival strategies of *Deinococcus radiodurans*. Annu Rev Microbiol 51:203–224
45. Anton J, Oren A, Benlloch S, Rodriguez-Valera F, Amann R, Rossello-Mora R (2002) *Salinibacter ruber* gen. nov., sp. nov., a novel, extremely halophilic member of the bacteria from saltern crystallizer ponds. Int J Syst Evol Microbiol 52:485–491
46. Takaichi S, Oh-Oka H, Maoka T, Jung DO, Madigan MT (2003) Novel carotenoid glucoside esters from alkaliphilic heliobacteria. Arch Microbiol 179:95–100
47. Yokoyama A, Sandmann G, Hoshino T, Adachi K, Sakai M, Shizuri Y (1995) Thermozeaxanthins, new carotenoid-glycoside-esters from thermophilic eubacterium Thermus thermophilus. Tetrahedron Lett 46: 4901–4904
48. Kametani K, Matsumura T (1983) Determination of 238U, 234U, 226Ra and 228Ra in spring waters of sanin district. Radioisotopes 32:18–21
49. Dufossé L (2009) Pigments, microbial. In: Moselio S (ed) Encyclopedia of microbiology. Academic, Oxford, pp 457–471
50. Tangeras A, Marøy T, Wahlstrøm S et al. (1989) Release of astaxanthin from the red yeast Phaffia rhodozyma by high pressure homogenization. in Program 1. Int. Mar. Biotechnol. Conf., IMB '89, Tokyo, Japan
51. Gentles A, Haard NF (1991) Pigmentation of rainbow trout with enzyme-treated and spray-dried *Phaffia rhodozyma*. Progr Fish-Cult 53:1–6
52. Johnson EA, Conklin DE, Lewis MJ (1977) The yeast *Phaffia rhodozyma* as a dietary pigment source for salmonids and crustaceans. J Fish Res Board Can 34:2417–2421
53. Johnson EA, Villa TG, Lewis MJ (1980) *Phaffia rhodozyma* as an astaxanthin source in salmonid diets. Aquaculture 20:123–134
54. Storebakken T, Sørensen M, Bjerkeng B, Harris J, Monahan P, Hiu S (2004) Stability ofastaxanthinfromredyeast,*Xanthophyllomyces dendrorhous*, during feed processing: effects of enzymatic cell wall disruption and extrusion temperature. Aquaculture 231:489–500
55. Colla E, Pereira A, Hernalsteens S, Maugeri F, Rodrigues M (2010) Optimization of trehalose production by *Rhodotorula dairenensis* following a sequential strategy of experimental design. Food Bioprocess Tech 3:265–275
56. Fontana JD, Czeczuga B, Bonfim TMB, Chociai MB, Oliveira BH, Guimarães MF et al (1996) Bioproduction of carotenoids: the comparative use of raw sugarcane juice and depolymerized bagasse by *Phaffia rhodozyma*. Bioresour Technol 58:121–125
57. Aguilera-Carbo A, Hernández J, Augur C, Prado-Barragan L, Favela-Torres E, Aguilar C (2009) Ellagic acid production from biodegradation of creosote bush ellagitannins by *Aspergillus niger* in solid state culture. Food Bioprocess Tech 2:208–212
58. Johnson EA, An GH (1991) Astaxanthin from Microbial Sources. Crit Rev Biotechnol 11:297–326
59. Mendes-Pinto MM, Raposo MFJ, Bowen J, Young AJ, Morais R (2001) Evaluation of different cell disruption processes on encysted cells of *Haematococcus pluvialis*: effects on astaxanthin recovery and implications for bioavailability. J Appl Phycol 13:19–24
60. An GH, Choi ES (2003) Preparation of the red yeast, *Xanthophyllomyces dendrorhous*, as feed additive with increased availability of astaxanthin. Biotechnol Lett 25:767–771
61. Valduga E, Valerio A, Tatsch PO, Treichel H, Furigo A, Di Luccio M (2009) Assessment of cell disruption and carotenoids extraction from *Sporidiobolus salmonicolor* (CBS 2636). Food Bioprocess Tech 2:234–238
62. Sedmak JJ, Weerasinghe DK, Jolly SO (1990) Extraction and quantification of astaxanthin from *Phaffia rhodozyma*. Biotechnol Tech 4:107–112
63. Asker D, Isaka K (2006) Production of astaxanthin by microorganisms. Japan Patent office 2006, 340676A
64. Ladislav F, Vera P, Karel S, Karel V (2005) Reliability of carotenoid analyses: a review. Curr Anal Chem 1:93–102
65. Khachik F (2009) Analysis of carotenoids in nutritional studies. Birkhäuser, Basel
66. Pupin AM, Dennis MJ, Toledo MCF (1999) HPLC analysis of carotenoids in orange juice. Food Chem 64:269–275
67. Scott KJ, Finglas PM, Seale R, Hart DJ, deFroidmontGortz I (1996) Interlaboratory studies of HPLC procedures for the analysis of carotenoids in foods. Food Chem 57:85–90
68. Hart DJ, Scott KJ (1995) Development and evaluation of an Hplc method for the analysis of carotenoids in foods, and the measurement of the carotenoid content of vegetables and fruits commonly consumed in the UK. Food Chem 54:101–111

69. Stackebrandt E, Goebel BM (1994) A place for DNA-DNA reassociation and 16s ribosomal-Rna sequence-analysis in the present species definition in bacteriology. Int J Syst Bacteriol 44:846–849
70. Schleifer KH, Ludwig W (1989) Phylogenetic relationships of bacteria. Elsevier Science Publishers B. V, Amsterdam
71. Woese CR (1987) Bacterial evolution. Microbiol Rev 51:221–271
72. Rossello-Mora R, Amann R (2001) The species concept for prokaryotes. FEMS Microbiol Rev 25:39–67
73. Egan S, Thomas T, Kjelleberg S (2008) Unlocking the diversity and biotechnological potential of marine surface associated microbial communities. Curr Opin Microbiol 11:219–225
74. Yokoyama A, Miki W, Izumida H, Shizuri Y (1996) New trihydroxy-keto-carotenoids isolated from an astaxanthin-producing marine bacterium. Biosci Biotech Biochem 60:200–203
75. Yokoyama A, Izumida H, Miki W (1994) Production of astaxanthin and 4-ketozeaxanthin by the marine bacterium, *Agrobacterium aurantiacum*. Biosci Biotech Biochem 58: 1842–1844
76. Berry A, Janssens D, Humbelin M, Jore JP, Hoste B, Cleenwerck I et al (2003) *Paracoccus zeaxanthinifaciens* sp. nov., a zeaxanthin-producing bacterium. Int J Syst Evol Microbiol 53:231–238
77. Silva C, Cabral JM, van Keulen F (2004) Isolation of a beta-carotene over-producing soil bacterium, *Sphingomonas* sp. Biotechnol Lett 26:257–262
78. Fukuoka S, Ajiki Y, Ohga T, Kawanami Y, Izumori K (2004) Production of dihydroxy C50-carotenoid by *Aureobacterium* sp. FERM P-18698. Biosci Biotechnol Biochem 68:2646–2648
79. Tao L, Yao H, Cheng Q (2007) Genes from a *Dietzia* sp. for synthesis of C(40) and C(50) beta-cyclic carotenoids. Gene 386:90–97
80. Lemee L, Peuchant E, Clerc M, Brunner M, Pfander H (1997) Deinoxanthin: a new carotenoid isolated from *Deinococcus radiodurans.* Tetrahedron 53:919–926
81. Duc LH, Fraser PD, Tam NK, Cutting SM (2006) Carotenoids present in halotolerant *Bacillus* spore formers. FEMS Microbiol Lett 255:215–224
82. Asker D, Awad TS, Beppu T, Ueda K (2008) *Deinococcus misasensis* and *Deinococcus roseus*, novel members of the genus *Deinococcus,* isolated from a radioactive site in Japan. Syst Appl Microbiol 31:43–49
83. Asker D, Awad TS, Beppu T, Ueda K (2009) *Deinococcus aquiradiocola* sp. nov., isolated from a radioactive site in Japan. Int J Syst Evol Microbiol 59:144–149
84. Asker D, Awad TS, McLandsborough L, Beppu T, Ueda K (2010) *Deinococcus depolymerans* sp. nov., a gamma- and UV-radiation-resistant bacterium, isolated from a radioactive site in Japan. Int J Syst Evol Microbiol 61: 1448–1453
85. Asker D, Beppu T, Ueda K (2007) *Sphingomonas astaxanthinifaciens* sp. nov., a novel astaxanthin-producing bacterium of the family *Sphingomonadaceae* isolated from Misasa, Tottori, Japan. FEMS Microbiol Lett 273:140–148
86. Asker D, Beppu T, Ueda K (2007) *Sphingomonas jaspsi* sp. nov., a novel carotenoid-producing bacterium isolated from Misasa, Tottori, Japan. Int J Syst Evol Microbiol 57:1435–1441
87. Asker D, Beppu T, Ueda K (2008) *Nubsella zeaxanthinifaciens* gen. nov., sp. nov., a zeaxanthin-producing bacterium of the family *Sphingobacteriaceae* isolated from freshwater. Int J Syst Evol Microbiol 58:601–606
88. Suzuki K, Collins MD, Iijima E, Komagata K (1988) Chemotaxonomic characterization of a radiotolerant bacterium, Arthorobacter radiotolerans: description of Rubrobacter radiotolerans gen. nov., comb. nov. FEMS Microbiol Lett 52:33–40
89. Nishimura Y, Ino T, Iizuka H (1988) *Acinetobacter radioresistens* sp. nov. isolated from cotton and soil. Int J Syst Bacteriol 38:209–211
90. Billi D, Friedmann EI, Hofer KG, Caiola MG, Ocampo-Friedmann R (2000) Ionizing-radiation resistance in the desiccation-tolerant cyanobacterium *Chroococcidiopsis.* Appl Environ Microbiol 66:1489–1492
91. Collins MD, Hutson RA, Grant IR, Patterson MF (2000) Phylogenetic characterization of a novel radiation-resistant bacterium from irradiated pork: description of *Hymenobacter actinosclerus* sp. nov. Int J Syst Evol Microbiol 50(2):731–734
92. Phillips RW, Wiegel J, Berry CJ, Fliermans C, Peacock AD, White DC et al (2002) *Kineococcus radiotolerans* sp. nov., a radiation-resistant, gram-positive bacterium. Int J Syst Evol Microbiol 52:933–938
93. Green P, Bousfield IJ (1983) Emendation of *Methylobacterium* Patt, Cole, and Hanson 1976; *Methylobacterium rhodinum* (Heumann 1962) comb. nov. corrig.; *Methylobacterium radiotolerans* (Ito and Iizuka 1971) comb. nov., corrig.; and *Methylobacterium mesophili-*

cum (Austin and Goodfellow 1979) comb. nov. Int J Syst Bacteriol 33:875–877

94. Shahmohammadi HR, Asgarani E, Terato H, Saito T, Ohyama Y, Gekko K et al (1998) Protective roles of bacterioruberin and intracellular KCl in the resistance of *Halobacterium salinarium* against DNA-damaging agents. J Radiat Res (Tokyo) 39:251–262
95. Minton KW (1994) DNA repair in the extremely radioresistant bacterium *Deinococcus radiodurans*. Mol Microbiol 13:9–15
96. Markillie LM, Varnum SM, Hradecky P, Wong KK (1999) Targeted mutagenesis by duplication insertion in the radioresistant bacterium *Deinococcus radiodurans*: radiation sensitivities of catalase (*katA*) and superoxide dismutase (*sodA*) mutants. J Bacteriol 181:666–669
97. Kobayashi I, Tamura T, Sghaier H, Narumi I, Yamaguchi S, Umeda K et al (2006) Characterization of monofunctional catalase KatA from radioresistant bacterium *Deinococcus radiodurans*. J Biosci Bioeng 101:315–321
98. Saito T, Miyabe Y, Ide H, Yamamoto O (1997) Hydroxyl radical scavenging ability of bacterioruberin. Radiat Phys Chem 50:267–269
99. Arrange AA, Phelps TJ, Benoit RE, Palumbo AV, White DC (1993) Bacterial sensitivity to UV light as a model for ionizing radiation resistance. J Microbiol Methods 18:127–136
100. Saito T, Terato H, Yamamoto O (1994) Pigments of *Rubrobacter radiotolerans*. Arch Microbiol 162:414–421
101. Shahmohammadi HR, Asgarani E, Terato H, Ide H, Yamamoto O (1997) Effects of 60Co gamma-rays, ultraviolet light, and mitomycin C on *Halobacterium salinarium* and *Thiobacillus intermedius*. J Radiat Res (Tokyo) 38:37–43
102. Li Y, Kawamura Y, Fujiwara N, Naka T, Liu H, Huang X et al (2004) *Sphingomonas yabuuchiae* sp. nov. and *Brevundimonas nasdae* sp. nov., isolated from the Russian space laboratory Mir. Int J Syst Evol Microbiol 54: 819–825
103. Ben-Amotz A, Katz A, Avron M (1982) Accumulation of β-carotene in halotolerant algae: purification and characterization of β-carotene rich globules from *Dunaliella bardawil* (*chlorophyceae*). J Phycol 18:529–537
104. Murillo FJ, Calderon IL, Lopez-Diaz I, Cerda-Olmedo E (1978) Carotene-superproducing strains of *Phycomyces*. Appl Environ Microbiol 36:639–642
105. Mehta BJ, Salgado LM, Bejarano ER, Cerda-Olmedo E (1997) New mutants of *Phycomyces blakesleeanus* for beta-carotene production. Appl Environ Microbiol 63:3657–3661
106. Pegklidou K, Mantzouridou F, Tsimidou MZ (2008) Lycopene production using *Blakeslea trispora* in the presence of 2-methyl imidazole: yield, selectivity, and safety aspects. J Agric Food Chem 56:4482–4490
107. Jones JD, Hohn TM, LeathersT D (2004) Genetically modified strains of *Fusarium sporotrichioides* for production of lycopene and β-Carotene. Society of Industrial Microbiology Annual Meeting, San Diego, USA, p 91
108. Del Rio E, Acien FG, Garcia-Malea MC, Rivas J, Molina-Grima E, Guerrero MG (2008) Efficiency assessment of the one-step production of astaxanthin by the microalga *Haematococcus pluvialis*. Biotechnol Bioeng 100:397–402
109. Lorenz RT, Cysewski GR (2000) Commercial potential for *Haematococcus microalgae* as a natural source of astaxanthin. Trends Biotechnol 18:160–167
110. Jacobson GK, Jolly SO, Sedmak JJ, Skatrud TJ, Wasileski JM (1995) Astaxanthin over-producing strains of *Phaffia rhodozyma*. United States Patent 5,466,599
111. Liu YS, Wu JY (2006) Hydrogen peroxide-induced astaxanthin biosynthesis and catalase activity in *Xanthophyllomyces dendrorhous*. Appl Microbiol Biotechnol 73:663–668
112. Calo P, Miguel TD, Sieiro C, Velazquez JB, Villa TG (1995) Ketocarotenoids in halobacteria: 3-hydroxy-echinenone and trans-astaxanthin. J Appl Bacteriol 79:282
113. Tsubokura, A., Yoneda, H. and Mizuta, H. (1999) *Paracoccus carotinifaciens* sp. nov., a new aerobic gram-negative astaxanthin-producing bacterium. *Int J Syst Bacteriol* 49 Pt 1, 277-282.
114. Yokoyama A, Miki W, Izumida H, Shizuri Y (1996) New trihydroxy-keto-carotenoids isolated from an astaxanthin-producing marine bacterium. Biosci Biotechnol Biochem 60:200–203
115. Yokoyama A, Izumide H, Miki W (1994) Production of astaxanthin and 4-ketozeaxanthin by the marine bacterium, *Agrobacterium aurantiacum*. Biosci Biotech Biochem 58:1842–1844
116. Osanjo GO, Muthike EW, Tsuma L, Okoth MW et al (2009) A salt lake extremophile, *Paracoccus bogoriensis* sp. nov., efficiently produces xanthophyll carotenoids. Afr J Microbiol Res 8:426–433
117. Shepherd D, Dasek J, Suzanne M, Carels C (1976) Production of zeaxanthin. US Patent, 3,951,743
118. Bhosale P, Larson AJ, Bernstein PS (2004) Factorial analysis of tricarboxylic acid cycle

intermediates for optimization of zeaxanthin production from *Flavobacterium multivorum*. J Appl Microbiol 96:623–629

119. Lagarde D, Beuf L, Vermaas W (2000) Increased production of zeaxanthin and other pigments by application of genetic engineering techniques to *Synechocystis* sp. strain PCC 6803. Appl Environ Microbiol 66:64–72
120. Hundle BS, O'Brien DA, Beyer P, Kleinig H, Hearst JE (1993) In vitro expression and activity of lycopene cyclase and beta-carotene hydroxylase from *Erwinia herbicola*. FEBS Lett 315:329–334
121. McDermott JC, Britton G, Goodwin TW (1973) Carotenoid biosynthesis in a *Flavobacterium* sp.: stereochemistry of hydrogen elimination in the desaturation of phytoene to lycopene, rubixanthin and zeaxanthin. Biochem J 134:1115–1117
122. Cooney JJ, Marks HW, Smith AM (1966) Isolation and identification of canthaxanthin from *Micrococcus roseus*. J Bacteriol 92: 342–345
123. Veiga-Crespo P, Blasco L, Rosa-Dos-Santos F, Poza M, Villa TG (2005) Influence of culture conditions of *Gordonia jacobaea* MV-26 on canthaxanthin production. Int Microbiol 8:55–58
124. de Miguel T, Sieiro C, Poza M, Villa TG (2001) Analysis of canthaxanthin and related pigments from *Gordonia jacobaea* mutants. J Agric Food Chem 49:1200–1202
125. Asker D, Ohta Y (2002) Production of canthaxanthin by *Haloferax alexandrinus* under non-aseptic conditions and a simple, rapid method for its extraction. Appl Microbiol Biotechnol 58:743–750
126. Lorquin J, Molouba F, Dreyfus BL (1997) Identification of the carotenoid pigment canthaxanthin from photosynthetic *bradyrhizobium* strains. Appl Environ Microbiol 63: 1151–1154
127. Guyomarch F, Binet A, Dufosse L (2000) Production of carotenoids by *Brevibacterium lines*: variation among strains, kinetic as pects and HPLC profiles. J Ind Microbiol Biotech 24:64–70
128. Del Campo JA, Rodriguez H, Moreno J, Vargas MA, Rivas J, Guerrero MG (2004) Accumulation of astaxanthin and lutein in *Chlorella zofingiensis* (*Chlorophyta*). Appl Microbiol Biotechnol 64:848–854
129. Shi XM, Liu HJ, Zhang XW, Chen F (1999) Production of biomass and lutein by *Chlorella protothecoides* at various glucose concentrations in heterotrophic cultures. Process Biochem 34:341–347
130. Del Campo JA, Rodriguez H, Moreno J, Vargas MA, Rivas J, Guerrero MG (2001) Lutein production by *Muriellopsis* sp. in an outdoor tubular photobioreactor. J Biotechnol 85:289–295
131. Thompson JD, Higgins DG, Gibson TJ (1994) CLUSTAL W: improving the sensitivity of progressive multiple sequence alignment through sequence weighting, position-specific gap penalties and weight matrix choice. Nucleic Acids Res 22:4673–4680

Chapter 4

A Novel Radio-Tolerant Astaxanthin-Producing Bacterium Reveals a New Astaxanthin Derivative: Astaxanthin Dirhamnoside

Dalal Asker, Tarek S. Awad, Teruhiko Beppu, and Kenji Ueda

Abstract

Astaxanthin is a red ketocarotenoid that exhibits extraordinary health-promoting activities such as antioxidant, anti-inflammatory, antitumor, and immune booster. The recent discovery of the beneficial roles of astaxanthin against many degenerative diseases such as cancers, heart diseases, and exercise-induced fatigue has raised its market demand as a nutraceutical and medicinal ingredient in aquaculture, food, and pharmaceutical industries. To satisfy the growing demand for this high-value nutraceuticals ingredient and consumer interest in natural products, many research efforts are being made to discover novel microbial producers with effective biotechnological production of astaxanthin. Using a rapid screening method based on 16S rRNA gene, and effective HPLC-Diodearray-MS methods for carotenoids analysis, we succeeded to isolate a unique astaxanthin-producing bacterium (strain TDMA-17^T) that belongs to the family *Sphingomonadaceae* (Asker et al., Appl Microbiol Biotechnol 77: 383–392, 2007). In this chapter, we provide a detailed description of effective HPLC-Diodearray-MS methods for rapid analysis and identification of the carotenoids produced by strain TDMA-17^T. We also describe the methods of isolation and identification for a novel bacterial carotenoid (astaxanthin derivative), a major carotenoid that is produced by strain TDMA-17^T. Finally, we describe the polyphasic taxonomic analysis of strain TDMA-17^T and the description of a novel species belonging to genus *Sphingomonas*.

Key words: Astaxanthin, Carotenoids, Antioxidant, *Sphingomonadaceae*, *Sphingomonas astaxanthinifaciens*, Screening, 16S rRNA, Phylogenetic, HPLC

1. Introduction

1.1. Structure, Function, and Importance of Astaxanthin

Astaxanthin, a natural red pigment, is the main carotenoid in marine biosystem (e.g., algae and aquatic animals). Astaxanthin like other carotenoids can be synthesized by photosynthetic organisms (plant and microalgae) and some nonphotosynthetic microorganisms such as yeast, fungi, and bacteria but animals cannot (1–3). Most wild aquatic animals (shrimp, crawfish, crabs, lobster, and salmon)

José-Luis Barredo (ed.), *Microbial Carotenoids from Bacteria and Microalgae: Methods and Protocols*, Methods in Molecular Biology, vol. 892, DOI 10.1007/978-1-61779-879-5_4, © Springer Science+Business Media, LLC 2012

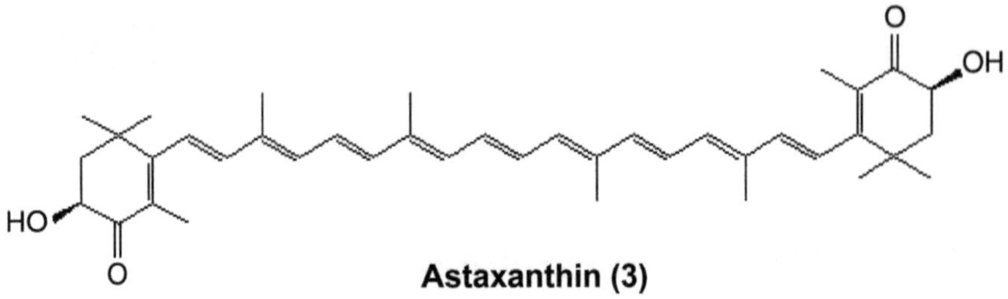

Astaxanthin (3)

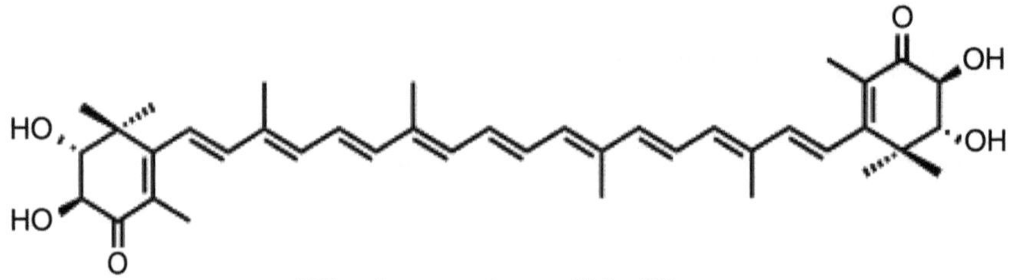

Dihydroxyastaxanthin (2)

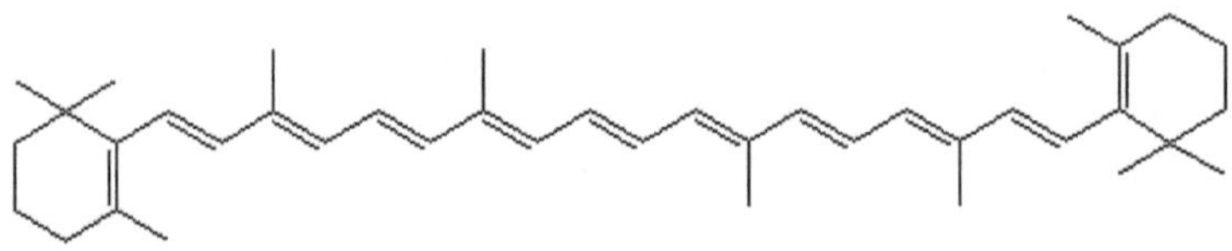

β-carotene

Fig. 1. Chemical structure of astaxanthin, dihydroxyastaxanthin, and β-carotene.

are able to accumulate astaxanthin from their food chains, which gives them the beautiful pink to red flesh (4–7). In the aquaculture industries, astaxanthin has been widely used as a feed supplement for proper growth and survival as well as to provide the favorable pigmentation in commercial fish and crustacean farms (8–11).

Astaxanthin (2) (3,3′-dihydroxy-β,β-carotene-4,4′-dione) is a diketo-dihydroxy-carotenoid that consists of polyene system with 11 conjugated ethylenic double bonds, the hydroxyl and keto endings on each ionone ring (Fig. 1). Owing to its unique chemical structure, astaxanthin is characterized by its extraordinary antioxidant activity, high polar configuration, and the ability to be esterified, which increases the solubility and stability of free astaxanthin in the cell and stabilizes cell membrane (12). Beyond the activities of other carotenoids (e.g., canthaxanthin, zeaxanthin, and β-carotene), astaxanthin is a super antioxidant that exhibits a strong free radical scavenging, singlet oxygen quenching and chain-breaking activities, which reduce oxidative damage in the cell. The photoprotective function of astaxanthin for many microorganisms has been well documented (13–16). It also protects against lipid peroxidation and oxidative damage of low-density lipoprotein cholesterol (17) and against UVA light-induced oxidative stress in cultured fibroblasts (18), prevents tumor formation, enhances in vitro antibody production (19, 20) and anti-inflammatory agent. Research in health and nutrition has indicated its beneficial roles against degenerative diseases, (21, 22) such as some kinds of cancers (23, 24), chronic inflammatory diseases (21), diabetes (25), diabetic nephropathy (26), cardiovascular diseases (27–29), gastrointestinal diseases (21), liver diseases (30), neurodegenerative diseases (21), eye diseases, skin diseases (31), exercise-induced fatigue (32, 33), male infertility (34), and mercuric chloride toxicity (35). The use of astaxanthin as a nutraceutical, supplement, and medicinal ingredient against degenerative diseases, such as cancer, skin-related illness, and heart diseases (21, 24, 36), has increased the demand and market of astaxanthin in aquaculture, medicinal, pharmaceutical, and food industries (21, 37).

1.2. Microbial Production of Astaxanthin

Astaxanthin is commercially produced by extraction from natural sources (shrimp processing waste) (38–40) or by chemical synthesis (41). Most commercial astaxanthin for aquaculture is produced synthetically. However, synthetic astaxanthin contains a mixture of stereoisomers, which is not eco-friendly and less favorable by consumer compared to natural astaxanthin. Due to consumer concerns about these undesirable effects, attempts have been made to produce astaxanthin from microbial sources. As listed in Table 1, the effective microbial production of astaxanthin was established using only *Haematococcus pluvialis* (12, 42) or *Xanthophyllomyces dendrorhous* (the teleomorphic state of *Phaffia rhodozyma*) (43). *H. pluvialis* is known for its high ability to accumulate (3S,3′S)-astaxanthin, up to 0.2 to 3.0% (on a dry mass basis) (12); however, there are few drawbacks in its biotechnological process. Because it grows at mild conditions, such as at 20–28°C, below 15 klux light intensity, and at pH 6.8–7.4, a possible contamination by bacteria, fungi, and protozoa may lead to a serious problem. Therefore, the outdoor cultivation of *Haematococcus* represents a challenge and requires control of contamination and environmental conditions such as light and temperature. On the other hand, process involving completely closed photobioreactors with artificial light is costly. In addition, *Haematococcus* accumulates astaxanthin in the aplanospore growth phase but their thick wall requires physical breakage before the astaxanthin can be extracted or be available to organisms consuming the algae (12).

X. dendrorhous is one of the best candidates for the commercial production of astaxanthin with the highest yield reported so far of almost over 6.6 mg/g cell dry weight of the product that is the (3R,3′R)-isomer astaxanthin (44). A major drawback in the use of *X. dendrorhous* is its rigid cell wall structure (43). For efficient intestinal absorption of the pigment, disruption of the cell wall of the yeast biomass is thus required before addition to an animal diet.

Attempts to isolate astaxanthin-producing bacteria have resulted in a few bacterial strains, most of which were from marine

Table 1
Microorganisms reported for astaxanthin production

Microorganism	Yield	Commercial producer	References
Microalgae			
Haematococcus pluvialis	1–3%	India, Japan, Sweden, U.S.	(12, 100)
Yeast			
Xanthophyllomyces dendrorhous (formerly *Phaffia rhodozyma*)	4–8.1 mg/L	China, Japan, U.S.	(44, 101)
Archaea			
Halobacterium salinarium	265 µg/g	–	(51)
Bacteria			
Paracoccus sp. N81106 (formerly *Agrobacterium aurantiacum* N81106)	0.3 mg/6 L	–	(102)
Paracoccus carotinifaciens E-396	–	–	(47)
Brevundimonas sp. SD212 (formerly *Pseudomonas* sp.)	1.5 mg/27 L	–	(49)
Paracoccus sp. strain PC1(formerly *Alcaligenes* sp. PC-1)	–	–	(48)
Paracoccus bogoriensis	0.4 mg/g of wet cells	–	(103)
Brevundimonas spp. (wild type)	27.6–365 µg/g	–	(50)
Brevundimonas sp. M7 (mutant)	1.3 mg/g	–	(50)

water, such as *Paracoccus* spp. (45–48), *Brevundimonas* sp. (49, 50), and *Halobacterium* sp. (51). Nevertheless, no commercial processes have been developed. Owing to its extraordinary antioxidant activity and its emerging application as nutraceuticals in food, medical, and cosmetic industries (22), and growing demand for natural carotenoids, it is important to maximize the commercial production of astaxanthin by seeking to isolate novel bacterial sources with unique biotechnological properties.

1.3. Isolation of Novel Astaxanthin-Producing Bacteria

For the screening of high-value carotenoids-producing bacteria, we used fresh water samples collected at Misasa (Tottori, Japan), a region known for its high natural radioactivity (52). Using a rapid method for screening, a comprehensive taxonomic characterization based on 16S rRNA gene sequences and effective HPLC-Diodearray-MS methods for carotenoids analysis, we obtained a pigmented bacterial library that contains 104 strains (53) (see also Chapter 3). Among these strains, strain TDMA-17^{T} was isolated

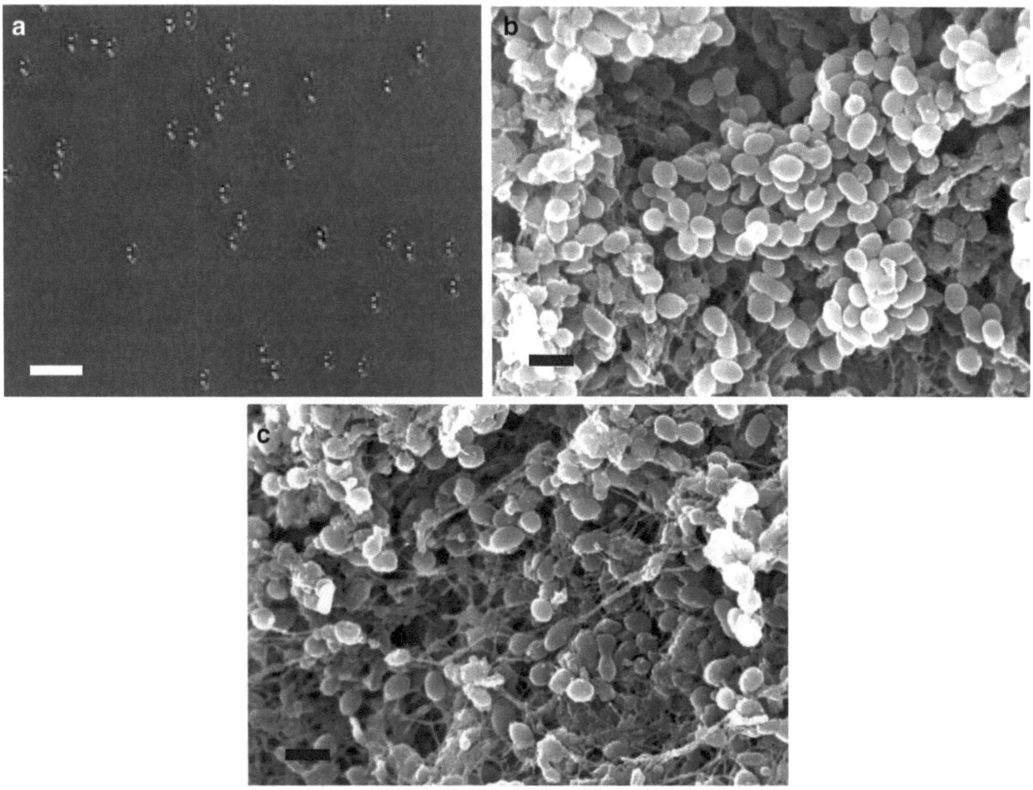

Fig. 2. Morphological characteristics of strain TDMA-17^T. Colony appearances (**a**) and cells observed under a scanning electron microscope (**b**, **c**) are shown. TDMA-17^T formed red-colored colonies. TDMA-17^T exhibited oval- or short rod-shaped cell morphology, and occasionally formed fiber network structure. Cells were grown for 2 days (**a** and **b**) and 4 days (**c**) at 40°C on Nutrient agar. The fiber structure was abundant in the latter condition. *Bars*, 1.0 mm (**a**) and 1.0 μm (**b** and **c**). Adapted from Asker et al. (54).

by its ability to produce red pigmented colonies on Nutrient agar plate after 2 days of cultivation at 40°C (Fig. 2a). Strain TDMA-17^T exhibited high accumulation of carotenoids including astaxanthin and its derivatives (53, 54). The 16S rRNA gene sequence revealed that this unique red-pigmented bacterium (strain TDMA-17^T) belongs to the family *Sphingomonadaceae*, and that the highest degree of sequence similarity was found with *Sphingomonas* spp (53, 54).

The genus *Sphingomonas* was first proposed by Yabuuchi et al. (55) and subsequently emended for description (56–58). *Sphingomonas* comprises Gram-negative, off-white-, yellow- or orange-pigmented, non-fermentative, non-sporulating, non-motile or motile, rod-shaped bacteria that are characterized by the presence of 2-hydroxymyristics acid (2-OH $C_{14:0}$), the absence of 3-hydroxy fatty acids and the content of ubiquinone 10 (Q10) as a major respiratory quinone (57, 58). Members of *Sphingomonas* are of biotechnological interest due to their degradation capabilities of various xenobiotic substances (59), as well as their potential to produce useful exopolysaccharides (60) and carotenoids such as β-carotene and nostoxanthin (61). This was the first description of a red-pigmented *Sphingomonas* spp. and of astaxanthin-producing *Sphingomonadaceae* bacterium (i.e., strain TDMA-17^T) (54).

1.4. Identification of Carotenoids Produced by Strain TDMA-17^T

In Chapter 3, we summarize rapid and effective HPLC methods with Diodearray or MS spectroscopy for the analysis and identification of carotenoids in their natural biological mixtures. These methods were also used for the analysis and identification of the carotenoids produced by strain TDMA-17^T. Advantageously, the extraction of carotenoids from the wet cells of strain TDMA-17^T was easier and efficient (100%) using methanol (54, 62). The amount of carotenoids produced by strain TDMA-17^T was significantly high (2.8 mg/g dry cells). HPLC analysis of the crude pigment isolated from the cells of strain TDMA-17^T using HPLC method A showed the presence of one broad peak with R_t (0.8–1 min). However, HPLC method B showed the presence of five pigments (corresponding to peak 1–5 in Fig. 3a). Table 2 summarizes the spectroscopic data for the HPLC five peaks of carotenoids produced by strain TDMA-17^T. The UV-visible absorption spectra of the first three peaks (Fig. 3b) were identical to the spectrum of astaxanthin; they exhibited a single absorption maximum at 476 nm. The astaxanthin and dihydroxyastaxanthin standards showed identical spectroscopic data as those for peak 3 and peak 2, respectively. Therefore, based on the spectroscopic data (Table 2), we identified the compounds corresponding to peak 2 (7.5% of the total carotenoids) and 3 (24.6% of the total carotenoids) as dihydroxyastaxanthin and astaxanthin (all-*trans*-astaxanthin), respectively.

Astaxanthin was further confirmed by HPLC-Mass (method C) retention time, R_t at 7.05 min; and molecular weight (M^++1), 597] (Table 2). The compounds corresponding to peak 4 and 5 (3.4 and 6.3% of the total carotenoids, respectively), which showed the asymmetrical spectra, were predicted to be 9-*cis*-astaxanthin and 13-*cis*-astaxanthin, respectively, by comparing their spectra to literature UV/VIS spectra (63). It is worthy to mention that we failed to identify the major constituent that corresponds to peak 1 (58.4% of the total carotenoids). It exhibited the same spectra of astaxanthin but with higher molecular weight and polarity than astaxanthin. Therefore, the unknown was primary identified as astaxanthin derivative.

1.5. Isolation and Purification of a Novel Astaxanthin Derivative

The discovery of the numerous health benefits of carotenoids has increased the demand to explore new carotenoids species with a novel property or activity. More than 700 identified naturally occurring carotenoids (64) differ in the length of the polyene chromophore, nature of the end groups and various substituents. The chemical structure of carotenoids determines how they interact with other molecules and integrate into membranes, which strongly

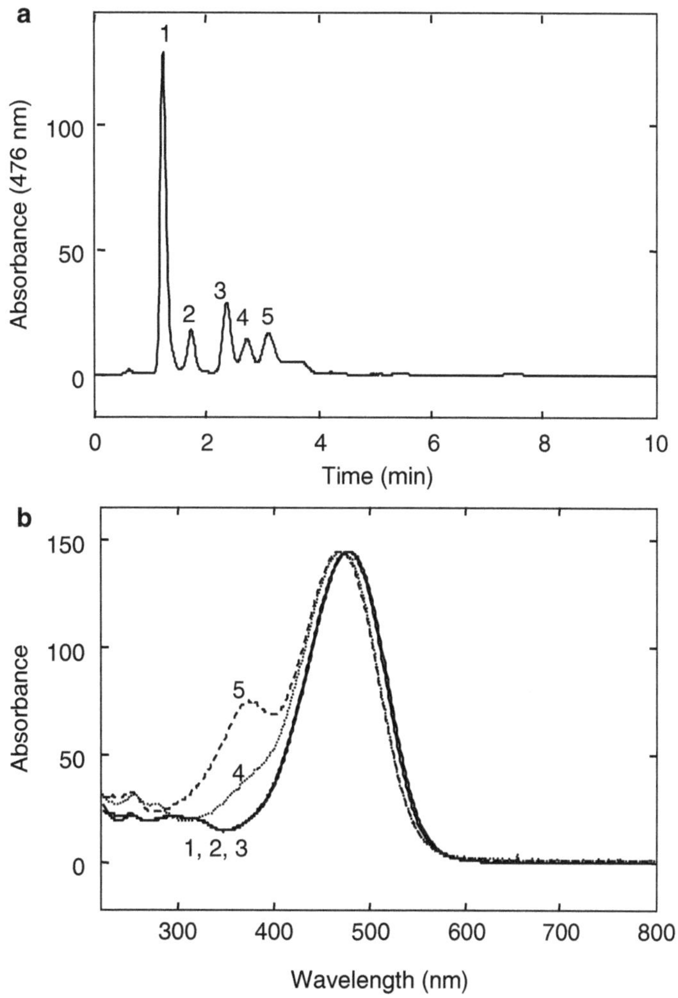

Fig. 3. Absorption profile of the carotenoids produced by strain TDMA-17^T. Elution profile (λ_{max} 476 nm) (**a**) and absorption spectrum of each peak (**b**) are shown. The carotenoid fraction was analyzed by reverse-phase liquid chromatography using method B (see Subheading 3.2). The fraction consisted of unidentified carotenoid (*peak 1*), dihydroxyastaxanthin (*peak 2*), all-*trans*-astaxanthin (*peak 3*), 9-*cis*-astaxanthin (*peak 4*), and 13-*cis*-astaxanthin (*peak 5*). Adapted from Asker et al. (54).

determine their antioxidant activity. Astaxanthin is a diketo and has two hydroxyl groups, which gives it an extraordinary antioxidant activity compared with β-carotene. About 12 astaxanthin derivative molecules were isolated and identified from marine bacterium, transformed *Escherichia coli* strains, salmon and microalgae (Table 3). Discovery of astaxanthin-like component that enclose astaxanthin with a modified functional group might increase significantly the polarity and thus may possess a higher antioxidant activity compared to astaxanthin, and high solubility or membrane integrity. All of them can increase or modify the biological function of astaxanthin.

Table 2
The spectroscopic data for the HPLC five peaks of carotenoids produced by strain TDMA-17^T

HPLC peak	Carotenoids	Absorption, (λ_{max}, nm)	HPLC, retention time (R_t, min) Method B	Method C	Mw (M^++1)
1	Unknown	476	1.24	3.4	889
2	Dihydroxyastaxanthin	476	1.34	4.21	629
3	Astaxanthin	476	2.31	7.05	597
4	9-*cis*-astaxanthin	350–370, 464	2.78	8.4	597
5	13-*cis*-astaxanthin	350–370, 462	3.05	11.0	597

Isolation, purification, and identification of novel (i.e., unknown) carotenoids are critical processes due to instability of carotenoids in the presence of light, heat, or oxygen. Therefore, all procedures must be performed under dim light. Identification of known carotenoid structure in biological samples can be simply done using HPLC–MS system, in which individual carotenoids can be identified by their respective retention time, absorption spectrum, and molecular weight. This requires the use of commercial carotenoid standards or known biological sources as references. On the other hand, identification of unknown carotenoids with novel structure is usually done using HPLC–UV–vis, NMR Mass, IR, and CD, in which a pure and relatively high amount (at least 10 mg) of the unknown carotenoid is required. In general, individual carotenoids are structurally highly similar and many compounds interfere in their isolation. It might be easy to isolate carotenoids such as β-carotene and astaxanthin in their natural mixture because of large difference in polarity, i.e., astaxanthin contains extra hydroxyl and keto endings on each ionone ring compared to β-carotene (Fig. 1). However, it is a challenge to separate carotenoids with slightly different polarities such as astaxanthin carotenoid 3 and the unknown carotenoid carotenoid 1, as shown in Fig. 3. Therefore, we use open column chromatography (OCC) followed by thin layer chromatography (TLC) for purification of carotenoid 1. In OCC, silica gel is used as the stationary phase while a mobile phase (single solvent or a mixture of solvents) is used to separate carotenoids' fractions based on polarity. It is also possible to monitor the fractions of carotenoids visually based on color differences.

Figure 4 illustrates a preparative scale for extraction, isolation, and purification of carotenoid 1 from wet cells of strain TDMA-17^T. The crude carotenoid extract of strain TDMA-17^T was subjected to fractionation using OCC with silica gel as the polar

Table 3
Reported natural astaxanthin derivatives

Common name	Formula	Mw	Sources	References
Astaxanthin/(3S,3′S)-Astaxanthin	$C_{40}H_{52}O_4$	596.84	Salmon, Shrimp, Krill, etc., (marine animals), *Adonis aestivalis* (flower petals of higher plant), *Haematococcus pluvialis* (green alga) *Agrobacterium aurantiacum* (bacteria)	(5, 47, 48, 104, 105)
(3R,3′R)-Astaxanthin	$C_{40}H_{52}O_4$	596.84	Shrimps, lobsters, etc. (marine animals) *Xanthophyllomyces dendrorhous* (Yeast)	(106, 107)
meso-Astaxanthin/(3R,3′S)-astaxanthin	$C_{40}H_{52}O_4$	596.838	Shrimps, lobsters, etc., (marine animals)	(108, 109)
Astaxanthin glucoside	$C_{46}H_{62}O_9$	758.98	Bacteria (*Agrobacterium aurantiacum)*	(110)
Astaxanthin diglucoside	$C_{52}H_{72}O_{14}$	921.12	Bacteria (*Escherichia coli* transformants)	(111)
7,8-Didehydro-astaxanthin	$C_{40}H_{50}O_4$	594.82	Marine animals	(109, 112)
7,8,7′,8′-Tetradehydro-astaxanthin	$C_{40}H_{48}O_4$	592.81	Marine animals	(109, 112)
Astaxanthin esters (mono-, diester	Average		Marine animals krill	(113, 114)
2-Hydroxy-astaxanthin	$C_{40}H_{52}O_5$	612.838	*Brevundimonas* sp. (Bacteria)	(49)
2,2′-Dihydroxy-astaxanthin	$C_{40}H_{52}O_6$	628.837	*Brevundimonas* sp. (Bacteria)	(49)
Bastaxanthin D/Bastaxanthol	$C_{40}H_{53}O_8SNa$	716.900	*Ianthella basta* (marine sponge)	(115)
Bastaxanthin C/Bastaxanthin	$C_{40}H_{51}O_8SNa$	714.884	*Ianthella basta* (marine sponge)	(115)

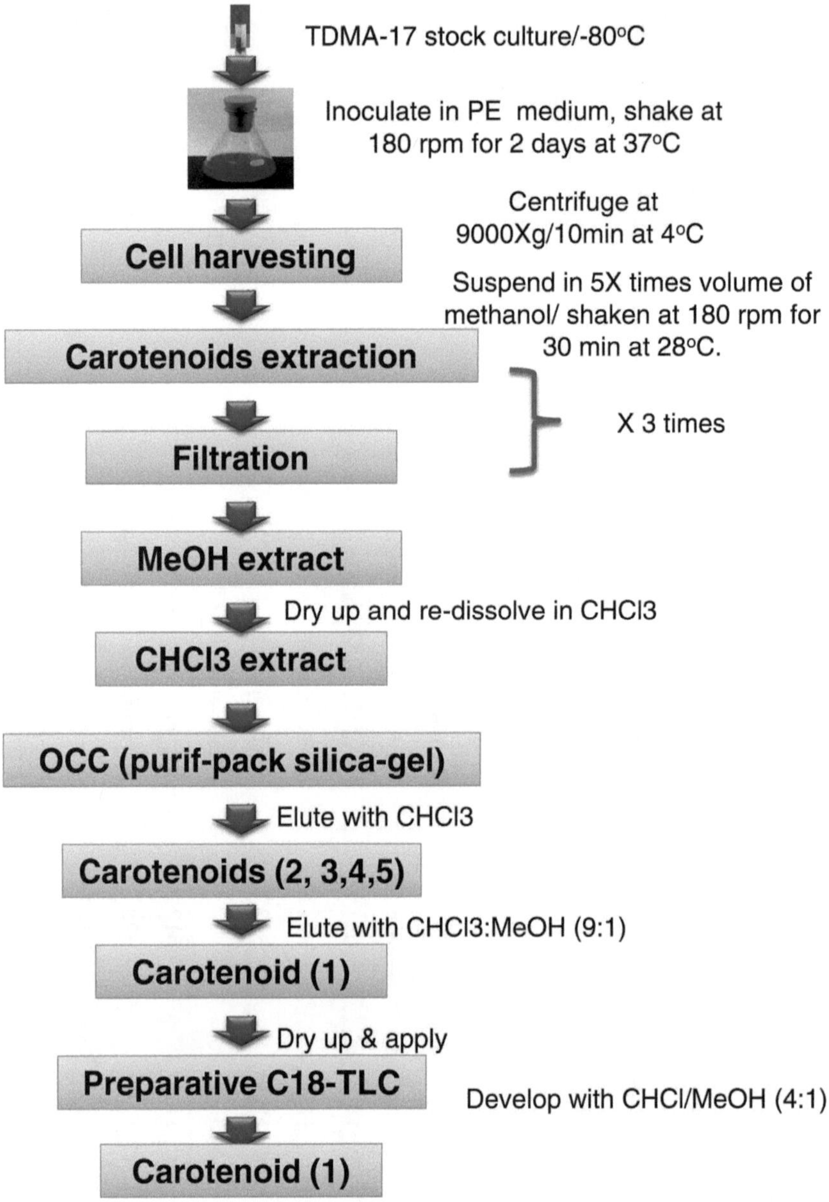

Fig. 4. Outline of isolation and purification process of the novel astaxanthin derivative carotenoid (53).

stationary phase (Fig. 4). First, astaxanthin carotenoid 3 was eluted with chloroform followed by carotenoid 1, which was eluted with $CHCl_3$/methanol (9:1). An additional purification step was done to purify carotenoid 1 using a preparative TLC (C18-TLC) that was developed with $CHCl_3$/methanol (4:1). By this process, ca 56 mg carotenoid 1 and 4.7 mg astaxanthin carotenoid 3 were obtained (Fig. 4). After obtaining a purified unknown carotenoid 1, a comprehensive spectroscopy analysis using UV–vis absorption, NMR, Mass, IR and, CD spectra were performed.

The purified carotenoid 1 was identified using HPLC–UV–vis, NMR Mass, IR, and CD. The analytical data of carotenoid 1 were as follows: Vis in ether/2-methylbutane/ethanol (5:5:1) λ_{max} (ε) 471 (130,000); CD in ether/2-methylbutane/ethanol (5:5:1) $\Delta\varepsilon_{381} = 1.5$, $\Delta\varepsilon_{351} = 0$, $\Delta\varepsilon_{310} = -18.1$, $\Delta\varepsilon_{282} = 0$, $\Delta\varepsilon_{274} = 6.5$, $\Delta\varepsilon_{253} = 0$, $\Delta\varepsilon_{245} = -15.0$; ^{1}H NMR (DMSO-$d_6$, 500 MHz) and ^{13}C NMR (125 MHz) (Table 4); ESI-TOFMS *m/z* 889 $(M+H)^+$, 911 $(M+Na)^+$; HRESI-TOFMS m/z 911.4878 $(M+Na)^+$ (calcd for $C_{52}H_{72}NaO_{12}$, 911.4921); IR (nujol) 3,337, 1,668, 1,039 cm^{-1}. The ^{1}H and ^{13}C NMR data of **1** in DMSO-d_6 (Table 4) indicated the presence of astaxanthin as the carotenoid moiety and two rhamnose residues. The values of $^3J_{H\text{-}1''(H\text{-}1'''),H\text{-}2''(H\text{-}2''')}$ (1 Hz)(65) and $^1J_{C\text{-}1''(C\text{-}1'''),H\text{-}1''(H\text{-}1''')}$ (171 Hz)(66) of rhamnose moieties indicated the α-glycoside.

The spectral data of authentic methyl 2,3,4-tribenzoyl-α-D-rhamnopyranoside: ESI-TOFMS *m/z* 513 $(M+Na)^+$; CD in acetonitrile $\Delta\varepsilon_{238} = 7.3$, $\Delta\varepsilon_{227} = 0$, $\Delta\varepsilon_{221} = -2.3$; δ_H ($CDCl_3$, 600 MHz): 7.2–8.2 (15 H, Bz), 5.92 (1H, d, *J*=4 Hz, H-2), 5.63 (1H, t, *J*=10 Hz, H-4), 5.55 (1H, dd, *J*=10, 4 Hz, H-3), 4.79 (1H, s, H-1), 3.85 (1H, m, H-5), 3.57 (3H, s, OCH_3), 1.45 (3H, d, *J*=6 Hz, H-6). Since the retention time on HPLC and ESI-TOFMS and CD spectra of the derivative from 1 were identical with those of authentic methyl 2,3,4-tribenzoyl-α-D-rhamnopyranoside, the configuration of the rhamnose moieties of 1 was determined as D. The CD spectrum of 2 isolated from the cells of TDMA-17^T indicated the (3*S*,3′*S*) configuration of 2, strongly suggesting the (3*S*,3′*S*) configuration of 1. This was confirmed by the CD spectrum of 1 having the minimum or maximum Cotton effects at 310 (Δε −18.1), 274 (Δε +6.5) and 245 nm (Δε −15.0), which was compatible with (3*S*,3′*S*)-**2** or (3*S*,3′*S*)-**2** monoglucoside [314 (Δε −18.5), 274 (Δε +10.1) and 242 nm (Δε −12.0)][10]. Therefore, the structure of the major carotenoid of strain TDMA 17^T was assigned as (3*S*,3′*S*)-3,3′-bis(α-D-rhamnopyranosyloxy)-β,β-carotene-4,4′-dione carotenoid 1 (Fig. 5).

1.6. Properties of the Novel Carotenoid

Antioxidant activity of compound 1 was measured by the inhibition of methyl linoleate hydroperoxides production due to the reaction of methyl linoleate with 2.2′-azobis(2,4-dimethylvaleronitrile) (AMVN) (67). The antioxidant activity was expressed as a percentage of inhibition of hydroperoxide production due to the addition of carotenoids. The result showed that 1 mM of astaxanthin, astaxanthin dirhamnoside, and β-carotene inhibited methyl linoleate hydroperoxides production at 27, 20, and 16% (a representative result; though the inhibitory values fluctuated, similar results were obtained in three-times-repeated analyses). This indicated that compound 1 exhibits a distinct antioxidant activity though it is slightly lower than that of astaxanthin. Astaxanthin-β-glucoside was reported to be the most active, followed by astaxanthin, zeaxanthin,

Table 4
[1]H and ^{13}C assignments of astaxanthin dirhamnoside[a]

C-No.	δ_C	δ_H
1, 1′	36.7	
2, 2′	44.26	1.89, 1.87
3, 3′	71.33	4.36 dd (13, 5.5)
4, 4′	197.80	
5, 5′	127.41	
6, 6′	159.95	
7, 7′	123.81	6.28 d (16)
8, 8′	141.58	6.50 d (16)
9, 9′	134.98	
10, 10′	134.92	6.42 d (12)
11, 11′	125.15	6.71 dd (15, 12)
12, 12′	139.29	6.49 d (15)
13, 13′	136.62	
14, 14′	133.76	6.40 dd (8, 3)
15, 15′	130.98	6.76 dd (8, 3)
16, 16′	26.06	1.28 s
17, 17′	30.09	1.17 s
18, 18′	13.78	1.79 s
19, 19′	12.37	1.99 s
20, 20′	12.62	1.96 s
1″, 1‴	100.29	4.92 d (1)
2″, 2‴	70.25	3.75 ($J_{2,3}$ = 3.6 Hz)[b]
2″-OH, 2‴-OH		4.84 d (4)
3″, 3‴	70.49	3.43 ($J_{3,4}$ = 9.5 Hz)[b]
3″-OH, 3‴-OH	70.4	4.58 d (6)
4″, 4‴	71.94	3.20 ($J_{4,5}$ = 9.5 Hz)[b]
4″-OH, 4‴-OH		4.78 d (5)
5″, 5‴	68.98	3.44
6″, 6‴	17.9	1.14 d (6)

[a]Spectra were obtained in DMSO-d_6
[b]Obtained by decoupling experiments
Coupling constants in Hertz are given in parentheses
Adapted from Asker et al. 2009 (62)

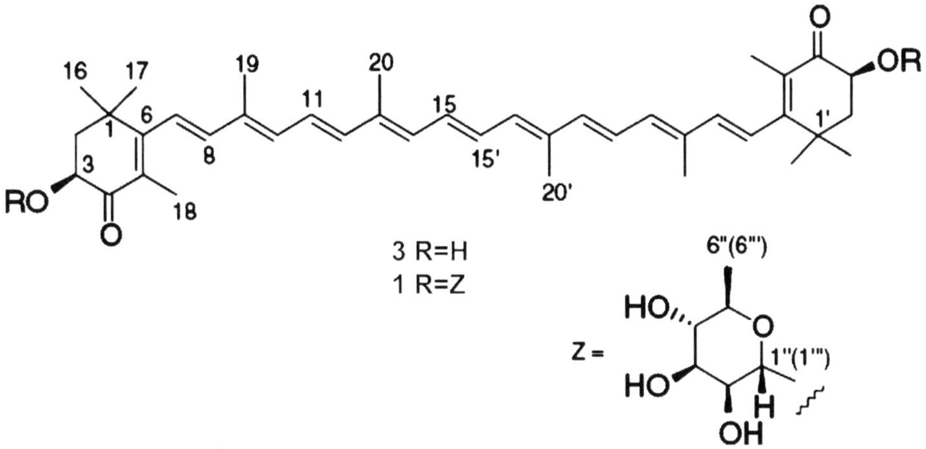

Fig. 5. Chemical structure of astaxanthin dirhamnoside carotenoid 1 and astaxanthin carotenoid 3. Adapted from Asker et al. (62).

and β-carotene in liposomes system, although there was no significant difference in the antioxidant activity of these carotenoids in chloroform solution (66).

Compound 1 was also studied for solubility in water-based solvent by a spectroscopic measurement. First, carotenoid 1 and astaxanthin were dissolved in 100% DMSO, acetone, and methanol at 20 mM. The solutions were then added with water at various ratios. After removing the insoluble pigment particles by filtration (0.1 μm pore size), each solution was measured for absorption at 475 nm. Similarly, the absorption of the above 20 mM solution and its diluted series was measured and used as standards to quantify the amount of solubilized pigment. Astaxanthin dirhamnoside was dissolved in solutions of H_2O–DMSO (99:1), H_2O–acetone (99:1), and H_2O–methanol (99:1) at 13, 6.2, and 5.4 μM, respectively. Solubility of astaxanthin in these water-containing solvents was below the detectable level (<0.1%). Appearance of various solutions is shown in Fig. 6.

The known carotenoid rhamnosides to date include myxocoxanthin rhamnoside from myxobacterium (68), and oscillol 2,2′-dirhamnoside (69) and myxol 2′-rhamnoside (70) from blue-green algae. While these known molecular species were produced as minor components, astaxanthin dirhamnoside carotenoid 1 identified in this study was a major carotenoid of strain TDMA-17^T. The unique properties of astaxanthin dirhamnoside may expand the application of this kind of carotenoids.

1.7. Identification of the Novel Astaxanthin-Producing Bacterium

Identification of unknown microorganism is a process that is used to determine whether the unknown belongs to one of the units (species, genus, etc.) and labeled (e.g., nomenclature) (71–73). This process is very important in academy (e.g., to establish novel taxa) as well as industry (e.g., for obtaining a patent) in order to

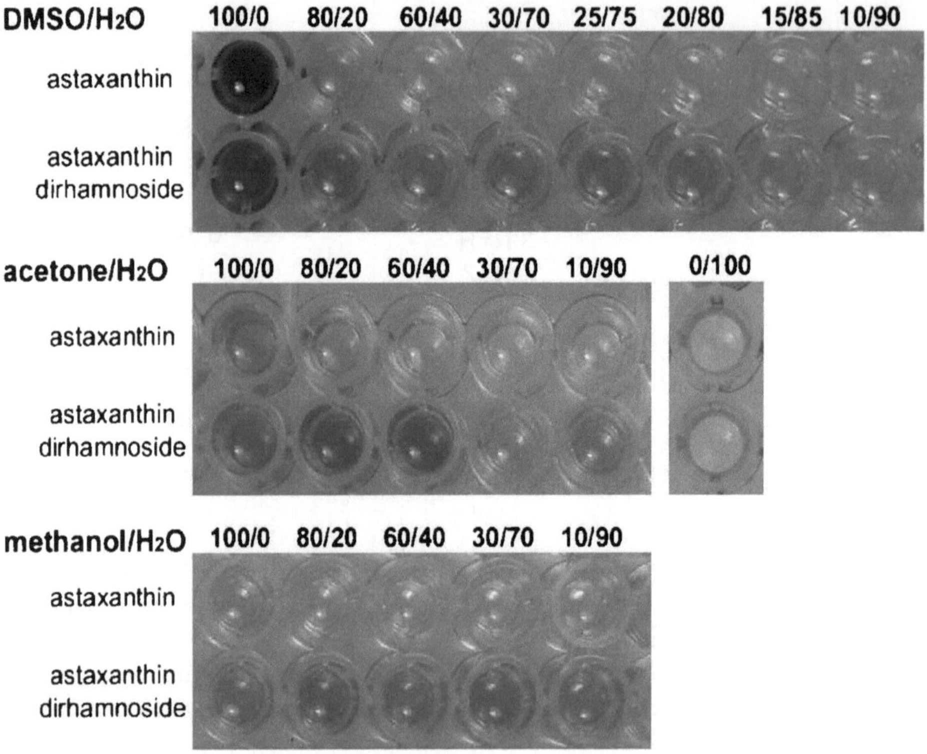

Fig. 6. Solubility of astaxanthin dirhamnoside in water-based solvents. Astaxanthin dirhamnoside was dissolved in water containing DMSO, acetone, and methanol at various ratios. Adapted from Asker et al. (62).

claim a novel producer for production of bioactive components at species or genus level. The identification of any taxon involves a polyphasic approach that includes a combination of phenotypic methods (e.g., biochemical testing, physiological analysis, and chemotaxonomic analysis) and genotypic testing methods (e.g., DNA–DNA hybridization, analysis of G+C content (in moles percent), 16S rDNA gene sequencing, and phylogenetic information) (73–75). The necessity of a careful selection and use of a wide variety of methodologies for describing a novel taxon has been stated (76). In our screening approach, 16S rRNA gene sequence was used as a fast and reliable molecular tool to determine the phylogenetic delineation (e.g., taxonomic position) of the interested unknown isolates. After classifying the unknown isolates into a group (e.g. species or genus, depending on 16S rRNA gene sequence), appropriate tests that provide all taxonomic information (e.g. phenotypic and genotypic) were carried out to distinguish between unknown isolate and its closely related taxa. Figure 7 summarizes the polyphasic taxonomy approach that we have used for identification of the unknown isolates. Nonetheless, specific tests can be added to the above identification approach depending on the taxa involved.

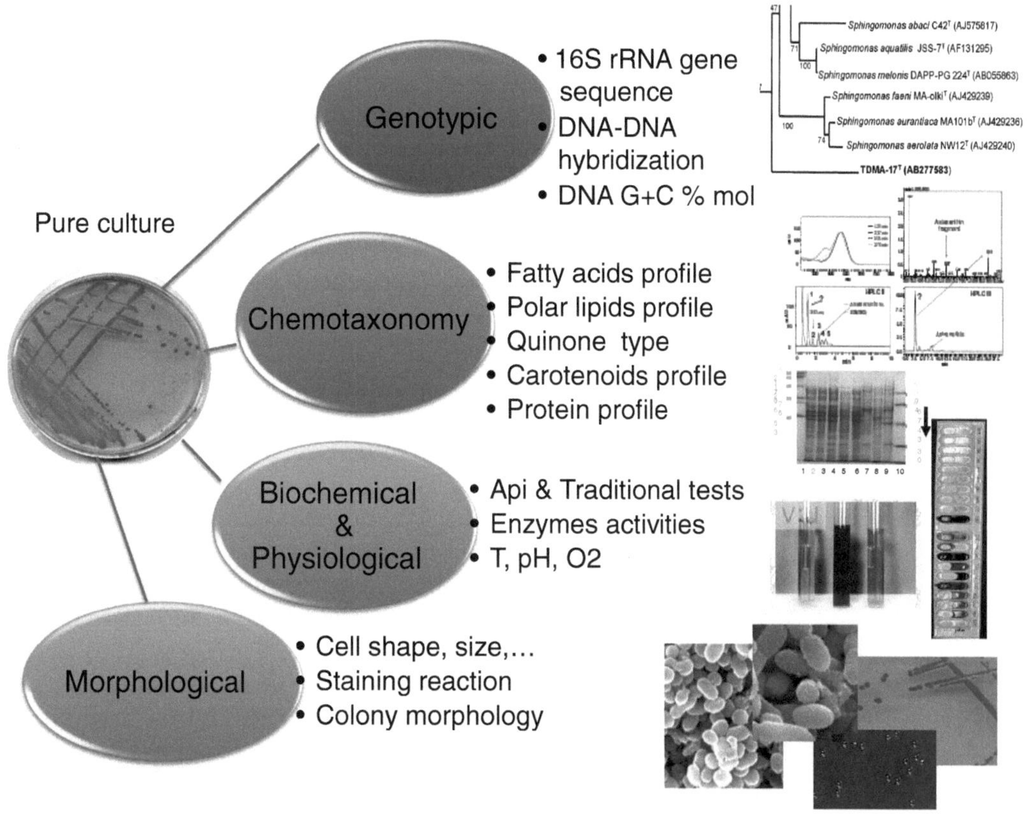

Fig. 7. Summary of the polyphasic taxonomy approach employed for identification of the unknown isolates.

1.8. Phylogenetic Characterization

16S rDNA gene sequencing is a powerful molecular technique presently used for bacterial species identification and to construct phylogenetic trees (77–81). It is recommended that new species descriptions should include an almost complete 16S rRNA gene sequence (>1,300 nt, <0.5% ambiguity) (79). When the 16S rRNA gene sequence similarity between an unidentified isolate and its closest phylogenetic neighbors is less than 97%, it indicates that a novel species has been isolated (76, 80, 82, 83). More than 97% similarity indicates a possibility that they belong to the same species, which requires the use of DNA–DNA hybridization, a standard method for defining species and subspecies, to differentiate between them (80). If the DNA–DNA hybridization between these strains is less than ~70%, it suggests that the two strains belonged to different species (76, 80, 84, 85). When the unknown isolate exhibits novel species properties, the phenotypic differences between the novel species and existing species within the genus are reported (76). At values lower than 95% 16S rRNA gene sequence similarity (over full pairwise comparisons), taxa should be tested by other methods to establish whether separate genera are present (76, 86). To claim a novel producer for production of astaxanthin, we targeted the predictable novel strains that exhibited low sequences homology in the range of 95–97% and 88–94% at species and genus levels, respectively.

A comparison of the 16S rRNA gene sequence of TDMA-17^T with those available in GenBank/EMBL/DDBJ nucleotide sequence database revealed that it belonged to the family *Sphingomonadaceae*. Highest similarity was found with sequences of established species of the genus *Sphingomonas* sensu stricto including *Sphingomonas aerolata* NW12^T (94.5%), *Sphingomonas aurantiaca* MA101b^T (94.0%), *Sphingomonas pruni* IFO 15498^T (94.4%), *Sphingomonas mali* IFO 15500^T (94.2%), *Sphingomonas dokdonensis* DS-4^T (94.2%), *Sphingomonas faeni* MA-olki^T (94.1%), *Sphingomonas melonis* DAPP-PG 224^T (94.0%), and *Sphingomonas abaci* C42^T (93.9%). These results clearly demonstrated that TDMA-17^T represents a novel species within the genus *Sphingomonas* sensu stricto. Phylogenetic analysis of nearly full-length 16S rRNA sequence of strain TDMA-17^T indicated that it is a member of the genus *Sphingomonas* sensu stricto (57), forming a distinct phylogenetic lineage (Fig. 8). The 16S rRNA gene sequence of strain TDMA-17^T retained all the signature nucleotides, namely, C:G at position 52:359, G at position 134, G at position 593, G:C at position 987:1218, and U:G at position 990:1215 (based on *E. coli* numbering) (87), that are conserved in the genus *Sphingomonas* sensu stricto (57).

1.9. Chemotaxonomic Characterization

Chemotaxonomy deals with various structural elements of the cell including the outer cell layers (peptidoglycan, techoic acids, mycolic acids, etc.), the cell membrane(s) (fatty acids, polar lipids, respiratory lipoquinones, pigments, etc.), or constituents of the cytoplasm (polyamines). In our taxonomy analysis, we analyzed polar lipids, fatty acid profile, respiratory lipoquinones, and carotenoids (Fig. 7).

Table 5 shows the fatty acid profile of strain TDMA-17^T. The major fatty acids were $C_{17:1}$ ω6c (33.0%), $C_{18:1}$ ω7c (20.8%), and Summed feature 3 (containing $C_{16:1}$ ω7c and/or 2-OH iso-$C_{15:0}$) (16.8%). The predominant hydroxy fatty acid was 2-OH $C_{18:1}$ (4.9%). Table 5 shows the detailed fatty acid composition of TDMA-17^T. This profile is in excellent agreement with those representative of the genus (Table 5) (58, 88–90) and its assignment to the genus *Sphingomonas* sensu stricto. The presence of 2-hydroxy fatty acids and the lack of 3-hydroxy fatty acids are important characteristics of members of the family *Sphingomonadaceae* (57, 88, 91). However, TDMA-17^T differed from the members of genus *Sphingomonas* sensu stricto both qualitatively and quantitatively with regard to certain fatty acids and by the presence of 2-OH $C_{18:1}$ as the major hydroxy fatty acid (Table 5). The quinone system of strain TDMA-17^T consisted exclusively of ubiquinone Q-10. This quinone profile is a characteristic of the majority of species within

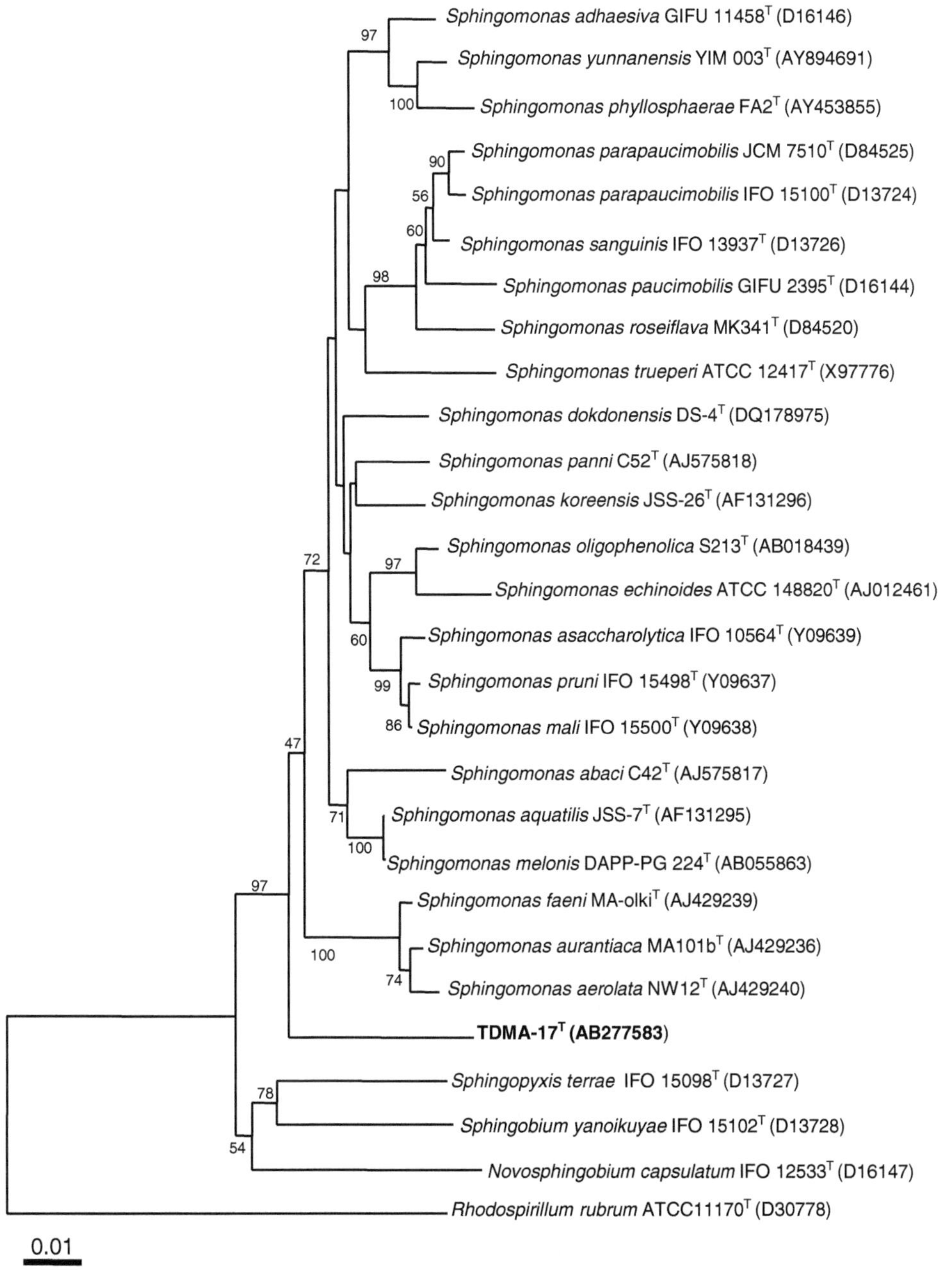

Fig. 8. A phylogenetic tree based on the 16S rRNA gene sequences of strain TDMA-17^{T} and representatives of some related taxa. The tree was constructed by the neighbor-joining method (99). GenBank accession numbers are shown in *parentheses*. *Numbers* at nodes indicate bootstrap values higher than 50%. *Rhodospirillum rubrum* ATCC 11170^{T} was used as an outgroup. *Bar*, 1 substitution per 100 nucleotide positions. Adapted from Asker et al. (54).

the class *Alphaproteobacteria* (92), including genus *Sphingomonas* sensu stricto (57, 88). Analysis of lipid extract by TLC revealed the presence of glycosphingolipids in TDMA-17^{T}. The DNA G+C content of TDMA-17^{T} was 67.7% (Table 6), in agreement with the values of 62–68% for the genus (57).

Table 5
Comparison of whole-cell fatty acid profiles (percentage content)

Fatty acid	1	2	3	4	5	6	7	8	9	10	11	12	13	14
Saturated fatty acids														
$C_{14:0}$	1.2	1.6–1.9	2.0	2.1–3.0	1.0	0.7		1.4	16.8		0.9	2.6	1.1	0.8
$C_{15:0}$	3.0	0.0–0.5		0.0–0.8	0.5			0.2		3.8			1.1	
$C_{16:0}$	7.4	12.0–15.8	11.2	12.2–15.6	10.9	18.4	12.4	10.9	16.9	7.5	10.0	13.0	17.8	19.0
$C_{17:0}$	1.2			0.0–0.2						8.9				
$C_{18:0}$	tr			0.0–0.3	0.9	1.2		0.5		0.4		0.8		
$C_{17:0}$ cyclo		0.0–0.4				0.4								
$C_{19:0}$ cyclo ω8*c*			0.8	0.4–0.7	0.5	0.4								
Unsaturated fatty acids														
$C_{15:1}$ ω6c														
$C_{16:1}$ ω5c	3.5	1.2–2.0	1.6	1.6–1.8	1.1	1.0					0.8	1.6	2.1	1.3
$C_{16:1}$ ω7c											2.9			
$C_{16:1}$								1.4						
$C_{17:1}$ ω6c	33.0	0.0–1.7	2.6	0.8–3.0	2.6	0.4	1.6					0.8	3.6	
$C_{17:1}$ ω8c	2.7													
$C_{17:1}$								1.7	4.9	27.5				
$C_{18:1}$ ω5c	1.7	0.6–0.8	1.0	0.6–1.0	2.0	1.5		1.8			2.7	1.4	0.7	
$C_{18:1}$ ω7c	20.8				63.0	60.3	64.8					62.5	51.3	45.7
$C_{18:1}$														
11 methyl $C_{18:1}$ ω7c				0.0–3.5	6.9	5.4						4.1		6.3

Hydroxy fatty acid														
2-OH $C_{12:0}$		0.3-0.6	0.6	0.4										
2-OH $C_{13:0}$														
2-OH $C_{14:0}$	tr	8.7-14.1	9.7	10.0-10.6	7.6	5.5	12.9	9.8		0.4	7.2	7.5	5.3	12.1
2-OH $C_{15:0}$	tr				0.5			0.3		1.0				
2-OH $C_{16:0}$	tr				0.8			0.8						
2-OH $C_{16:1}$		0.0-0.5												
2-OH $C_{18:1}$	4.9					0.6						0.8		
Summed features														
4	16.8	24.6-26.8	29.2	25.7-28.0	1.4	4.3	8.3	2.1	2.5		2.45	5.0	17.1	14.8
7		38.5-46.0	41.4	40.0-40.5				69.2	58.4	50.2	73.1			

1, *Sphingomonas astaxanthinifaciens* sp. nov. TDMA-17T; 2, *Sphingomonas aerolata* NW12T (58); 3, *Sphingomonas faeni* MA-olkiT (58); 4, *Sphingomonas aurantiaca* MA101bT (58); 5, *Sphingomonas melonis* DAPP-PG 224T (89); 6, *Sphingomonas aquatilis* JSS-7T (89); 7, *Sphingomonas abaci* C42T (89); 8, *Sphingomonas mali* IFO 15500T (90); 9, *Sphingomonas pruni* IFO 15498T (90); 10, *Sphingomonas asaccharolytica* IFO 10564T (90); 11, *Sphingomonas echinoides* ATCC 148820T (116); 12, *Sphingomonas koreensis* JSS-26T (89); 13, *Sphingomonas panni* C52T (89); 14, *Sphingomonas dokdonensis* DS-4T (117).

Summed feature 4 contains 2-OH iso-C15:0 and/or C16:1 ω7c. Summed feature 7 contains C18:1ω7*c*, C18:1ω9*t* and/or C18:1ω12*t*. Values are percentage of total fatty acids.

Adapted from Asker et al. (54)

Table 6
Differential characteristics of *Sphingomonas astaxanthinifaciens* sp. nov. TDMA-17^T and related type strains

Characteristic	1	2	3	4	5	6	7	8	9	10	11	12	13	14
Color of colonies	DR	O	O	O	DY	Y	O	LY	LY	LY	Y	Y	Y	Y
Astaxanthin production	+	–	–	–	–	–	–	–	–	–	–	–	–	–
Cell shape	o/sr	r	r	r	r	r	r	r	r	r	r	r	r	r
Motility	+	+	+	+	–	+	–	+	+	+	+	+	–	+
Growth at/in														
45°C	+	–	–	–	–	–	–	–	–	–	–	–	–	–
3.0 % NaCl	–	ND	ND	ND	+	ND	ND	+	+	+	ND	ND	ND	+
Assimilation of:														
Glucose	–	+	+	+	+	–	–	+	+	–	(+)	(+)	–	–
L-Arabinose	–	+	+	+	+	+	–	w	+	w	+	–	–	–
D-Mannose	–	+	+	+	+	–	–	+	+	w	–	–	(+)	+
N-Acetyl-D-glucosamine	–	+	+	+	+	–	–	+	+	+	+	+	–	–
Maltose	–	+	+	+	+	+	–	+	+	+	(+)	+	(+)	–
Gluconate	–	+	+	+	–	–	(+)	+	+	–	–	–	–	–
Adipate	–	+	+	+	–	–	–	+	–	–	–	–	–	+
DL-Malic acid	–	v	–	+	+	–	+	–	w	–	+	+	+	–
Citrate	–	v	+	+	–	–	–	–	–	–	–	–	+	–
Nitrate reduction	–	–	–	–	–	–	ND	–	–	–	+	–	ND	–
β-Galactosidase	–	+	(+)	(+)	+	+	ND	+	+	+	+	+	ND	–
Oxidase activity	–	+	+	+	+	+	–	+	+	+	+	+	–	+
Hydrolysis of:														
Aesculin	+	+	+	+	+	–	+	+	+	+	+	–	+	+
Gelatin	+	–	–	–	–	–	ND	–	–	–	–	–	ND	+
Starch	+	–	–	–	–	ND	ND	–	+	+	ND	ND	ND	+
Tween 80	+	ND	ND	ND	–	ND	ND	+	+	+	ND	ND	ND	+
DNA	NG	(+)	(+)	v	NG	ND	ND	+	NG	NG	ND	ND	ND	ND
DNA G+C content (mol%)	63.3	65.4	63.1	64.7	65.0	63	ND	65.4-65.9	65.4	64.8	65.8	66	ND	66.9

1, *Sphingomonas astaxanthinifaciens* sp. nov. TDMA-17^T; 2, *Sphingomonas aerolata* NW12^T (58), 3, *Sphingomonas faeni* MA-olki^T (58); 4, *Sphingomonas aurantiaca* MA101b^T (58), 5, *Sphingomonas melonis* DAPP-PG 224^T (118); 6, *Sphingomonas aquatilis* JSS-7^T (119); 7, *Sphingomonas abaci* C42^T (89); 8, *Sphingomonas mali* IFO 15500^T (57); 9, *Sphingomonas pruni* IFO 15498^T (57); 10, *Sphingomonas asaccharolytica* IFO 10564^T (57); 11, *Sphingomonas echinoides* ATCC 148820^T (116); 12, *Sphingomonas koreensis* JSS-26^T (119); 13, *Sphingomonas panni* C52^T (89); 14, *Sphingomonas dokdonensis* DS-4^T (117)
+, positive; (+), weakly positive; –, negative; ND, not determined; NG, no growth; Y, yellow; LY, light yellow; DY, deep yellow; o/sr, oval- or short rod; r, rods. All strains were negative for urease activity, indole production, glucose fermentation, arginine dihydrolase and assimilation of D-mannitol, caprate and acetate
Adapted from Asker et al. 54

1.10. Morphological, Physiological, and Biochemical Characterization

Examinations of morphological (e.g., cell shape and size, spore formation, and motility), physiological (e.g., requirement of carbon or nitrogen sources, growth at different temperature, pH, salt concentration oxygen, etc.) and biochemical (e.g., activity of various enzyme, metabolization of carbohydrates) features are the oldest tools for the characterization and classification of prokaryotes. Most of these methods are however time consuming and increase the work load. Commercial galleries of biochemical tests, such as API strips, biolog microplate, BBL crystal ID system, are fast, reduce the work load, and increase the standardization between different laboratories. We used a combination of traditional biochemical tests and API galleries combined with traditional methods to gather the physiological and biochemical information (Fig. 7).

For identification of a novel taxon, it is necessary to start with a pure culture. To insure the purity of the culture, isolation of a single colony from a plate that contains nonselected media is preferred (72). In addition, one should use a fresh culture and standardized conditions because biochemical characteristics and morphology depend on the age of culture, the medium used and other growth conditions.

Cells of strain TDMA-17^T were Gram-negative-staining motile ovals or short rods (0.3–0.5 μm in diameter and 0.5–0.9 μm in length) (Fig. 2b, c), which grew singly or in pairs. The cells occasionally formed fiber structures; the fibers were clearly observed in 4-day culture (Fig. 2c). Spore was not observed. Strain TDMA-17^T was a strictly aerobic chemoheterotroph. TDMA-17^T was slightly tolerant against gamma-ray irradiation, showing 2.1% survival rate after exposure to a dose of 2.3 kGy. At this dose, the survival rate of *E. coli* C600 was 0%. Other phenotypic properties of strain TDMA-17^T are given in the species description and Table 6.

1.11. Differential Characteristics of Strain TDMA-17^T

Table 6 summarizes the characteristics of strain TDMA-17^T that differentiate it from related members of the genus *Sphingomonas* sensu strict. TDMA-17^T differed from all related species by (1) having red pigmentation, (2) being able to produce astaxanthin, and (3) being more mesophilic. Unlike *Sphingomonas mali* IFO 15500^T, *Sphingomonas pruni* IFO 15498^T, *Sphingomonas asaccharolytica* IFO 10564^T, *Sphingomonas melonis* DAPP-PG 224^T, and *Sphingomonas dokdonensis* DS-4^T, TDMA-17^T showed severe growth arrest in the presence of >0.25% NaCl (Table 6). TDMA-17^T differed from *Sphingomonas aerolata* NW12^T, *Sphingomonas faeni* MA-olki^T, and *Sphingomonas aurantiaca* MA101b^T by (1) being unable to assimilate glucose, L-arabinose, D-mannose, *N*-acetyl-D-glucosamine, maltose, gluconate, adipate, and citrate, (2) being able to hydrolyse gelatine and starch, (3) being unable to hydrolyse DNA, and (4) being unable to produce β-galactosidase and oxidase. TDMA-17^T differed from *Sphingomonas melonis* DAPP-PG 224^T by (1) being unable to assimilate glucose, L-arabinose,

D-mannose, *N*-Acetyl-D-glucosamine, and malate, (2) being able to hydrolyse gelatine and starch, and (3) being unable to produce β-galactosidase and oxidase. Additionally, TDMA-17^T was distinguished from other closely related species by the traits detailed in Table 6. Interestingly, colonies of strain TDMA-17^T showed a characteristic reddish color probably due to astaxanthin production. Red pigmentation has never been given in the description of the genus *Sphingomonas* sensu stricto. It has been described that the genus *Sphingomonas* encompasses orange-, yellow-, or off-white-pigmented species (57, 58)

1.12. Taxonomic Position of Strain

The results of the analysis of 16S rRNA gene sequence, quinone system, fatty acids, the presence of *Sphingomonadaceae*-specific sphingoglycolipid in the polar lipid profile, and DNA G+C content thus proved that strain TDMA-17^T is a member of the genus *Sphingomonas* sensu stricto. The strain can be distinguished from its close relatives by its 16S rRNA gene sequence, fatty acid composition, and numerous physiological characteristics. On the basis of these results, we conclude that strain TDMA-17^T represents a novel species of the genus *Sphingomonas* sensu stricto; we propose the name *Sphingomonas astaxanthinifaciens* sp. nov. for this new astaxanthin-producing bacterium.

1.13. Description of Sphingomonas astaxanthinifaciens sp. nov

Sphingomonas astaxanthinifaciens (as.ta.xan.thi.ni.fa'ci.ens. N.L. neut. n. *astaxanthinum* astaxanthin; L. part. pres. *faciens* making/producing; N.L. adj. *astaxanthinifaciens* astaxanthin-producing).

Gram-negative, non-spore-forming oval or short motile rods. Strictly aerobic. Catalase positive and oxidase negative. Cells are 0.3–0.5 μm in diameter and 0.5–0.9 μm in length. Net-like structures that connected the cells may be observed. On nutrient agar, it forms red, circular, small colonies (0.5–1.0 mm in diameter) within 48 h. It contains red carotenoid pigments with high concentration of astaxanthin. Growth occurs between 20 and 45°C (optimum, 40°C) and between pH 5.5 and 11.0 (optimum, 7.0–8.0). Optimal growth occurs in the absence of NaCl; growth is inhibited by the presence of >0.25% (w/v) NaCl. Anaerobic growth does not occur. Aesculin, Tween 80, gelatin, casein, and starch are hydrolysed, while agar, DNA, cellulose, and chitin are not. Negative reactions are observed for glucose fermentation and acid production from all tested sugars. Indole and H_2S are not produced. Nitrate is not reduced. Catalase is positive. β-galactosidase, oxidase, tryptophan deaminase, lysine decarboxylase, ornithine decarboxylase, tryptophan deaminase, and urease activities are negative. Glucose, arabinose, mannose, mannitol, maltose, *N*-acetyl-glucosamine, gluconate, capric acid, adipic acid, malate, trisodium citrate, and phenylacetic acid are not assimilated. The major fatty acids are $C_{17:1}$ω6c (33.0%), $C_{18:1}$ω7c (20.8%) and Summed feature 4 (containing $C_{16:1}$ω7c and/or 2-OH iso-$C_{15:0}$) (16.8%).

The predominant hydroxy fatty acid is 2-OH $C_{18:1}$ (4.9%). Ubiquinone is predominantly Q-10. Glycosphingolipids are present. The DNA G+C content of the type strain is 67.7%.

The type strain is TDMA-17^T = NBRC 102146 = CCUG 53608 that was isolated from a fresh water sample collected at Misasa (Tottori, Japan).

1.14. Biotechnological Properties of Strain TDMA-17^T

Strain TDMA-17^T is a promising biotechnological source of astaxanthin. Its unique features are: (1) a high accumulation of carotenoids (2.8 mg g^{-1} dry cells), which has never been reported for a wild bacterial species before; (2) the procedure for extraction and purification of carotenoids from TDMA-17^T seem to be simpler than those from other microbial sources such as microalgae and yeasts as its carotenoids could be totally extracted using methanol (100%) and without any cell disrupting treatments. The presence of chlorophyll and glycerides in microalgae cells can complicate the purification process of astaxanthin, which is not the case with TDMA-17^T. In conclusion, strain TDMA-17^T is a promising source of astaxanthin. Improved production can be expected by obtaining a mutant with higher productivity or selectivity of astaxanthin, in which accumulates astaxanthin as end product (50).

2. Materials

2.1. Media and Buffers

1. Nutrient agar: 3 g/L beef extract, 5 g/L peptone, and 15 g/L agar (Difco Chemicals, Detroit, MI, USA).
2. PE medium: 4 g/L Bactopeptone (Difco Chemicals, Detroit, MI, USA), 2 g/L Bactoyeast extract (Difco Chemicals, Detroit, MI, USA), 5 g/L NaCl, 2 g/L fish meat extract (Kyokuto, Tokyo, Japan), 2 g/L $MgSO_4$, and 10 g/L maltose (pH 7.2).

2.2. Preparing Media and Cultivation

1. Analytical balance.
2. Flasks for cultivation.
3. Disposable Petri dishes.
4. Micropipettes and sterile tips (0–20 μL, 20–200 μL, and 100–1,000 μL).
5. Autoclave.
6. Incubators (28, 30, 37, 40, 50°C).
7. Sterile loops, sterile toothpicks, and sterile cotton swabs.
8. Clean bench equipped with UV light.

2.3. Carotenoids Analysis

1. HPLC grade solvents including dimethyl sulphoxide (DMSO), methanol, acetonitrile, tetrahydrofuran (THF).
2. Methanol:water (9:1, v/v).

3. Butylhydroxytoluene (BHT) (Sigma, St. Louis, MO, USA) (see Note 1).
4. Dim light (see Note 1).
5. Astaxanthin (Sigma, St. Louis, MO, USA).
6. Eppendorf tubes.
7. Racks.
8. Balance.
9. Eppendorf table top vial shaker.
10. Spectrophotometer.
11. 0.45 μm hydrophobic Teflon membrane syringe filter (Advantec, Tokyo, Japan) and syringe (1 mL).
12. 96-well plate.
13. HPLC automated injector with 96-well plate adaptor Class-VP (Shimadzu, Kyoto, Japan) with PDA detector.
14. HPLC/MS system LCMS-2010EV (Shimadzu, Kyoto, Japan).
15. HPLC column CAPCELLPAC C18 (35 mm × 4.6 mm; particle size, 5 μm) (Shiseido, Tokyo, Japan).
16. HPLC column Shim-Pack FC-ODS (150 mm × 4.6 mm; particle size, 5 μm) (Shimadzu, Kyoto, Japan).

2.4. Purification and Identification of a Novel Astaxanthin Derivative

1. Strain TDMA-17^T (53).
2. PE medium (see Subheading 2.1).
3. Shaker.
4. Incubators (28, 37, 40, 50°C).
5. HPLC grade solvents, including methanol, acetonitrile, tetrahydrofuran, chloroform, ethyl acetate, and hexane.
6. *n*-Hexane-ethyl acetate (95:5).
7. Purif-pack silica-gel chromatography (Moritex, Tokyo, Japan).
8. Preparative C18-TLC (Whatman, Piscataway, NJ, USA).
9. α-D-rhamnose.
10. HPLC Class-VP with photodiode array detector (PDA) Shimadzu SPD-M10AVP (Shimadzu, Kyoto, Japan).
11. Capcell-Pak C18 analytical column (20 mm i.d. × 250 mm).
12. NMR Varian Unity INOVA 500 (Agilent, Santa Clara, CA, USA).
13. Spectrometer JEOL JMN-600 (JEOL, Tokyo, Japan).
14. Spectrometer ESI-TOFMS JEOL JMS-T100LC AccuTOF (JEOL, Tokyo, Japan).
15. Spectrometer FT/IR 4000 (JASCO, Tokyo, Japan).
16. Spectropolarimeter J-720 (JASCO, Tokyo, Japan).

17. Lyophilizer.
18. OCC Dowex-1 (OH-type).
19. Silica gel column Wakogel C-200 (40 g).
20. HCl 4 N.

2.5. Identification of Bacterial Isolates: 16S rRNA Gene Sequences

1. Eppendorf tubes, toothpicks, and deionized water (all sterilized).
2. DNA purification kit (Edge BioSystems, Gaithersburg, MD, USA).
3. Primer B8F: 5′-AGAGTTATCCTGGCTCAG-3′.
4. Primer B1492R: 5′-GGTTACCTTGTTACGACTT-3′.
5. Thermocycler.
6. Ex Taq polymerase (Takara-shuzo, Kyoto, Japan).
7. Agarose.
8. 10× TBE buffer: 108 g Tris base, 55 g boric acid, and 40 mL 0.5 M EDTA (pH 8.0).
9. 1× TBE buffer: 100 mL of 10× TBE buffer and 900 mL of deionized water.
10. 6× blue loading dye: 30% glycerol in distilled water and 0.25% bromophenol blue.
11. Ethidium bromide (see Note 2).
12. Agarose gel electrophoresis system.
13. UV safety goggles and UV box.
14. 0.2-mL thin-walled PCR tubes.
15. BigDye Terminator v3.1 Cycle Sequencing Kit (Applied Biosystems, Foster City, CA, USA).
16. Pipettes and tips (0–10, 20–200, 100–1,000 μL).
17. Gene Clean kit II (Funakoshi, Tokyo, Japan).

2.6. Morphological, Physiological, and Biochemical Characterization

1. Nutrient agar plates and Nutrient broth (5 mL).
2. Incubators (4, 10, 20, 30, 37, 40 and 45°C).
3. BD™ Gram staining kit (BD, NJ, USA).
4. Zeiss Axioskop 2 microscope (Carl Zeiss MicroImaging, LLC, USA).
5. Scanning electron microscope (model VE-8800; Keyence, Tokyo, Japan).
6. GasPak anaerobic system (BBL).
7. Substrates for enzymatic activity testing: casein, tween 80, salmon DNA, starch, chitin, carboxymethyl cellulose, agar.
8. Cytochrome oxidase test strips from Fluka Biochemika (Sigma-Aldrich GmbH, Taufkirchen, Germany).

9. Carboxymethyl cellulose (high viscosity) (Sigma-Aldrich GmbH, Taufkirchen, Germany).
10. API 20E, API 20NE, and API 50CH (bioMérieux, Craponne, France).
11. *E. coli* C600.
12. Irradiation at 2.35 kGy (13 Gy min^{-1}, ^{137}Cs source).
13. 0.1 M HCl.
14. 0.1 M NaOH.

2.7. Chemotaxonomic Characterization

1. Sherlock Microbial Identification System Version 5.0 (MIDI, Inc. Newark, DE, USA).
2. HPLC automated injector with 96-well plate adaptor Class-VP with PDA detector (Shimadzu, Kyoto, Japan).
3. *Sphingopyxis alaskensis* DSMZ 13593^T (DSMZ culture collection, Braunschweig, Germany) and *Sphingomonas asaccharolytica* NBRC 15499^T (NBRC, Chiba, Japan) as a reference for Q-10.
4. TLC plates size 20 × 10 cm (Merck, NJ, USA).

3. Methods

3.1. Isolation of the Astaxanthin-Producing Bacterium

1. Collect water samples from the surface water layer (our samples were collected at Misasa radioactive site in Tottori, Japan).
2. Do serial dilutions with sterile water.
3. Spread 0.1-mL aliquots on Nutrient agar.
4. Incubate at 40°C for up to 3 days.
5. Do screening of colonies based on its color (the colonies of TDMA-17^T were isolated based on their red color).

3.2. Identification of Carotenoids

1. Extract the red pigments by agitating the bacterial cells in methanol on a rotary shaker (100 rpm) at 50°C in the dark until the cells are bleached (see Note 1).
2. Centrifuge at 5,000 × *g* for 5 min.
3. Scan the resultant supernatant for absorbance in the wavelength range ($\lambda = 260–700$ nm) at room temperature on a spectrophotometer.
4. Identify each carotenoid on a high-performance liquid chromatography (HPLC) system containing a photodiode array detector using two methods (B and C) (see Note 3).
5. Perform HPLC method B using a CAPCELLPAC C18 column (35 × 4.6 mm; particle size, 5 μm; column temperature,

35°C) and elute with methanol:water (9:1, v/v) at a flow rate of 1.0 mL/min.

6. Perform HPLC method C on a LC/MS system to determine the molecular weight of carotenoids. Apply each 10-μL sample onto a Shim-Pack FC-ODS (150×4.6 mm, particle size, 5 μm; column temperature, 35°C) and elute with 95% methanol at a flow rate 0.8 mL/min. Wash the column regularly (see Note 4).
7. Identify astaxanthin by retention time, absorption spectrum, and molecular weight, using commercial β-carotene and astaxanthin as standards. Use an extract of *Brevundimonas* sp. SD-212 (NBRC 101024) as a reference for dihydroxyastaxanthin(2,3,2′,3′- tetrahydroxy-β, β-carotene-4,4′-dione) (49).

3.3. Purification and Identification of a Novel Astaxanthin Derivative

All this procedure must be performed under dim light.

1. To isolate the major carotenoid pigment, culture strain TDMA-17^T in PE medium is shaken at 180 rpm for 48 h at 37°C.
2. Harvest the cells of TDMA-17^T by centrifugation at 9,000×*g*, and suspend them in five-times volume of methanol.
3. Shake at 180 rpm for 30 min at 28°C and filtrate.
4. Repeat this extraction process for three times (Fig. 4).
5. Combine the filtrates and dry up by evaporation.
6. Dissolve in $CHCl_3$.
7. Apply the $CHCl_3$ solution to Purif-pack silica-gel chromatography.
8. Wash the column with $CHCl_3$, and elute the pigment with $CHCl_3$/methanol (9:1).
9. Dry up the eluate, and dissolve it in an adequate volume of $CHCl_3$/methanol (9:1).
10. Apply onto a preparative C18-TLC developed with $CHCl_3$/methanol (4:1) to isolate the final principle. By this process, ca 56 mg of carotenoid 1 and 4.7 mg of astaxanthin can be obtained (Fig. 4).
11. Identify the purified carotenoid 1 by HPLC method B as described (Subheading 3.2).
12. Identify the UV–vis absorption spectra of carotenoid 1 in the HPLC eluent of MeOH using Shimadzu SPD-M10AVP photodiode array spectrophotometer.
13. Measure the NMR spectra of the purified carotenoid 1 a Varian Unity INOVA 500 or JEOL JMN-600 spectrometer.
14. Measure the ESI-TOFMS spectra of the purified carotenoid 1 using a JEOL JMS-T100LC AccuTOF.
15. Measure IR and CD spectra of the purified carotenoid 1 using a FT/IR 4000 spectrometer and J-720 spectropolarimeter of JASCO, respectively.

16. Determine the absolute configuration of rhamnose moieties of purified carotenoid 1:
    (a) Hydrolyze astaxanthin dirhamnoside 1 (7.2 mg) in a sealed tube with 4 M HCl at 105°C for 3 h.
    (b) Evaporate the solution to remove HCl, lyophilize the residue, and determine its yield.
    (c) Treat the obtained residue with 5% HCl-methanol (3 mL) at room temperature for 1 h.
    (d) Apply the reaction solution on a Dowex-1 (O^{H-} type) column.
    (e) Evaporate the passed-through solution from the column.
    (f) Benzoylate with benzoyl chloride (0.5 mL) in dry pyridine (4 mL) containing 4-dimethylaminopyridine (2 mg).
    (g) Add water into the reaction solution and extract with ethyl acetate.
    (h) Wash the ethyl acetate solution with saturated NaCl, saturated $NaHCO_3$, and water successively; dry with anhydrous Na_2SO_4; and evaporate.
    (i) Load the residue on a silica gel column (Wakogel C-200, 40 g) packed with *n*-hexane and elute stepwise with *n*-hexane-ethyl acetate (95 : 5, 1.5 L) and ethyl acetate (500 mL).
    (j) Wash ethyl acetate fraction (26.5 mg) with saturated NaCl, saturated $NaHCO_3$, and water successively and evaporate.
    (k) Purify the obtained residue by HPLC on a 250×20 mm i.d. Capcell-Pak $_{C18}$ column with a gradient of 60–100% acetonitrile in water in 15 min and then isocratic elution of 100% acetonitrile at flow rate of 5 mL/min to obtain methyl 2,3,4-tribenzoyl – α-rhamnopyranoside (retention time: 26.7 min; yield: 40 μg).
    (l) Prepare the authentic methyl 2,3,4-tribenzoyl-α-D-rhamnopyranoside from α-D-rhamnose (25.0 mg) by 5% HCl–methanol treatment and benzoylation using the same method mentioned above.
    (m) Compare CD spectrum of methyl 2,3,4-tribenzoyl-α-rhamnopyranoside and with that of authentic D-rhamnose.

3.4. Properties of the Novel Carotenoid

Antioxidant activity of compound 1 is measured by the inhibition of methyl linoleate hydroperoxides production due to the reaction of methyl linoleate with 2.2′-azobis(2,4-dimethylvaleronitrile) (AMVN) (67).

1. Prepare a reaction mixture containing: 50 μL methyl linoleate, 450 μL ethanol, 125 μL 0.15 mM AMVN, and 125 μL 1 mM carotenoid in $CHCl_3$.

2. Incubate at 37°C for 4 h with reciprocal shaking at 80 rpm under dark condition.
3. Analyze the reactant by normal phase HPLC using a Shodex SIL 5B column (4.6×250 mm) [mobile phase, *n*-hexane/2-propanol (99:1); flow rate, 1.0 mL/min; λ=235 nm), and measure for the content of the four kinds of hydroperoxides.
4. Express the antioxidant activity as a percentage of inhibition of hydroperoxide production due to the addition of carotenoids.
5. Measure solubility of compound 1 in water-based solvent by a spectroscopic measurement:
   (a) Dissolve carotenoid 1 and astaxanthin in 100% DMSO, acetone, and methanol at 20 mM.
   (b) Add water in each solution at various ratios.
   (c) Remove the insoluble pigment particles by filtration (pore size, 0.1 μm).
6. Measure the absorption of each solution at 475 nm.

3.5. Phylogenetic Characterization of the Astaxanthin-Producing Bacterium

1. Carefully transfer 50 mg wet cells from bacterial isolate into eppendorf tubes.
2. Extract genomic DNA using the bacterial genomic DNA purification kit.
3. Check the quality of the extracted genomic DNA by agarose gel electrophoresis (steps 4–9).
4. Prepare 1% gel solution by melting 1 g of agarose in 100 mL of 0.5× TBE buffer in a microwave for approximately 2 min. Allow to cool for 5 min then add 2.5 μL ethidium bromide, stir to mix (see Note 2).
5. Cast the gel using the supplied tray and comb. Allow the gel to set for a minimum of 30–40 min at room temperature.
6. Load into separate wells: 10 μL of 1-kb ladder, diluted samples (5 μL sample + 5 μL water + 2 μL 6× loading dye).
7. Run the gel for 30 min at 100 V.
8. Expose the gel to UV light and photograph (wear UV goggles).
9. Confirm DNA quality.
10. Set up PCR reaction mixtures in a final volume of 50 μL: 10 μL 10×PCR-buffer, 1 μL PCR Nucleotide Mix, 50 U Taq DNA polymerase, 1 μL (10 μM) primer B8F, 1 μL (10 μM) primer B1500R, 2 μL DNA extract, and 33 μL deionised and sterilized water.
11. Perform PCR using a protocol that includes an initial denaturation for a period of 5 min at 94°C; 30 cycles of 94°C for 1 min, 55°C for 1 min, and 72°C for 1 min; 72°C for 3 min; and incubation at 4°C until further processing.

12. Purify PCR amplicons using agarose gel electrophoresis (steps 13–19).
13. Prepare 2% agarose gel, and let it to solidify at room temperature for 30–40 min.
14. Add 10 μL 6× blue loading dye to each PCR product (50 μL).
15. Carefully load all 60 μL into wide gel well.
16. Run the gel for 30 min at 100 V.
17. Wear UV goggles. Remove gel and place on UV box to identify region with PCR products (see Note 2).
18. Cut DNA bands from agarose gel and transfer into sterilized eppendorf tubes.
19. Purify PCR products from the gel using the Gene Clean kit.
20. Determine the 16S rRNA gene sequences using the BigDye Terminator v3.1 cycle sequencing kit.

3.6. Phylogenetic Analysis

1. Compare the nearly complete 16S rRNA gene (about 1,396-1, 450 bp) sequences of isolates to those available in GenBank/EMBL/DDBJ nucleotide sequence database using the BLASTN program (http://www.ncbi.nlm.nih.gov/BLAST/).
2. Using ClustalW, align the obtained sequences with those of closed relatives.
3. Construct a neighbor-joining phylogenetic tree (see Fig. 8).
4. Use NJ plot to draw and edit the obtained tree.

3.7. Morphological, Physiological, and Biochemical Characterization

1. Cultivate TDMA-17^{T} cells grown on Nutrient agar for 48 h at 40°C unless specified otherwise (see Note 5).
2. Perform Gram staining according to the method described by Smibert & Krieg (93), and observe the bacterial cells under a microscope.
3. Observe the bacterial cells under scanning electron microscope according to the manufacturer's instructions. The specimen for scanning electron microscopy was prepared using the osmium fixation protocol and critical point drying (93).
4. Test the anaerobic growth of strain TDMA-17^{T} by incubation of inoculated nutrient agar plates in a GasPak anaerobic system.
5. Perform the following tests as described in each reference: casein hydrolysis (94); catalase activity, spore formation, and hydrolysis of Tween 80, DNA, and starch (93); and hydrolysis of chitin, carboxymethyl cellulose, and agar (95).
6. Test oxidase activity by using commercial cytochrome oxidase test strips.

7. Examine the other enzyme activities, growth on carbohydrates, acid production from carbohydrates, nitrate reduction, and the production of H_2S, indole, and acetoin using the commercial systems API 20E, API 20NE, and API 50CH, according to the manufacturer's instructions.
8. Determine the optimal growth temperature: cultivate the strain on NA at 4, 10, 20, 30, 37, 40, and 45°C.
9. Test the NaCl tolerance using Nutrient broth supplemented with different concentrations of NaCl.
10. Evaluate the growth at various pH values (pH 3.0–10.0) in NB adjusted with HCl or NaOH.
11. Determine the survival rate after exposure to gamma radiation, grow cells of TDMA-17^T on NA plate and suspended in NB (5 mL). Prepare aliquots of the cell suspension (1 mL) and irradiate them at 2.35 kGy (13 Gy min^{-1}, ^{137}Cs source), dilute serially, and plate onto Nutrient agar. Aerobically incubate the plates incubated at 37°C. Similarly treat *E. coli* C600. Determine the percentage of survival as by comparing with unirradiated cultures.

3.8. Chemotaxonomic Characterization

1. Determine the cellular fatty acids of strain TDMA-17^T:
   (a) Grow the cells aerobically on Nutrient agar at 40°C (see Note 5).
   (b) Extract and analyse the fatty acid methyl esters of TDMA-17^T according to the standard protocol of the Sherlock Microbial Identification System (Version 5·0).
2. Determine the respiratory quinone by an HPLC method (96) using cell extracts of *Sphingopyxis alaskensis* DSMZ 13593^T and *Sphingomonas asaccharolytica* NBRC 15499^T as a reference for Q-10.
3. Analyse glycosphingolipids using TLC as described previously (57).
4. Determine the DNA G+C content by an HPLC method (97).

4. Notes

1. Carotenoids are highly unstable and sensitive to light, oxygen, and temperature, which induce their oxidation. To minimize the oxidation of carotenoids, the extraction must be quickly done without exposure to light, oxygen, and high temperatures. Therefore, all the experimental procedures should be carried out in a dark room. Alternatively, dark tubes or tubes foiled with aluminum foil can be used. Nitrogen gas should be

also used for flushing the pigment extracts before fridge storage at –20°C to avoid oxidation. In addition, the use of an antioxidant such as butylhydroxytoluene (BHT) (98) can significantly reduce the oxidation of carotenoids during extraction.

2. Ethidium bromide is mutagenic and moderately toxic. You must wear gloves and handle the chemical with care.
3. For details, see HPLC methods B and C in Chapter 3.
4. Usually, DMSO–MeOH carotenoid extracts contain impurities such as salts, lipids, fatty compounds, proteins, and other cell compounds. These contaminants usually are adsorbed on the stationary phase of reversed-phase HPLC column, which may become contaminated by the repeated injection of samples. This causes shifts in retention times (0.5–1 min) and also tailing may occur. In addition, this may cause the column backpressure to build up to intolerably high levels, which overpressures the pump and can cause the column to settle and create a void, depending upon where the blockage occurs. To prevent this problem, we clean the columns on a regular basis throughout our work. Washing the HPLC columns with 100% acetonitrile and then with 100% methanol is sufficient to clean and regenerate the column. We strongly recommend using a pre-column.
5. For more details see the cited references.

References

1. Johnson EA, An GH (1991) Astaxanthin from microbial sources. Crit Rev Biotechnol 11:297–326
2. Nelis HJ, Deleenheer AP (1991) Microbial sources of carotenoid-pigments used in foods and feeds. J Appl Bacteriol 70:181–191
3. Britton G, Liaaen-Jensen S, Pfander H (1995) Carotenoids today and challenges for the future. Birkhauser, Basel
4. Rufer CE, Moeseneder J, Briviba K, Rechkemmer G, Bub A (2008) Bioavailability of astaxanthin stereoisomers from wild (*Oncorhynchus* spp.) and aquacultured (*Salmo salar*) salmon in healthy men: a randomised, double-blind study. Br J Nutr 99:1048–1054
5. Page GI, Davies SJ (2006) Tissue astaxanthin and canthaxanthin distribution in rainbow trout (*Oncorhynchus mykiss*) and Atlantic salmon (*Salmo salar*). Comp Biochem Physiol A Mol Integr Physiol 143:125–132
6. Fujita T, Satake M, Watanabe T, Kitajima C, Miki W, Yamaguchi K et al (1983) Pigmentation of cultured red sea bream with astaxanthin diester purified from krill oil. Bull Jpn Soc Sci Fish 49:1855–1865
7. Torrissen OJ (1989) Pigmentation of salmonids – interactions of astaxanthin and canthaxanthin on pigment deposition in Rainbow-Trout. Aquaculture 79:363–374
8. Bjerkeng B, Berge GM (2000) Apparent digestibility coefficients and accumulation of astaxanthin E/Z isomers in Atlantic salmon (*Salmo salar* L.) and Atlantic halibut (*Hippoglossus hippoglossus* L.). Comp Biochem Phys B 127:423–432
9. An GH, Choi ES (2003) Preparation of the red yeast, *Xanthophyllomyces dendrorhous*, as feed additive with increased availability of astaxanthin. Biotechnol Lett 25:767–771
10. Christiansen R, Glette J, Lie O, Torrissen OJ, Waagbo R (1995) Antioxidant status and immunity in Atlantic salmon, *Salmo salar* L, fed semi-purified diets with and without astaxanthin supplementation. J Fish Dis 18: 317–328
11. Christiansen R, Torrissen OJ (1997) Effects of dietary astaxanthin supplementation on

fertilization and egg survival in Atlantic salmon (*Salmo salar* L). Aquaculture 153:51–62
12. Lorenz RT, Cysewski GR (2000) Commercial potential for *Haematococcus microalgae* as a natural source of astaxanthin. Trends Biotechnol 18:160–167
13. Gorton HL, Williams WE, Vogelmann TC (2001) The light environment and cellular optics of the snow alga *Chlamydomonas nivalis* (Bauer) Wille. Photochem Photobiol 73: 611–620
14. Gorton HL, Vogelmann TC (2003) Ultraviolet radiation and the snow alga *Chlamydomonas nivalis* (Bauer) Wille. Photochem Photobiol 77:608–615
15. Kim JH, Chang HI (2006) High-level production of astaxanthin by *Xanthophyllomyces dendrorhous* mutant JH1, using chemical and light induction. J Microbiol Biotechnol 16: 381–385
16. Fan L, Vonshak A, Zarka A, Boussiba S (1998) Does astaxanthin protect *Haematococcus* against light damage? Z Naturforsch C 53: 93–100
17. Iwamoto T, Hosoda K, Hirano R, Kurata H, Matsumoto A, Miki W et al (2000) Inhibition of low-density lipoprotein oxidation by astaxanthin. J Atheroscler Thromb 7:216–222
18. O'Connor I, O'Brien N (1998) Modulation of UVA light-induced oxidative stress by beta-carotene, lutein and astaxanthin in cultured fibroblasts. J Dermatol Sci 16:226–230
19. Wang X, Willen R, Wadstrom T (2000) Astaxanthin-rich algal meal and vitamin C inhibit Helicobacter pylori infection in BALB/cA mice. Antimicrob Agents Chemother 44:2452–2457
20. Jyonouchi H, Sun S, Gross M (1995) Effect of carotenoids on in vitro immunoglobulin production by human peripheral blood mononuclear cells: astaxanthin, a carotenoid without vitamin A activity, enhances in vitro immunoglobulin production in response to a T-dependent stimulant and antigen. Nutr Cancer 23:171–183
21. Guerin M, Huntley ME, Olaizola M (2003) *Haematococcus* astaxanthin: applications for human health and nutrition. Trends Biotechnol 21:210–216
22. Yuan JP, Peng J, Yin K, Wang JH (2010) Potential health-promoting effects of astaxanthin: A high-value carotenoid mostly from microalgae. Mol Nutr Food Res 55:150–165
23. Tanaka T, Makita H, Ohnishi M, Mori H, Satoh K, Hara A (1995) Chemoprevention of rat oral carcinogenesis by naturally-occurring xanthophylls, astaxanthin and canthaxarathin. Cancer Res 55:4059–4064
24. Chew BP, Park JS, Wong MW, Wong TS (1999) A comparison of the anticancer activities of dietary beta-carotene, canthaxanthin and astaxanthin in mice in vivo. Anticancer Res 19:1849–1853
25. Uchiyama K, Naito Y, Hasegawa G, Katada K, Sakamoto N, Isozaki H et al (2002) Beneficial effects of astaxanthin in type 2 diabetes model of db/db mouse. Free Radical Bio Med 33:S211–S211
26. Naito Y, Uchiyama K, Aoi W, Hasegawa G, Nakamura N, Yoshida N et al (2004) Prevention of diabetic nephropathy by treatment with astaxanthin in diabetic db/db mice. Biofactors 20:49–59
27. Monroy-Ruiz J, Sevilla MA, Carron R, Montero MJ (2010) Astaxanthin-enriched-diet reduces blood pressure and improves cardiovascular parameters in spontaneously hypertensive rats. Pharmacol Res 63:44–50
28. Fassett RG, Coombes JS (2009) Astaxanthin, oxidative stress, inflammation and cardiovascular disease. Future Cardiol 5:333–342
29. Pashkow FJ, Watumull DG, Campbell CL (2008) Astaxanthin: a novel potential treatment for oxidative stress and inflammation in cardiovascular disease. Am J Cardiol 101:58D–68D
30. Gradelet S, Astorg P, Leclerc J, Chevalier J, Vernevaut MF, Siess MH (1996) Effects of canthaxanthin, astaxanthin, lycopene and lutein on liver xenobiotic-metabolizing enzymes in the rat. Xenobiotica 26:49–63
31. Suganuma K, Nakajima H, Ohtsuki M, Imokawa G (2010) Astaxanthin attenuates the UVA-induced up-regulation of matrix-metalloproteinase-1 and skin fibroblast elastase in human dermal fibroblasts. J Dermatol Sci 58: 136–142
32. Ikeuchi M, Koyama T, Takahashi J, Yazawa K (2006) Effects of astaxanthin supplementation on exercise-induced fatigue in mice. Biol Pharm Bull 29:2106–2110
33. Aoi W, Naito Y, Sakuma K, Kuchide M, Tokuda H, Maoka T et al (2003) Astaxanthin limits exercise-induced skeletal and cardiac muscle damage in mice. Antioxid Redox Signal 5:139–144
34. Comhaire FH, El Garem Y, Mahmoud A, Eertmans F, Schoonjans F (2005) Combined conventional/antioxidant "Astaxanthin" treatment for male infertility: a double blind, randomized trial. Asian J Androl 7:257–262
35. Augusti PR, Conterato GMM, Somacal S, Sobieski R, Spohr PR, Torres JV et al (2008) Effect of astaxanthin on kidney function

impairment and oxidative stress induced by mercuric chloride in rats. Food Chem Toxicol 46:212–219

36. Mayne ST (1996) Beta-carotene, carotenoids, and disease prevention in humans. FASEB J 10:690–701
37. Mortensen A (2009) Supplements. Nutrition and health. In: Britton G, Liaaen-Jensen S, Pfander H (eds) Vol 5. Birkhäuser Verlag, Basel, Boston, Berlin, pp 68–82
38. Bi W, Tian M, Zhou J, Row KH (2010) Task-specific ionic liquid-assisted extraction and separation of astaxanthin from shrimp waste. J Chromatogr B Analyt Technol Biomed Life Sci 878:2243–2248
39. Franco-Zavaleta ME, Jimenez-Pichardo R, Tomasini-Campocosio A, Guerrero-Legarreta I (2010) Astaxanthin extraction from shrimp wastes and its stability in 2 model systems. J Food Sci 75:C394–399
40. Pacheco N, Garnica-Gonzalez M, Ramirez-Hernandez JY, Flores-Albino B, Gimeno M, Barzana E et al (2009) Effect of temperature on chitin and astaxanthin recoveries from shrimp waste using lactic acid bacteria. Bioresour Technol 100:2849–2854
41. An GH, Song JY, Chang KS, Lee BD, Chae HS, Jang BC (2004) Pigmentation and delayed oxidation of broiler chickens by the red carotenoid, astaxanthin, from chemical synthesis and the yeast, Xanthophyllomyces dendrorhous. Asian Austral J Anim 17: 1309–1314
42. Bubrick P (1991) Production of astaxanthin from *Haematococcus*. Bioresour Technol 38: 237–239
43. Miller MW, Yoneyama M, Soneda M (1976) *Phaffia*, a new yeast genus in the *Deuteromycotina* (*Blastomycetes*). Int J Syst Bacteriol 26:286–291
44. Jacobson GK, Jolly SO, Sedmak JJ, Skatrud TJ, Wasileski JM (1999) Astaxanthin over-producing strains of *Phaffia rhodozyma*, methods for their cultivation, and their use in animal feeds. Patent US6015684
45. Harker M, Hirschberg J, Oren A (1998) *Paracoccus marcusii* sp. nov., an orange gram-negative coccus. Int J Syst Bacteriol 48(Pt 2): 543–548
46. Lee JH, Kim YS, Choi TJ, Lee WJ, Kim YT (2004) *Paracoccus haeundaensis* sp. nov., a Gram-negative, halophilic, astaxanthin-producing bacterium. Int J Syst Evol Microbiol 54:1699–1702
47. Tsubokura A, Yoneda H, Mizuta H (1999) *Paracoccus carotinifaciens* sp. nov., a new aerobic gram-negative astaxanthin-producing bacterium. Int J Syst Bacteriol 49(Pt 1): 277–282
48. Yokoyama A, Izumide H, Miki W (1994) Production of astaxanthin and 4-ketozeaxanthin by the marine bacterium, *Agrobacterium aurantiacum*. Biosci Biotech Biochem 58: 1842–1844
49. Yokoyama A, Miki W, Izumida H, Shizuri Y (1996) New Trihydroxy-keto-carotenoids isolated from an astaxanthin-producing marine bacterium. Biosci Biotechnol Biochem 60:200–203
50. Asker D, Isaka K (2005) Production of astaxanthin by microorganisms. Japan patent No. 2005–170393 (X1050492)
51. Calo P, Miguel TD, Sieiro C, Velazquez JB, Villa TG (1995) Ketocarotenoids in halobacteria: 3-hydroxy-echinenone and trans-astaxanthin. J Appl Bacteriol 79:282
52. Kametani K, Matsumura T (1983) Determination of ^{238}U, ^{234}U, ^{226}Ra and ^{228}Ra in spring waters of sanin district. Radioisotopes 32:18–21
53. Asker D, Beppu T, Ueda K (2007) Unique diversity of carotenoid-producing bacteria isolated from Misasa, a radioactive site in Japan. Appl Microbiol Biotechnol 77:383–392
54. Asker D, Beppu T, Ueda K (2007) *Sphingomonas astaxanthinifaciens* sp. nov., a novel astaxanthin-producing bacterium of the family *Sphingomonadaceae* isolated from Misasa, Tottori, Japan. FEMS Microbiol Lett 273:140–148
55. Yabuuchi E, Yano I, Oyaizu H, Hashimoto Y, Ezaki T, Yamamoto H (1990) Proposals of *Sphingomonas paucimobilis* gen. nov. and comb. nov., *Sphingomonas parapaucimobilis* sp. nov., *Sphingomonas yanoikuyae* sp. nov., *Sphingomonas adhaesiva* sp. nov., *Sphingomonas capsulata* comb. nov., and two genospecies of the genus *Sphingomonas*. Microbiol Immunol 34:99–119
56. Yabuuchi E, Kosako Y, Fujiwara N, Naka T, Matsunaga I, Ogura H et al (2002) Emendation of the genus *Sphingomonas* Yabuuchi et al. 1990 and junior objective synonymy of the species of three genera, *Sphingobium*, *Novosphingobium* and *Sphingopyxis*, in conjunction with *Blastomonas ursincola*. Int J Syst Evol Microbiol 52:1485–1496
57. Takeuchi M, Hamana K, Hiraishi A (2001) Proposal of the genus *Sphingomonas sensu stricto* and three new genera, *Sphingobium*, *Novosphingobium* and *Sphingopyxis*, on the basis of phylogenetic and chemotaxonomic analyses. Int J Syst Evol Microbiol 51: 1405–1417
58. Busse HJ, Denner EB, Buczolits S, Salkinoja-Salonen M, Bennasar A, Kämpfer P (2003)

Sphingomonas aurantiaca sp. nov., *Sphingomonas aerolata* sp. nov. and *Sphingomonas faeni* sp. nov., air- and dust-borne and Antarctic, orange-pigmented, psychrotolerant bacteria, and emended description of the genus *Sphingomonas*. Int J Syst Evol Microbiol 53:1253–1260

59. Ederer MM, Crawford RL, Herwig RP, Orser CS (1997) PCP degradation is mediated by closely related strains of the genus *Sphingomonas*. Mol Ecol 6:39–49
60. White DC, Sutton SD, Ringelberg DB (1996) The genus *Sphingomonas*: physiology and ecology. Curr Opin Biotechnol 7:301–306
61. Silva C, Cabral JM, Van Keulen F (2004) Isolation of a beta-carotene over-producing soil bacterium, *Sphingomonas* sp. Biotechnol Lett 26:257–262
62. Asker D, Amano S, Morita K, Tamura K, Sakuda S, Kikuchi N et al (2009) Astaxanthin dirhamnoside, a new astaxanthin derivative produced by a radio-tolerant bacterium, *Sphingomonas astaxanthinifaciens*. J Antibiot 62:397–399
63. Nelis HJ, De Leenheer AP (1989) Profiling and quantitation of bacterial carotenoids by liquid chromatography and photodiode array detection. Appl Environ Microbiol 55: 3065–3071
64. Britton G, Liaaen-Jensen S, Pfander H (2004) Carotenoids handbook. Birkhäuser, Basel, Switzerland
65. Yang SW, Zhou BN, Wisse JH, Evans R, van der Werff H, Miller JS et al (1998) Three new ellagic acid derivatives from the bark of Eschweilera coriacea from the Suriname rainforest. J Nat Prod 61:901–906
66. Bock K, Pedersen C (1974) A study of 13CH coupling constants in hexopyranoses. J Chem Soc Perkin Trans II; 293–297
67. Terao J (1989) Antioxidant activity of beta-carotene-related carotenoids in solution. Lipids 24:659–661
68. Kleinig H, Reichenbach H, Achenbach H, Stadler J (1971) Carotenoid pigments of Sorangium compositum (Myxobacterales) including two new carotenoid glucoside esters and two new carotenoid rhamnosides. Arch Mikrobiol 78:224–233
69. Hertzberg S, Liaaen-Jensen S (1969) Carotenids of blue-green algae: 5. The structure of oscillaxanthin Phytochemistry 8: 1281–1292
70. Halfen LN, Francis GW (1972) The influence of culture temperature on the carotenoid composition of the blue-green alga, Anacystis nidulans. Arch Mikrobiol 81:25–35
71. Cowan ST (1968) A dictionary of microbial taxonomic usage. Oliver & Boyd, Edinburgh
72. Staley JT, Krieg NJ (1984) Classification of prokaryotic organisms:an overview. The Williams & Wilkins Co., Baltimore, pp 1–3
73. Vandamme P, Pot B, Gillis M, de Vos P, Kersters K, Swings J (1996) Polyphasic taxonomy, a consensus approach to bacterial systematics. Microbiol Rev 60: 407–438
74. Colwell RR (1970) Polyphasic taxonomy of the genus Vibrio: numerical taxonomy of *Vibrio cholerae*, *Vibrio parahaemolyticus*, and related *Vibrio* species. J Bacteriol 104: 410–433
75. Murray RGE, Brenner DJ, Colwell RR, Devos P, Goodfellow M, Grimont PAD et al (1990) Report of the Ad-Hoc committee on approaches to taxonomy within the *Proteobacteria*. Int J Syst Bacteriol 40: 213–215
76. Tindall BJ, Rossello-Mora R, Busse HJ, Ludwig W, Kampfer P (2010) Notes on the characterization of prokaryote strains for taxonomic purposes. Int J Syst Evol Microbiol 60:249–266
77. Egan S, Thomas T, Kjelleberg S (2008) Unlocking the diversity and biotechnological potential of marine surface associated microbial communities. Curr Opin Microbiol 11: 219–225
78. Schleifer KH, Ludwig W (1989) Phylogenetic relationships of bacteria. Elsevier Science Publishers B. V, Amsterdam
79. Stackebrandt E, Frederiksen W, Garrity GM, Grimont PAD, Kämpfer P, Maiden MCJ et al (2002) Report of the ad hoc committee for the re-evaluation of the species definition in bacteriology. Int J Syst Evol Microbiol 52:1043–1047
80. Stackebrandt E, Goebel BM (1994) A place for DNA-DNA reassociation and 16s ribosomal-Rna sequence-analysis in the present species definition in bacteriology. Int J Syst Bacteriol 44:846–849
81. Woese CR (1987) Bacterial evolution. Microbiol Rev 51:221–271
82. Fox GE, Wisotzkey JD, Jurtshuk P JR (1992) 16S rRNA sequence identity may not be sufficient to guarantee species identity. Int J Syst Bacteriol 42:166–170
83. Martinez-Murcia AJ, Collins MD (1990) A phylogenetic analysis of the genus Leuconostoc based on reverse transcriptase sequencing of 16S rRNA. FEMS Microbiol Lett 70:73–83

84. Wayne LG, Brenner DJ, Colwell RR, Grimont PAD, Kandler O, Krichevsky MI et al (1987) Report of the Ad Hoc committee on reconciliation of approaches to bacterial systematics. Int J Syst Bacteriol 37:463–464
85. Johnson JL (1984) Nucleic acids in bacterial classification. Williams & Wilkins, Baltimore
86. Yarza P, Richter M, Peplies J, Euzeby J, Amann R, Schleifer K-H et al (2008) The all-species living tree project: A 16 S rRNA-based phylogenetic tree of all sequenced type strains. Syst Appl Microbiol 31:241–250
87. Brosius J, Palmer ML, Kennedy PJ, Noller HF (1978) Complete nucleotide sequence of a 16 S ribosomal RNA gene from *Escherichia coli*. Proc Natl Acad Sci U S A 75: 4801–4805
88. Busse HJ, Kämpfer P, Denner EB (1999) Chemotaxonomic characterisation of *Sphingomonas*. J Ind Microbiol Biotechnol 23:242–251
89. Busse HJ, Hauser E, Kämpfer P (2005) Description of two novel species. *Sphingomonas abaci* sp. nov. and *Sphingomonas panni* sp. nov. Int J Syst Evol Microbiol 55:2565–2569
90. Yang DC, Im WT, Kim MK, Ohta H, Lee ST (2006) *Sphingomonas soli* sp. nov., a beta-glucosidase-producing bacterium in the family *Sphingomonadaceae* in the alpha-4 subgroup of the *Proteobacteria*. Int J Syst Evol Microbiol 56:703–707
91. Takeuchi M, Hiraishi A (2001) Taxonomic significance of 2-hydroxy fatty acid profiles of the species in the genus *Sphingomonas* and related taxa. IFO Res Commun 20:72–82
92. Collins MD, Jones D (1981) Distribution of isoprenoid quinone structural types in bacteria and their taxonomic implication. Microbiol Rev 45:316–354
93. Smibert RM, Krieg NR (1994) Phenotypic characterization. In: Gerhardt P, Murray RGE, Wood WA, Krieg NR (eds) Methods for general and molecular bacteriology. American Society for Microbiology, Washington, DC, pp 607–654
94. Norris JR, Ribbons DW, Varma AK (1985) Methods in microbiology. Academic, London
95. Cowan ST, Steel KJ (1993) Manual for the identification of medical bacteria, 3rd ed. Cambridge University Press, London
96. Collins MD (1994) Isoprenoid quinones. In: O'Donnell MGAG (ed) Chemical methods in prokaryotic systematics. Wiley, Chichester, pp 265–310
97. Mesbah M, Whitman WB (1989) Measurement of deoxyguanosine/thymidine ratios in complex mixtures by high-performance liquid chromatography for determination of the mole percentage guanine + cytosine of DNA. J Chromatogr 479:297–306
98. Meyers SP, Bligh D (1981) Characterization of astaxanthin pigments from heat-processed crawfish waste. J Agric Food Chem 29: 505–508
99. Saitou N, Nei M (1987) The neighbor-joining method: a new method for reconstructing phylogenetic trees. Mol Biol Evol 4:406–425
100. Del Rio E, Acien FG, Garcia-Malea MC, Rivas J, Molina-Grima E, Guerrero MG (2008) Efficiency assessment of the one-step production of astaxanthin by the microalga *Haematococcus pluvialis*. Biotechnol Bioeng 100:397–402
101. Liu YS, Wu JY (2006) Hydrogen peroxide-induced astaxanthin biosynthesis and catalase activity in *Xanthophyllomyces dendrorhous*. Appl Microbiol Biotechnol 73:663–668
102. Yokoyama A, Izumida H, Miki W (1994) Production of astaxanthin and 4-ketozeaxanthin by the marine bacterium, *Agrobacterium aurantiacum*. Biosci Biotech Biochem 58:1842–1844
103. Osanjo GO, Muthike EW, Tsuma L, Okoth MW et al (2009) A salt lake extremophile, *Paracoccus bogoriensis* sp nov, efficiently produces xanthophyll carotenoids. Afr J Microbiol Res 3(8):426–433
104. Takaichi S, Matsui K, Nakamura M, Muramatsu M, Hanada S (2003) Fatty acids of astaxanthin esters in krill determined by mild mass spectrometry. Comp Biochem Physiol B Biochem Mol Biol 136:317–322
105. Breithaupt DE (2004) Identification and quantification of astaxanthin esters in shrimp (*Pandalus borealis*) and in a microalga (*Haematococcus pluvialis*) by liquid chromatography-mass spectrometry using negative ion atmospheric pressure chemical ionization. J Agric Food Chem 52:3870–3875
106. Matsuno T, Maoka T, Katsuyama M, Ookubo M, Katagiri K, Jimura H (1984) The occurence of enantiomeric and meso-astaxanthin in aquatic animals. Bull Jap Soc Sci Fish 50:1589–1592
107. Johnson EA, Schroeder W (1995) Astaxanthin from the yeast *Phaffia rhodozyma*. Stud Mycol. 81–90
108. Rønneberg H, Renstrom B, Aareskjold K, LiaaenJensen S, Veechi M, Leuenberger FJ et al (1980) Natural occurrence of enatiomeric and meso-astaxanthin 1. Ex Lobster Eggs (Homarus gammarus). Helv Chim Acta 63:711–715
109. Bernhard K, Englert G, Meister W, Vecchi M, Renstrom B, Liaaen-Jensen S (1982)

Carotenoids of the carotenoprotein asteriarubin. Optical purity of asterinic acid. Helv Chim Acta 65:2224–2229

110. Yokoyama A, Adachi K, Shizuri Y (1995) New carotenoid glucosides, astaxanthin glucoside and adonixanthin glucoside, isolated from the Astaxanthin-producing marine bacterium, *Agrobacterium aurantiacum*. J Nat Prod 58:1929–1933
111. Misawa N, Shimada H (1998) Metabolic engineering for the production of carotenoids in non-carotenogenic bacteria and yeasts. J Biotechnol 59:169–181
112. Maoka T, Tsushima M, Matsuno T (1989) New acetylenic carotenoids from the starfishes *Asterina pectinifera* and *Asterias amurensis*. Comp Biochem Physiol 93:829–834
113. Maoka T, Katsuyama M, Kaneko N, Matsuno T (1985) Stereochemical investigation of the carotenoids in the antarctic krill *Euphausia superba*. Bull Jap Soc Sci Fish 51:1671–1673
114. Takaichi S, Matsui K, Nakamura M, Muramatsu M, Hanada S (2003) Fatty acids of astaxanthin esters in krill determined by mild mass spectrometry. Comp Biochem Physiol Part B: Biochem Mol Biol 136: 317–322
115. Hertzberg S, Ramdahl T, Johansen J, Liaaen-Jensen S (1983) Carotenoid sulfates: 2. Structural elucidation of bastaxanthin. Acta Chem Scand 37:267–280
116. Denner EB, Kämpfer P, Busse HJ, Moore ER (1999) Reclassification of *Pseudomonas echinoides* Heumann 1962, 343AL, in the genus *Sphingomonas* as *Sphingomonas echinoides* comb. nov. Int J Syst Bacteriol 49(Pt 3): 1103–1109
117. Yoon JH, Lee MH, Kang SJ, Lee SY, Oh TK (2006) *Sphingomonas dokdonensis* sp. nov., isolated from soil. Int J Syst Evol Microbiol 56:2165–2169
118. Buonaurio R, Stravato VM, Kosako Y, Fujiwara N, Naka T, Kobayashi K et al (2002) *Sphingomonas melonis* sp. nov., a novel pathogen that causes brown spots on yellow Spanish melon fruits. Int J Syst Evol Microbiol 52:2081–2087
119. Lee JS, Shin YK, Yoon JH, Takeuchi M, Pyun YR, Park YH (2001) *Sphingomonas aquatilis* sp. nov., *Sphingomonas koreensis* sp. nov., and *Sphingomonas taejonensis* sp. nov., yellow-pigmented bacteria isolated from natural mineral water. Int J Syst Evol Microbiol 51: 1491–1498

Chapter 5

Novel Zeaxanthin-Producing Bacteria Isolated from a Radioactive Hot Spring Water

Dalal Asker, Tarek S. Awad, Teruhiko Beppu, and Kenji Ueda

Abstract

Zeaxanthin is a powerful antioxidant that is widely found in vegetables and fruits. Epidemiological evidences suggest that increasing the consumption of zeaxanthin in the diet is associated with a lower risk of age-related macular degeneration, helps prevent glaucoma and cataracts, and supports normal eye health. Zeaxanthin is a promising nutraceutical with many applications in the feed, food, and pharmaceutical industries. Currently, the commercial production of zeaxanthin is still dependant on synthetic routes with limitation for the biological one. Nevertheless, the biotechnological production of zeaxanthin is emerging due to its safety, potential large-scale production, and consumers' demand and preference for natural additives. Using a rapid screening method based on 16S rRNA gene and effective high-performance liquid chromatography (HPLC)-Diodearray-MS methods for carotenoids' analysis, we isolated effective zeaxanthin-producing bacteria (strain TDMA-5^{T} and -16^{T}) that belong to the family *Sphingobacteriaceae* and *Sphingomonadaceae*, respectively. In this chapter, we provide a detailed description of the HPLC-Diodearray-MS methods used for rapid analysis and identification of the carotenoids produced by both strains. In addition, the polyphasic taxonomic analysis of both novel strains and the description of a novel species and genus are described.

Key words: Zeaxanthin, Carotenoids, Antioxidant, *Sphingobacteriaceae*, *Sphingomonadaceae*, Screening, 16S rRNA, Phylogenetic, HPLC

1. Introduction

1.1. Structure, Function, and Importance of Zeaxanthin

Zeaxanthin is a natural yellow pigment and the main oxygenated carotenoid (xanthophyll) found in the common yellow maize corn, from which its name (pronounced zee-uh-zan'-thin) is derived. It is a phytochemical that is naturally found in egg yolks and some of the green, orange, and yellow vegetables and fruits, such as spinach, collard greens, lettuce, broccoli, zucchini, kiwifruit, garden peas,

José-Luis Barredo (ed.), *Microbial Carotenoids from Bacteria and Microalgae: Methods and Protocols*, Methods in Molecular Biology, vol. 892, DOI 10.1007/978-1-61779-879-5_5, © Springer Science+Business Media, LLC 2012

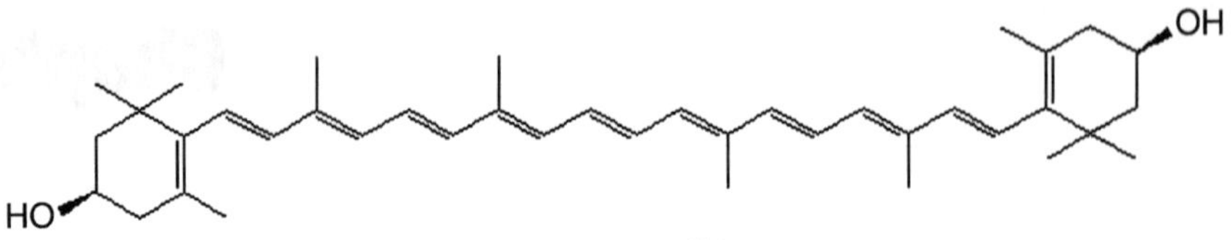

zeaxanthin

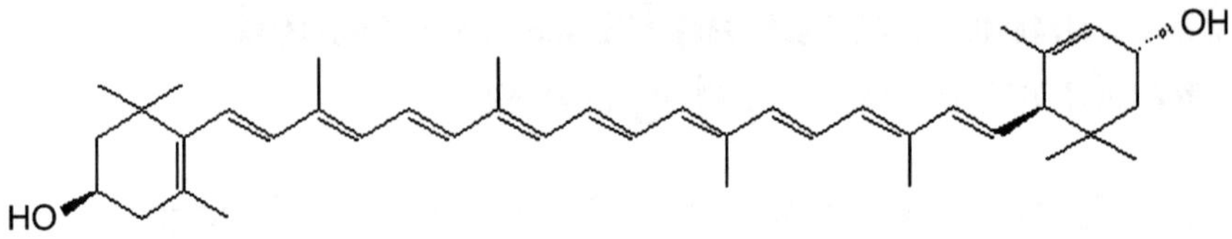

lutein

Fig. 1. Chemical structure of zeaxanthin and lutein.

alfalfa, and marigold flowers (1–6). Zeaxanthin (3,3′-dihydroxy-β-carotene) is a dihydroxy-carotenoid that has a chemical formula of $C_{40}H_{56}O_2$ and a molecular weight of 568.88 Da. Its chemical structure includes 40 carbon atoms with 11 conjugated ethylenic double bonds and 2 rings on both (Fig. 1). Zeaxanthin and lutein (another carotenoid) are isomers that differ from each other just by the shift of a single double bond.

Zeaxanthin exhibits no vitamin A activity but has powerful antioxidant protective benefits. Like other carotenoids, zeaxanthin can be synthesized by photosynthetic organisms (e.g., plant and microalgae) and some nonphotosynthetic microorganisms (e.g., yeast, fungi, and bacteria) (7–11), where it serves as an accessory light-gathering pigment as well as a protectant against the toxic effects of ultraviolet radiation and oxygen radicals. Animals and human cannot synthesize zeaxanthin and thus must obtain from the diet. To enhance the pigmentation of animal flesh and egg yolk, zeaxanthin is supplemented as an ingredient in feed for fish, birds, swine, and poultry (5). It is also suitable for use as a colorant in the cosmetics and food industries (12). Zeaxanthin and its isomer (lutein) are important antioxidants that are naturally condensed in the retina of the eye, where they act as UV filters to protect the eye against photooxidative damage from UV light from the sun, which reduces the risk of age-related macular degeneration (AMD) and cataracts, two of the leading causes of blindness in the world (13–20). The macular pigments can be increased in primates either by increasing the intake of foods rich in lutein and zeaxanthin or supplementation. In addition, epidemiological studies have evidenced that zeaxanthin has antioxidant (i.e., inhibition of LDL oxidation) (21) and cancer-preventive properties (22). These important health benefits multiplied the number of industrial applications of zeaxanthin, which thus increased the commercial demand for its production.

1.2. Microbial Production of Zeaxanthin

At present, the commercial production of zeaxanthin is mostly carried out either by the extraction from plant tissues such as marigolds and alfalfa (23) or chemical synthesis. Synthetic zeaxanthin is prepared by the Wittig reaction, which yields 96–98% of trans-(3R,3R)-zeaxanthin and minor quantities of impurities (24). A biological route of zeaxanthin extraction is from marigolds (*Tagetes erecta*) flowers; however, the produced zeaxanthin has low purity as the extract comprises only about 20% of trans-zeaxanthin, various amounts of other carotenoids, fats, oils, waxes, and other organic solvent extractable, mainly unsaponifiable compounds originating from plant material (24). A comparative study has shown that the bioavailability of zeaxanthin produced from a microbial source (*Flavobacterium multivorum*) is two to three times higher than that produced from marigold, which also had low stability (25). Different groups of microorganisms produce zeaxanthin including microalgae, yeast, and bacteria (Table 1). However, microbial production at the commercial level has not yet been established. A demand for natural zeaxanthin is now emerging in the pharmaceutical industry owing for its health benefits as antioxidant for reducing the risk of AMD and cataracts, and cancer-preventive properties.

1.3. Screening of Novel Zeaxanthin-Producing Bacteria

We screened for high-value carotenoid-producing bacteria from freshwater samples that were collected at Misasa (Tottori, Japan), a region known for the high natural radioactivity content (26) (see Chapter 3). Using a rapid screening method, comprehensive taxonomic characterization based on 16S rRNA gene sequences, and rapid and effective high-performance liquid chromatography (HPLC)-Diodearray-MS methods for carotenoids' analysis, we discovered a unique diversity of carotenoid producers including six yellow pigmented strains (TDMA-2, TDMA-5, TDMA-12, TDMA-14, TDMA-15, and TDMA-16) that produce zeaxanthin and its derivatives (27). From 16S rRNA gene sequences and phylogenetic analysis, we found that these strains belong to three different bacterial classes (α-*Proteobacteria*, *Flavobacteria*, and *Sphingobacteria*) and may represent novel taxa at the species or genus level (27). Importantly, the HPLC analysis of carotenoids revealed that the strains with noteworthy production efficiency of zeaxanthin were TDMA-5 (*Flavobacteria*) and TDMA-16 (*Sphingobacteria*), which represented novel taxa that belong to the family *Sphingobacteriaceae* and family *Sphingomonadaceae*, respectively (27). Both strains were further studied for their biotechnological interest (28, 29). In this chapter, we summarize their potential as novel bacteria for the production of zeaxanthin.

The family *Sphingobacteriaceae* belongs to the phylum *Bacteroidetes* was first proposed by Steyn et al. (30), accommodates the genera *Sphingobacterium* and *Pedobacter*. Two additional genera, *Olivibacter* and *Parapedobacter*, have been recently

Table 1
Reported microorganisms for production of zeaxanthin

Microorganism	Yield	References
Cyanobacteria		
Microcystis aeruginosa		(74)
Spirulina sp.		(75)
Synechococcus sp. PCC 7942 (formally *Anacystis nidulans*)		(76)
Prochloron sp.		(77)
Green algae		
Neospongiococcum excentricum	0.65%	(78)
Dunaliella salina zea1 (mutant)	6 mg/g	(79)
Dunaliella parva		(80)
Dunaliella bardawil		(81)
Blue-green algae		
Cyanophora paradoxa and *Glaucocystis nostochinearum*		(82)
Yellow-green algae		
Pleurochloris commutata		(82)
Red algae		
Corallina officinalis, *C. elongate* and *Jania* sp.		(83)
Yeast		
Phaffia rhodozyma (recombinant)		(10)
Bacteria		
Flavobacterium sp. (ATCC 21588)	16 mg/g	(84)
Sphingobacterium (formerly *Flavobacterium*) *multivorum* ATCC 55238	10.6 μg/ml	(85)
Synechocystis sp. PCC 6803	0.98 μg/ml	(86)
Erwinia herbicola		(87)
Paracoccus zeaxanthinifaciens (formerly *Flavobacterium* sp. R-1 512)		(9, 88)
E. coli (metabolic engineered strain)	1.6 mg/g	(89)
Zeaxanthinibacter enoshimensis		(7)
Mesoflavibacter zeaxanthinifaciens		(8)

proposed by Ntougias et al. (31) and Kim et al. (32). The genus *Sphingobacterium*, the type genus of this family, was originally described by Yabuuchi et al. (33) and at present comprises *Sphingobacterium antarcticum*, *S. faecium*, *S. multivorum*, *S. spiritivorum*, *S. thalpophilum*, *S. mizutaii*, and *S. daejeonense* (30, 33–36). Steyn et al. (30) proposed the genus *Pedobacter* initially to accommodate four heparinolytic species *Pedobacter africanus*, *P. heparinus*, *P. piscium*, and *P. saltans*; nine further species, *Pedobacter caeni*, *P. cryoconitis*, *P. himalayensis*, *P. sandarakinus*, *P. aquatilis*, *P. ginsengisoli*, *P. suwonensis*, *P. panaciterrae*, and *P. roseus* have since been added (30, 37–45).

The family *Sphingomonadaceae* that belongs to the phylum *Alphaproteobacteria* was first proposed by Koasko et al. (46). Subsequently, it was emended for description (47, 48), and included in the second edition of Bergey's Manual of systematic Bacteriology (46). Currently, the family comprises the following genera: *Sphingomonas*, *Porphyrobacter*, *Erythrobacter*, *Erythromicrobium*, *Zymomonas*, *Sandaracinobacter*, and *Sphingosinicella*. The genus *Sphingomonas*, the type genus of this family that was originally described by Yabuuchi et al. (49), has recently been proposed for the dissection into four genera including *Sphingomonas sensu stricto* and three new genera, *Sphingobium*, *Novosphingobium*, and *Sphingopyxis* (47). *Sphingomonads* are strictly aerobic, chemo-heterotrophic, Gram-negative bacteria that are of biotechnological interest due to their degradation capabilities for various xenobiotic substances (50–52) and also their potential to produce useful exopolysaccharides (53) and carotenoids (54).

This chapter deals with identification of carotenoids produced by these organisms and their taxonomic affiliation based on the results of polyphasic taxonomic study of the isolated zeaxanthin-producing strains (28, 29).

1.4. Identification of Zeaxanthin Production by Novel Isolates

In Chapter 3, we summarized rapid and effective HPLC methods with Diodearray or MS spectroscopy for the analysis and identification of carotenoids in their natural biological mixtures. These methods were also used for the analysis and identification of the carotenoids produced by strains TDMA-5^T and TDMA-16^T (28, 29). Extraction of carotenoids from the wet cells of both strains was done as described in Chapter 3. The yellow pigments of strains TDMA-5^T and TDMA-16^T were identified as carotenoids. The amount of carotenoids produced by strains TDMA-5^T was slightly high (0.8 mg/g dry cells of zeaxanthin). Analysis of the crude pigment isolated from the cells of strain TDMA-5^T using HPLC method A showed the presence of one sharp peak at $R_t = 0.98$ min, which is identical to zeaxanthin standard (Fig. 2a). In addition, the UV-visible absorption spectrum of this peak was also identical to that of zeaxanthin, which exhibited absorption maximum at 451 and 476 nm (Fig. 2b). These results, thus, indicated that zeaxanthin was the predominant carotenoid produced by TDMA-5^T.

TDMA-16^T produced significantly high amount of carotenoids (1.7 mg/g dry cells), which comprised a mixture of zeaxanthin, nostoxanthin, and an unknown carotenoid. Analysis of its crude pigment using HPLC (method A) showed the presence of one broad peak at $R_t = 0.8–1.1$ min. However, HPLC method B showed the presence of four pigments (corresponding to peaks 1–4 in Fig. 3a). Table 2 summarizes the spectroscopic data for the HPLC four peaks of carotenoids produced by strain TDMA-16^T. The UV-visible absorption spectra of the first and third peaks (Fig. 3b) were identical to the spectrum of zeaxanthin standard as they

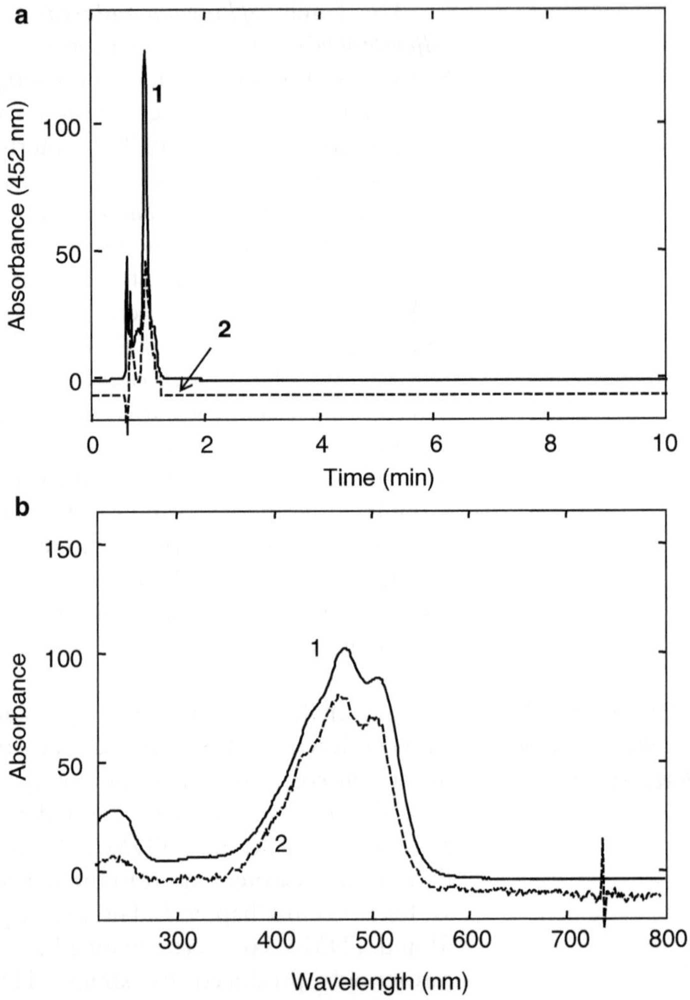

Fig. 2. Absorption profile of the carotenoids produced by strain TDMA-5[T]. Elution profile (λ_{max} 452 nm) (**a**) and absorption spectrum for each peak (**b**) are shown. The carotenoid fraction was analyzed by reverse-phase liquid chromatography using method A (see Subheadings 2 and 3). The fraction consisted of zeaxanthin as a major carotenoid.

exhibited absorption maxima at 451 and 476 nm. The zeaxanthin standard showed identical R_t and spectroscopic data as those for peak 3. Therefore, based on the spectroscopic data (Table 2), we identified the compound corresponding to peak 3 (27.5% of the total carotenoids) as zeaxanthin. The UV-visible absorption spectrum of peak 2 (2.5% of the total carotenoids) (Fig. 3b) was at 449 and 475 nm, which is identical to nostoxanthin that was extracted from cells *Sphingomonas asaccharolytica* NBRC 15499[T]. Zeaxanthin was further confirmed by HPLC-Mass (method C) [retention time, R_t at 10.3 min; and molecular weight ($M^+ + 1$), 569] (Table 2). The compounds corresponding to peak 4 (15.7% of the total carotenoids), which showed asymmetrical spectra, were predicted to be 13-cis-zeaxanthin by comparing their UV/VIS spectra to those

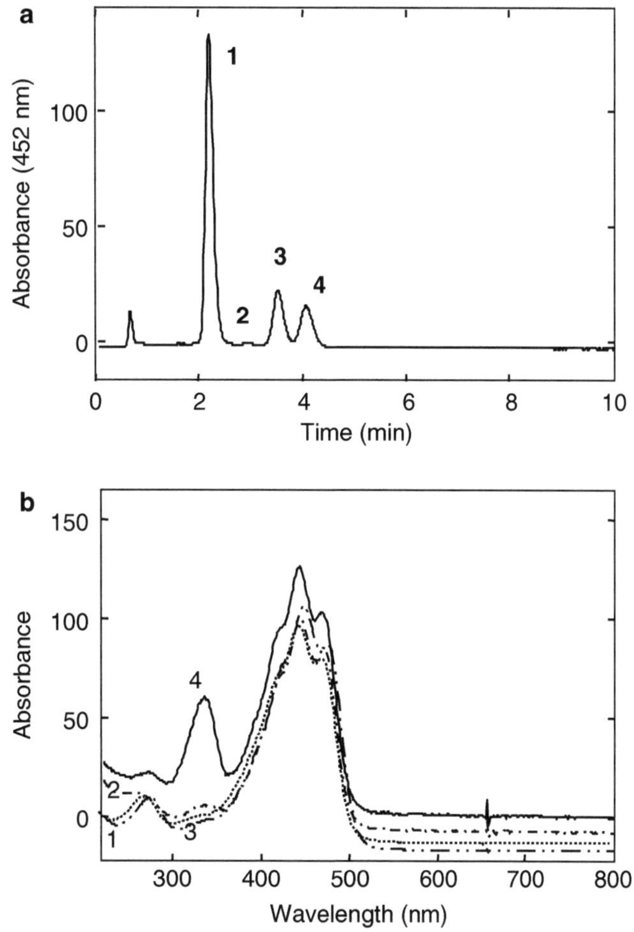

Fig. 3. Absorption profile of the carotenoids produced by strain TDMA-16^T. Elution profile (λ_{max} 452 nm) (**a**) and absorption spectrum of each peak (**b**) are shown. The carotenoid fraction was analyzed by reverse-phase liquid chromatography using method B (see Subheadings 2 and 3). The fraction consisted of unidentified carotenoid (peak 1), dihydroxyzeaxanthin (peak 2), zeaxanthin (peak 3), 9-*cis*-zeaxanthin (peak 4), and 13-*cis*-zeaxanthin (peak 5).

Table 2
The spectroscopic data for the HPLC five peaks of carotenoids produced by strain TDMA-16^T

HPLC peak	Carotenoids	Absorption (l_{max}, nm)	HPLC, retention time (R_t, min) Method B	Method C	Mw ($M^+ + 1$)
1.	Unknown	451–476	1.24	3.4	861
2.	Nostoxanthin (dihydroxyzeaxanthin)	449–475	2.29	6.5	601
3.	Zeaxanthin	451–476	3.85	10.3	569
4.	13-*cis*-zeaxsnthin	350–370, 451–476	4.27	11.04	569

published (55). It is noteworthy to mention that we failed to identify the major constituent that corresponds to peak 1 (54.3% of the total carotenoids). It exhibited the same spectra of zeaxanthin but with higher molecular weight and polarity. Therefore, this unknown was primarily identified as zeaxanthin derivative. In addition, the presence of nostoxanthin in TDMA-16^T is consistent with its presence in *S. paucimobilis* and *S. echinoides* (56, 57).

1.5. 16S rRNA Gene-Based Phylogenetic Analysis by Novel Isolates

The phylogenetic position of the isolates was studied by the standard analysis based on 16S rRNA gene sequence. A sequence similarity search in the GenBank/EMBL/DDBJ nucleotide sequence database performed using the BLASTN program (http://www.ncbi.nlm.nih.gov/BLAST/) revealed that TDMA-5^T belonged to the family *Sphingobacteriaceae*. The phylogenetic analysis revealed that strain TDMA-5^T formed a clade together with several uncharacterized bacterial strains with sequence similarities of 96–97%. This clade did not cluster robustly with any recognized species or genus (Fig. 3). TDMA-5^T was nearest to the members of the genus *Pedobacter* (except *P. saltans*), with 16S rRNA gene sequence similarity ranging from 90.3 to 93.4%. TDMA-5^T exhibited low 16S rRNA gene sequence similarity values to *P. saltans* (89.4%). Sequence similarities between TDMA-5^T and other relatives ranged from 87.6 to 89.8%; they include *Sphingobacterium* spp. (87.6–88.9%), *Olivibacter sitiensis* AW-6^T (89.8%), and *Parapedobacter koreensis* Jip14^T (89.3%).

It is evident from the phylogenetic tree shown in Fig. 4 that strain TDMA-5^T belonged to a distinct lineage in the family *Sphingobacteriaceae*, and *Pedobacter* species except *P. saltans*, which clearly formed a monophyletic clade. *P. saltans* exhibited low 16S rRNA gene sequence similarity values to other *Pedobacter* species (89.4–91.4%), and joined the cluster comprising *Sphingobacterium* spp., *Olivibacter sitiensis* AW-6^T, and *Parapedobacter koreensis* Jip14^T with a low bootstrap resampling value. These results agree with previous phylogenetic studies based on 16S rRNA gene sequences (31, 32, 38, 42–45). Accordingly, *P. saltans* may have to be reclassified as a member of a new genus. Trees based on maximum-parsimony and maximum likelihood methods showed essentially the same topology (data is not shown). The results of phylogenetic analysis, thus, suggest that TDMA-5^T represents a novel genus within the family *Sphingobacteriaceae*.

The phylogenetic analysis revealed that TDMA-16^T belongs to the family *Sphingomonadaceae*. Highest similarity was found with sequences of established species of the genus *Sphingomonas sensu stricto* including *Sphingomonas mali* IFO 15500^T (95.1%), *Sphingomonas aquatilis* JSS-7^T (95.0%), *Sphingomonas pruni* IFO 15498^T (94.9%), *Sphingomonas melonis* DAPP-PG 224^T (94.9%), *Sphingomonas asaccharolytica* IFO 10564^T (94.5%), *Sphingomonas faeni* MA-olki^T (94.3%), *Sphingomonas aerolata* NW12^T (94.2%),

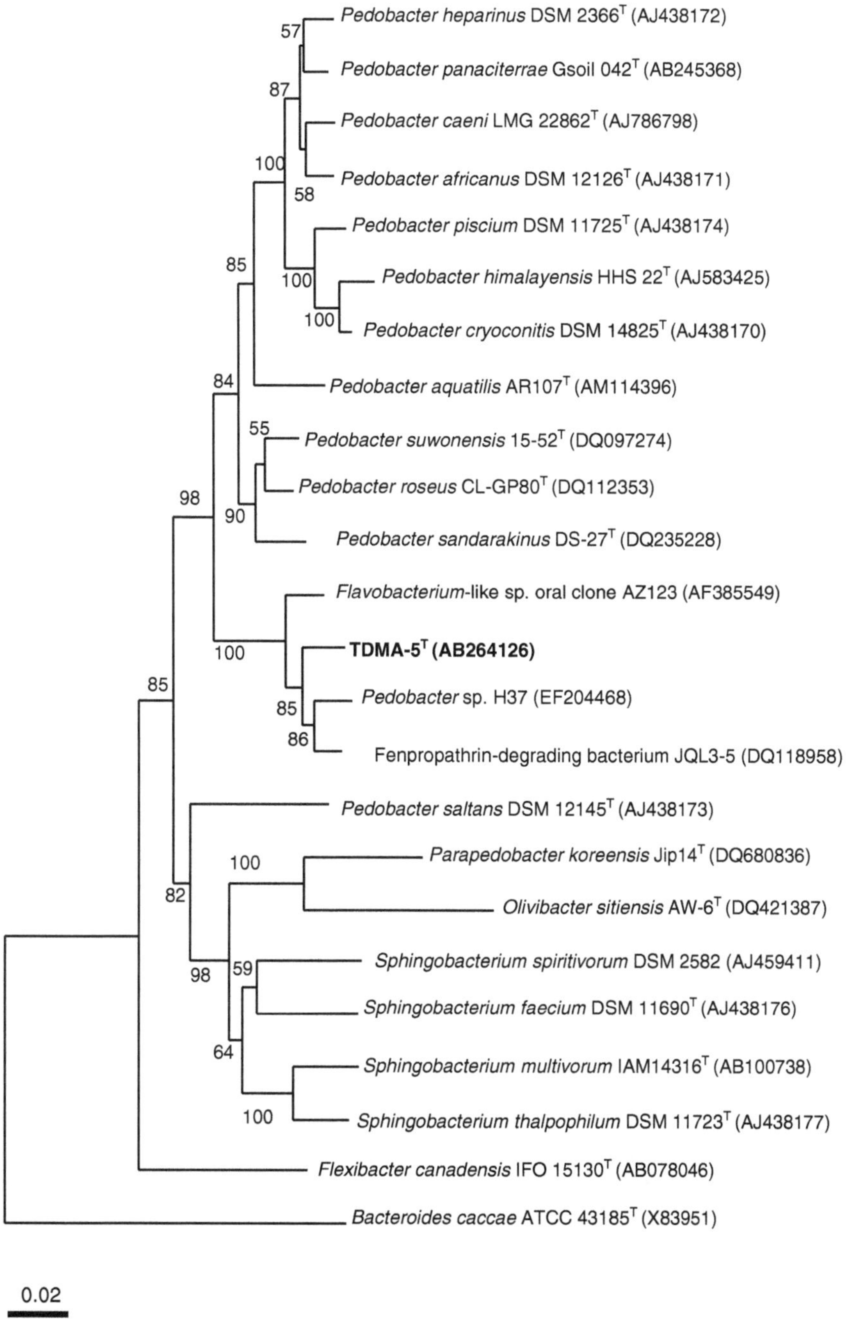

Fig. 4. A phylogenetic tree based on the 16S rRNA gene sequences of strain TDMA-5^T and representative members of related genera of the family *Sphingobacteriaceae*. The tree was constructed by the neighbor-joining method (94). Bar, 2 substitutions per 100 nucleotide positions. Bootstrap values, expressed as a percentage of 1,000 replicates, are given at the branching points. Values above 50% are shown. 16S rRNA gene sequence from *Bacteroides caccae* ATCC 43185^T (AB078046) was used as outgroup. GenBank accession numbers are shown in parentheses. Adapted from Asker et al. 29.

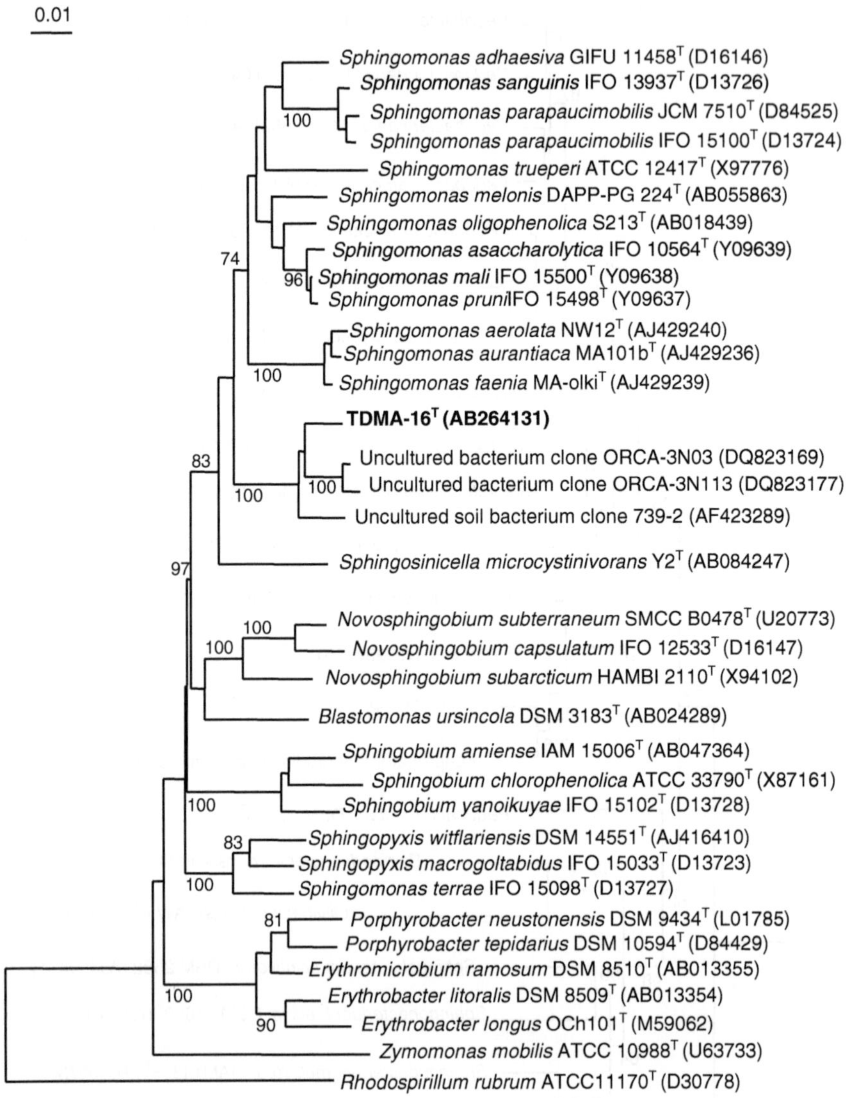

Fig. 5. A phylogenetic tree based on the 16S rRNA gene sequences of strain TDMA-16^{T} and representative members of related species of the genus *Sphingomonas sensu stricto*. The tree was constructed by the neighbor-joining method (94). Bar, 1 substitution per 100 nucleotide positions. Bootstrap values, expressed as a percentage of 1,000 replicates, are given at the branching points. Values above 50% are shown. GenBank accession numbers are shown in parentheses. Adapted from Asker et al. 28.

Sphingomonas abaci C42^{T} (94.2%), *Sphingomonas echinoides* ATCC 148820^{T} (94.2%), *Sphingomonas panni* C52^{T} (94.0%), and *Sphingomonas aurantiaca* MA101b^{T} (93.9%). These results clearly demonstrated that TDMA-16^{T} represents a novel species within the genus *Sphingomonas sensu stricto*.

Phylogenetic analysis of nearly full-length 16S rRNA sequence of strain TDMA-16^{T} indicated that it is a member of the genus *Sphingomonas sensu stricto* (47), forming a distinct phylogenetic lineage (Fig. 5). The 16S rRNA gene sequence of strain TDMA-16^{T}

retained all the signature nucleotides, namely, C:G at position 52:359, G at position 134, G at position 593, G:C at position 987:1218, and U:G at position 990:1215 (based on *E. coli* numbering) (58), that are conserved in the genus *Sphingomonas sensu stricto* (47).

1.6. Chemotaxonomic Characterization of Zeaxanthin-Producing Isolates

Table 3 shows the fatty acid profile of strain TDMA-5^T. The predominant cellular fatty acids were iso-$C_{15:0}$ (29.4%), iso-$C_{17:0}$ 3-OH (15.9%), and summed feature 4 (iso-$C_{15:0}$ 2-OH and/or $C_{16:1}\omega7c$) (20.1%). These fatty acids represented 65.4% of the total fatty acids of TDMA-5^T (Table 3). The major fatty acids of all members of the family *Sphingobacteriaceae* are iso-$C_{15:0}$, iso-$C_{17:0}$ 3-OH, iso-$C_{15:0}$

Table 3
Whole-cell fatty acid profile of strain TDMA-5^T

Fatty acid	TDMA-5^T
Straight-chain saturated	
$C_{14:0}$	0.3
$C_{15:0}$	3.9
$C_{16:0}$	0.4
$C_{17:0}$	0.1
$C_{15:0}$ 2-OH	1.0
$C_{15:0}$ 3-OH	1.3
$C_{16:0}$ 3-OH	0.6
$C_{17:0}$ 2-OH	0.6
$C_{17:0}$ 3-OH	0.6
Branched saturated	
iso-$C_{13:0}$	0.6
iso-$C_{14:0}$	0.3
iso-$C_{15:0}$	**29.4**
iso-$C_{16:0}$	0.3
iso-$C_{17:0}$	0.3
iso-$C_{15:0}$ 3-OH	3.2
iso-$C_{16:0}$ 3-OH	1.4
iso-$C_{17:0}$ 3-OH	**15.9**
anteiso-$C_{15:0}$	2.3
Summed feature 4[a]	**20.1**
Monounsaturated	
$C_{15:1}\ \omega6c$	3.6
$C_{17:1}\ \omega8c$	1.2
$C_{17:1}\ \omega6c$	0.8
iso-$C_{17:1}\ \omega9c$	5.1
Unknown[b]	
ECL 13.566	3.1
ECL 16.580	1.4

[a]Summed feature 4 contained iso-$C_{15:0}$ 2-OH and/or $C_{16:1}\omega7c$
[b]*ECL* equivalent chain length. Adapted from Asker et al. 29

2-OH, and $C_{16:1}$ω7c (30–32, 36–39, 43). However, TDMA-5^T differed from *Pedobacter* spp., *Sphingobacterium* spp., *Olivibacter sitiensis* AW-6^T, and *Parapedobacter koreensis* Jip14^T by the presence of minor fatty acid components including $C_{17:0}$, $C_{15:0}$ 3-OH (except for *Parapedobacter koreensis* Jip14^T), $C_{17:0}$ 3-OH, iso-$C_{13:0}$, iso-$C_{14:0}$, and $C_{17:1}$ω6c (Table 3). TDMA-5^T differed from members of all related genera by the absence of $C_{16:1}$ω5c. Menaquinone-7 (MK-7) was the major respiratory quinone. MK-7 is a characteristic of the majority of species within the family *Sphingobacteriaceae*. The DNA G+C content of TDMA-5^T was 38.6% (Table 5), in agreement with the values of 36.9–44.2% for the family *Sphingobacteriaceae* (30).

Table 4 shows the detailed fatty acid composition of TDMA-16^T. The major fatty acids were $C_{17:1}$ ω6c (34.5%) and $C_{18:1}$ ω7c (29.3%). This profile is in excellent agreement with those of representative of the genus (Table 4) (59–62), and its assignment to the genus *Sphingomonas sensu stricto*. The presence of 2-hydroxy fatty acids and the lack of 3-hydroxy fatty acids are important characteristics of members of the family *Sphingomonadaceae* (47, 61, 63). However, TDMA-16^T differed from the members of genus *Sphingomonas sensu stricto* both qualitatively and quantitatively with regard to certain fatty acids and by the presence of 2-OH $C_{18:1}$ as the major hydroxy fatty acid (Table 4).

The quinone system of strain TDMA-16^T consisted exclusively of ubiquinone Q-10. This quinone profile is a characteristic of the majority of species within the class "*Alphaproteobacteria*" (64), including genus *Sphingomonas sensu stricto* (47, 61). The polar lipid profile of strain TDMA-16^T was characterized by the presence of phosphatidyl ethanolamine, phosphatidyl glycerol, diphosphatidyl glycerol, sphingoglycolipd, and unidentified glycolipid lipid as the predominant lipids and moderate amounts of two unidentified glycolipids lipids. Additionally, TDMA-16^T contained minor amounts of various unidentified lipids, including an aminophospholipid, and two glycolipids. Phosphatidylmonomethylethanolamine and phosphatidyldimethylethanolamine were not detected in strain TDMA-16^T. The polar lipid profile of strain TDMA-16^T was distinguishable from *Sphingomonas mali*, *Sphingomonas pruni*, and *Sphingomonas asaccharolytica* in that phosphatidylmonomethylethanolamine and phosphatidyldimethylethanolamine were absent (61, 65). The DNA G+C content of TDMA-16^T was 63.3% (Table 6), in agreement with the values of 62–68% for the genus (47).

1.7. Morphological, Physiological, and Biochemical Characterization

TDMA-5^T formed deep yellow-pigmented, circular, convex, shiny, opaque, and slimy colonies (2–3 mm in diameter) that are slightly adherent to agar, with the entire margins after 3 days of incubation at 30°C on LB agar plates. Cells of strain TDMA-5^T were Gram-negative rods (0.2–0.3 μm × 1.0–5.0 μm) (Fig. 6). Spores were not observed. TDMA-5^T formed deep yellow-pigmented, circular, convex, shiny, opaque, and slimy colonies (2–3 mm in diameter)

Table 4
Comparison of whole-cell fatty acid profiles (percentage content)

Fatty acid	1	2	3	4	5	6	7	8	9	10	11	12	13	14	15
Saturated fatty acids															
$C_{14:0}$	0.4	1.4	16.8		0.7	1.0		1.6–1.9	2.0	2.1–3.0	0.9		2.6	1.1	0.8
$C_{15:0}$	2.1	0.2		3.8		0.5		0.0–0.5		0.0–0.8		0.8		1.1	
$C_{16:0}$	4.2	10.9	16.9	7.5	18.4	10.9	12.4	12.0–15.8	11.2	12.2–15.6	10.0	14.3	13.0	17.8	19.0
$C_{17:0}$	3.8			8.9						0.0–0.2					
$C_{18:0}$	0.2	0.5		0.4	1.2	0.9				0.0–0.3		0.5	0.8		
$C_{17:0}$ cyclo					0.4			0.0–0.4							
$C_{19:0}$ cyclo ω8*c*					0.4	0.5			0.8	0.4–0.7					
Unsaturated fatty acids															
$C_{15:1}$ ω6c	0.2														
$C_{16:1}$ ω5c	2.0				1.0	1.1		1.2–2.0	1.6	1.6–1.8	0.8		1.6	2.1	1.3
$C_{16:1}$ ω7c											2.9				
$C_{16:1}$		1.4													
$C_{17:1}$ ω6c	34.5				0.4	2.6	1.6	0.0–1.7	2.6	0.8–3.0			0.8	3.6	
$C_{17:1}$		1.7	4.9	27.5								9.6			
$C_{18:1}$ ω5c	2.1	1.8			1.5	2.0		0.6–0.8	1.0	0.6–1.0	2.7		1.4	0.7	
$C_{18:1}$ ω7c	29.3				60.3	63.0	64.8						62.5	51.3	45.7
$C_{18:1}$															
11 methyl $C_{18:1}$ ω7c					5.4	6.9				0.0–3.5			4.1		6.3
Hydroxy fatty acid															
2-OH $C_{12:0}$								0.3–0.6	0.6	0.4					
2-OH $C_{13:0}$	0.1														
2-OH $C_{14:0}$	0.7	9.8		0.4	5.5	7.6	12.9	8.7–14.1	9.7	10.0–10.6	7.2	3.4	7.5	5.3	12.1
2-OH $C_{15:0}$	0.5	0.3		1.0		0.5						0.9			

(continued)

Table 4 (continued)

Fatty acid	1	2	3	4	5	6	7	8	9	10	11	12	13	14	15
2-OH $C_{16:0}$	0.4	0.8				0.8									
2-OH $C_{16:1}$								0.0–0.5							
2-OH $C_{18:1}$	6.1				0.6								0.8		
Summed features															
4	8.1	2.1	2.5		4.3	1.4	8.3	24.6–26.8	29.2	25.7–28.0	2.45	3.9	5.0	17.1	14.8
7	0.5	69.2	58.4	50.2				38.5–46.0	41.4	40.0–40.5	73.1	65.5			

1, *Sphingomonas jaspsi* sp. nov. TDMA-16^T; 2, *Sphingomonas mali* IFO 15500^T (62); 3, *Sphingomonas pruni* IFO 15498^T (42); 4, *Sphingomonas asaccharolytica* IFO 15499^T (62); 5, *Sphingomonas aquatilis* JSS7^T (60); 6, *Sphingomonas melonis* DSM 14444^T (60); 7, *Sphingomonas abaci* C42^T (60); 8, *Sphingomonas aerolata* NW12^T (59); 9, *Sphingomonas faeni* MA-olki^T (59); 10, *Sphingomonas aurantiaca* MA101b^T (59); 11, *Sphingomonas echinoides* ATCC 14820^T (90); 12, *Sphingomonas soli* T5-04^T (62); 13, *Sphingomonas koreensis* JSS26^T (60); 14, *Sphingomonas panni* C52^T (60); 15, *Sphingomonas dokdonensis* DS-4^T (91)
Summed feature 4 contains 2-OH iso-$C_{15:0}$ and/or $C_{16:1}$ ω7c. Summed feature 7 contains $C_{18:1}$ω7*c*, $C_{18:1}$ω9*t*, and/or $C_{18:1}$ω12*t*. Values are percentage of total fatty acids. Adapted from Asker et al. 28

Table 5
Differentiation of *Nubsella zeaxanthinifaciens* gen. nov., sp. nov. TDMA-5[T] from related genera of the family *Sphingobacteriaceae*

Characteristic	1	2	3	4	5
Growth at:					
5°C	–	+[a]	v	+	–
40°C	+	–	v	+	+
42°C	–	–	v	+	+
Glucose fermentation	–	v	+	ND	+
Nitrate reduction	–	–	v	ND	–
Arginine dihydrolase	–	v[b]	ND	ND	–
Lipase	+	v	v	–[c]	–
Hydrolysis of:					
DNA	+	v	v	ND	ND
Gelatin	+	v	v	ND	–
Starch	+	v	v	–	ND
Aesculin	+	v	v	ND	+
Urea	–	–	v	ND	–
Assimilation of:					
Glucose	+	v	+	+	+
Arabinose	–	v	v	ND	–
Mannitol	–	v	v	–	+
Maltose	+	v	+	–	+
Malate	+	–	–	–	–
N-Acetyl-D-glucosamine	+	+	ND	ND	+
Detection of fatty acids:					
$C_{17:0}$	+ (0.1)	–	–	–	–
$C_{15:0}$ 3-OH	+ (1.3)	–	–	–	+ (1.7)
$C_{17:0}$ 3-OH	+ (0.6)	–	–	–	–
iso-$C_{13:0}$	+ (0.6)	–	–	–	–
iso-$C_{14:0}$	+ (0.3)	–	–	–	–
$C_{16:1}$ ω5c	–	v (tr-3.0)	v(tr-1.5)	+ (1.3)	–
$C_{17:1}$ ω6c	+ (0.8)	–	–	–	–
DNA G+C content (mol%)	38.6	36.9–44.2	37.7–44.2	45.6	45.6

Taxa: 1, strain TDMA-5[T]; 2, *Pedobacter* (13 species) (data from Gallego et al. 2006; Hwang et al. 2006; Kwon et al. 2007; Margesin et al. 2003; Shivaji et al. 2005; Steyn et al. 1998; Ten et al. 2006; Vanparys et al. 2005; Yoon et al. 2006; Yoon et al. 2007); 3, *Sphingobacterium* (7 species) (Kim et al. 2006; Shivaji et al. 1992; Steyn et al. 1998; Takeuchi and Yokota 1992); 4, *Olivibacter sitiensis* AW-6[T] (Ntougias et al. 2007); 5, *Parapedobacter koreensis* Jip14[T] (Kim et al. 2007). In all studies, cells for fatty acid methyl ester analysis were grown in peptone-yeast extract media at 28–30°C

Symbols: +, positive; –, negative; v, variable; ND, no data available. Adapted from Asker et al. 29

[a]Not done for *P. caeni*, *P. africanus*, and *P. saltans*

[b]Not done for *P. caeni*, *P. africanus*, and *P. saltans*

[c]Tween 20 is not hydrolyzed

Table 6
Differential characteristics of *Sphingomonas jaspsi* sp. nov. TDMA-16ᵀ and related type strains

Characteristic	1	2	3	4	5	6	7	8	9	10	11	12	13	14	15
Color of colonies	DY	LY	LY	LY	Y	DY	O	O	O	O	Y	Y	Y	Y	Y
Cell shape	PSR	R	R	R	R	R	R	R	R	R	R	R	R	R	R
Motility	+	+	+	+	+	–	–	+	+	+	+	–	+	–	+
Budding formation	+	–	–	–	–	–	–	–	–	–	–	–	–	–	–
Growth in the presence of 3% NaCl	–	+	+	+	ND	+	ND	ND	ND	ND	ND	ND	ND	ND	+
Assimilation of:															
Glucose	(+)	+	+	–	–	+	–	+	+	+	(+)	+	(+)	–	–
L-Arabinose	–	(+)	+	(+)	+	+	–	+	+	+	+	+	–	–	–
D-Mannose	–	+	+	(+)	–	+	–	+	+	+	–	–	–	(+)	+
D-Mannitol	–	–	–	–	–	–	–	–	–	–	–	–	–	–	–
N-Acetyl-D-glucosamine	–	+	+	+	–	+	–	+	+	+	+	–	+	–	–
Maltose	(+)	+	+	+	+	+	–	+	+	+	(+)	+	+	(+)	–
Gluconate	–	+	+	–	–	–	(+)	+	+	+	–	–	–	–	–
Adipate	–	+	–	–	–	–	–	+	+	+	–	–	–	–	+
DL-Malic acid	–	–	(+)	–	–	+	+	v	–	+	+	–	+	+	–
Citrate	–	–	–	–	–	–	–	v	+	+	–	–	–	+	–
Nitrate reduction	–	–	–	–	–	–	ND	–	–	–	+	–	c	ND	–
β-Galactosidase	+	+	+	+	+	+	ND	+	(+)	(+)	+	–	+	ND	–
Oxidase activity	+	+	+	+	+	+	–	+	+	+	+	+	+	–	+

Hydrolysis of:															
Aesculin	–	+	+	+	–	+	+	+	+	+	+	+	–	+	+
Gelatin	(+)	–	–	–	–	–	ND	–	–	–	–	+	–	ND	+
Starch	+	–	+	+	ND	–	ND	–	–	–	ND	ND	ND	ND	+
Tween 80	+	+	+	+	ND	–	ND	ND	ND	ND	ND	ND	ND	ND	+
DNA	NG	+	NG	NG	ND	NG	ND	(+)	(+)	v	ND	ND	ND	ND	ND
DNA G+C content (mol%)	63.3	65.4–65.9	65.4	64.8	63	65.0	ND	65.4	63.1	64.7	65.8	63.9	66	ND	66.9

1, *Sphingomonas jaspsi* sp. nov. TDMA-16^T; 2, *Sphingomonas mali* IFO 15500^T (47); 3, *Sphingomonas pruni* IFO 15498^T (47); 4, *Sphingomonas asaccharolytica* IFO 15499^T (47); 5, *Sphingomonas aquatilis* JSS7^T (92); 6, *Sphingomonas melonis* DSM 14444^T (60, 93); 7, *Sphingomonas abaci* C42^T (60); 8, *Sphingomonas aerolata* NW12^T (59), 9, *Sphingomonas faeni* MA-olki^T (59); 10, *Sphingomonas aurantiaca* MA101b^T (59), 11, *Sphingomonas echinoides* ATCC 14820^T (90); 12, *Sphingomonas soli* T5-04^T (62); 13, *Sphingomonas koreensis* JSS26^T (92); 14, *Sphingomonas panni* C52^T (60); 15, *Sphingomonas dokdonensis* DS-4^T (91) +, positive; (+), weakly positive; –, negative; v, variable; ND, not determined; NG, no growth; Y, yellow; LY, light yellow; DY, deep yellow; PSR, pleomorphic short rods; R, rods. All strains were negative for urease activity, indole production, glucose fermentation, arginine dihydrolase, and assimilation of D-mannitol, caprate, and acetate. Adapted from Asker et al. 28

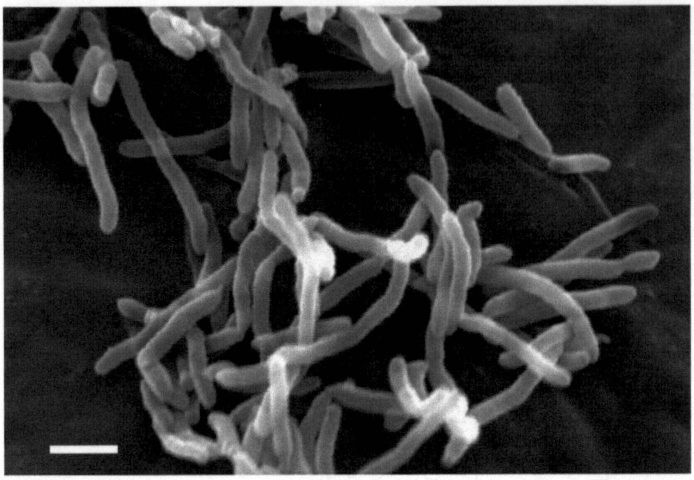

Fig. 6. Scanning electron micrograph of strain TDMA-5^T grown on LB agar at 30°C. Cells were long rods after 3 days' cultivation. Bar, 1 μm. Adapted from Asker et al. 29.

that are slightly adherent to agar, with the entire margins after 3 days of incubation at 30°C on LB agar plates. The temperature range for the growth of TDMA-5^T was between 18 and 40°C (optimum, 30°C). No growth occurred at 45°C. The pH range for the growth was between 6.0 and 9.0. TDMA-5^T was strictly aerobic and chemo-organotrophic. The strain exhibited positive reactions for oxidase and catalase. TDMA-5^T was sensitive to gamma-ray irradiation. Other phenotypic properties of strain TDMA-5^T are given in the species description and those characteristics that differentiate strain TDMA-5^T from related members of the family *Sphingobacteriaceae* are listed in Table 5.

Strain TDMA-16^T was isolated from a freshwater sample collected at Misasa, a region known for its high natural radioactivity, by its ability to produce yellow pigment. TDMA-16^T formed yellow-pigmented, convex, circular colonies (1–2 mm in diameter) after 48 h of incubation at 37°C.

Cells of strain TDMA-16^T were Gram-negative, pleomorphic motile, short rods (0.4–0.5 μm × 0.7–1.0 μm). Spores were not observed. Buds formed on the mature cells (Fig. 7). TDMA-16^T formed yellow-pigmented, convex, circular colonies (1–2 mm in diameter) after 48 h of incubation at 37°C. The temperature range for the growth of TDMA-16^T was between 20 and 40°C (optimum, 35–37°C). No growth occurred at 45°C. The pH range for the growth was between 6.0 and 9.0. TDMA-16^T was strictly aerobic and chemo-organotrophic. The strain exhibited positive reactions for oxidase and catalase but negative reactions in the oxidation/fermentation test and gas/acid production test with glucose. TDMA-16^T was slightly tolerant against gamma-ray irradiation, showing 0.45% survival rate after exposure to a dose of 2.3 kGy

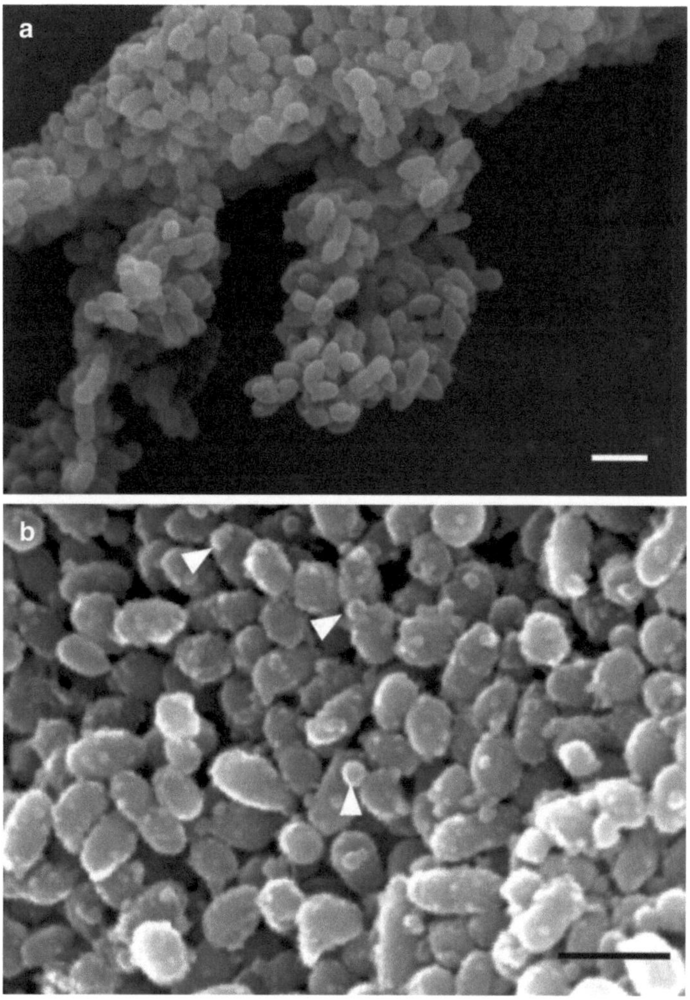

Fig. 7. Scanning electron micrographs of the TDMA-16^T cells. Pleomorphic-shaped morphology and budding formation (indicated by *arrow heads*) on mature cells were observed. Cells were grown for 2 days (**a**) and 4 days (**b**) at 37°C on nutrient agar. Bar, 1.0 μm. Adapted from Asker et al. 28.

gamma irradiation (13 Gy/min, ^{137}Cs source). At this dose, the survival rate of *E. coli* C600 was 0%. Other phenotypic properties of strain TDMA-16^T are given in the species description and those characteristics that differentiate strain TDMA-16^T from related members of the genus *Sphingomonas sensu stricto* are listed in Table 6.

1.8. Differential Characteristics of Novel Zeaxanthin-Producing Isolates

Strain TDMA-5^T differed from the members of *Pedobacter* (Table 5) by (1) having different minor fatty acids, (2) being more mesophilic, (3) being less psychrotolerant (no data was available for *P. caeni*, *P. africanus*, and *P. saltans*), and (4) being able to assimilate malate. TDMA-5^T differed from the members of *Sphingobacterium* by (1) having different minor fatty acids, (2) being less psychrotolerant, (3) being able to assimilate malate, (4) being unable to ferment glucose, and (5) being unable to hydrolyze urea. TDMA-5^T differed from *Olivibacter sitiensis* AW-6^T by (1) having different minor fatty acids, (2) having a narrower temperature growth range, (3) being able to assimilate malate and maltose, (4) being able to hydrolyze starch, and (5) having a low DNA G+C content (mol %). TDMA-5^T differed from *Parapedobacter koreensis* Jip14^T by (1) having different minor fatty acids, (2) being able to assimilate malate and not able to assimilate mannitol, (3) being able to hydrolyze gelatin, (4) being unable to ferment glucose, and (5) having a low DNA G+C content (mol %). Additionally, TDMA-5^T was distinguished from other closely related species by the traits detailed in Table 5.

Unlike *Sphingomonas mali* IFO 15500^T, *Sphingomonas pruni* IFO 15498^T, *Sphingomonas asaccharolytica* IFO 10564^T, *Sphingomonas melonis* DAPP-PG 224^T, and *Sphingomonas dokdonensis* DS-4^T, TDMA-16^T showed severe growth arrest in the presence of >0.25% NaCl. TDMA-16^T differed from *Sphingomonas mali* IFO 15500^T, *Sphingomonas aerolata* NW12^T, *Sphingomonas faeni* MA-olki^T, and *Sphingomonas aurantiaca* MA101b^T by (1) being unable to assimilate L-arabinose, D-mannose, *N*-Acetyl-D-glucosamine, gluconate, and adipate; (2) being able to hydrolyze gelatine and starch; and (3) being unable to hydrolyze aesculin and DNA. TDMA-16^T differed from *Sphingomonas pruni* IFO 15498^T and *Sphingomonas asaccharolytica* IFO 10564^T by (1) being unable to assimilate L-arabinose, D-mannose, and *N*-Acetyl-D-glucosamine; (2) being able to hydrolyze gelatine; and (3) being unable to hydrolyze aesculin. TDMA-16^T differed from *Sphingomonas melonis* DAPP-PG 224^T by (1) being able to motile; (2) being unable to assimilate L-arabinose, D-mannose, *N*-Acetyl-D-glucosamine, and DL-malic acid; (3) being able to hydrolyze gelatine and starch; and (4) being unable to hydrolyze aesculin. TDMA-16^T differed from *Sphingomonas aquatilis* JSS-7^T by (1) being unable to assimilate L-arabinose, (2) being able to assimilate glucose, and (3) being able to hydrolyze gelatine. Additionally, TDMA-16^T was distinguished from other closely related species by the traits detailed in Table 6.

1.9. Taxonomic Positions of Novel Zeaxanthin-Producing Isolates

Based on the result of 16S rRNA gene-based phylogenetic analysis and on the above-mentioned phenotypic features, we conclude that TDMA-5^T is a novel genus within the family *Sphingobacteriaceae* for which we propose the name *Nubsella zeaxanthinifaciens*.

The results of the analysis of 16S rRNA gene sequence, quinone system, fatty acids, polar lipids, and DNA G+C content, thus, proved that strain TDMA-16^T is a member of the genus *Sphingomonas sensu stricto*. The strain can be distinguished from its close relatives by its 16S rRNA gene sequence, fatty acid composition, and numerous physiological characteristics. On the basis of these results, we conclude that strain TDMA-16^{T} represents a novel species of the genus *Sphingomonas sensu stricto*, for which we propose the name *Sphingomonas jaspsi* sp. nov. for this new carotenoid-producing bacterium.

1.10. Description of Nubsella gen. nov.

Nubsella (Nu.b.sel'la N.L. fem. dim. n. *Nubsella* derived from the acronym NUBS for Nihon University College of Bioresource Sciences, where the type species was isolated).

Gram-negative, non-spore-forming rods. Strictly aerobic. Catalase and oxidase positive. MK-7 is the only respiratory quinone. Carotenoid pigments are produced. As determined by 16S rRNA gene sequence analysis, the genus *Nubsella* is a member of the family *Sphingobacteriaceae*. The G + C content of the type strain of the type species is 38.6%. The type species of the genus is *Nubsella zeaxanthinifaciens*.

1.11. Description of Nubsella zeaxanthinifaciens sp. nov.

Nubsella zeaxanthinifaciens (ze.a.xan.thi.ni.fa'ci.ens. N.L.neut. n. *zeaxanthinum* zeaxanthin; L. part. pres. *faciens* making/producing; N.L. part. adj. *zeaxanthinifaciens* zeaxanthin-producing): Displays the following characteristics in addition to those given in the genus description.

Cells are Gram-negative-staining rods (0.2–0.3 μm × 1.0–5.0 μm) that are strictly aerobic, motile, and non-spore forming. Forms deep yellow, circular, convex, shiny, opaque, and slimy colonies that are slightly adherent to agar, with the entire margins and a diameter of 2–3 mm on LB agar plates after 3 days' incubation at 30°C. TDMA-5^{T} contains nondiffusible yellow carotenoid pigments with zeaxanthin as the major component. Growth occurs between 18.0 and 40.0°C (optimum, 30.0°C) and between pH 6.0 and 9.0 (optimum, 6.0–7.0). Anaerobic growth does not occur. Tween 80, starch, gelatin, casein, aesculin, and DNA are hydrolyzed while agar is not. Indole and H_2S are not produced. Nitrate is not reduced. Catalase and oxidase positive. Glucose, mannose, *N*-acetyl-glucosamine, maltose, and malate are assimilated, but arabinose, mannitol, gluconate, capric acid, adipic acid, trisodium citrate, and phenylacetic acid are not. In API 20E tests, β-galactosidase and the Voges–Proskauer tests are positive; tests for arginine dihydrolase, lysine decarboxylase, ornithine decarboxylase, tryptophan deaminase, hydrogen sulfide, indole production, and urease are negative. Acid is not produced from D-glucose, D-melibiose, amygdalin, L-arabinose, D-mannitol, inositol, D-sorbitol, L-rhamnose, and sucrose. MK-7 is the predominant menaquinone, and the major fatty acids are iso-$C_{15:0}$, iso-$C_{17:0}$ 3-OH, and summed feature 4 (iso-$C_{15:0}$ 2-OH, and/or $C_{16:1}\omega7c$). The DNA G + C content of the type strain is 38.6%. The type strain is TDMA-5^{T} = NBRC 102579 = CCUG 54348 that was isolated from a freshwater sample collected at Misasa (Tottori, Japan).

1.12. Description of *Sphingomonas jaspsi* sp. nov.

Sphingomonas jaspsi (ja.s.p.si. N.L. n. *jaspsum* arbitrary name derived from the acronym JSPS (Japan Society for the Promotion of Science); N.L. gen. n. jaspsi, of JSPS, the organization that supported this study).

Cells are Gram-negative-staining pleomorphic short rods that are strictly aerobic, motile, and non-spore forming. They are 0.4–0.5 μm in width and 0.7–1.0 μm in length. Budding morphology may be observed. On nutrient agar, it forms yellow, circular, dry colonies (1–2 mm in diameter) within 48 h. Carotenoids, such as zeaxanthin and nostaxanthin, are produced. Growth occurs between 20.0 and 40.0°C (optimum, 35.0–37.0°C) and between pH 6.0 and 9.0 (optimum, 6.0–7.0). Optimal growth occurs in the absence of NaCl; growth is inhibited by the presence of >0.25% (w/v) NaCl. Anaerobic growth does not occur. Tween 80, gelatin, and starch are hydrolyzed while aesculin, agar, DNA, casein, cellulose, and chitin are not. Negative reactions are observed for glucose fermentation and acid production from all tested sugars. Indole and H_2S are not produced. Nitrate is not reduced. Catalase and oxidase positive. β-Galactosidase positive. Tryptophan deaminase, lysine decarboxylase, ornithine decarboxylase, tryptophan deaminase, and urease activities are negative. Glucose and maltose are assimilated, but arabinose, mannose, mannitol, *N*-acetyl-glucosamine, gluconate, capric acid, adipic acid, malate, trisodium citrate, and phenylacetic acid are not. The major fatty acids are $C_{17:1}\omega6c$ (34.5%), $C_{18:1}\omega7c$ (29.3%), and Summed feature 3 (containing $C_{16:1}\omega7c$ and/or 2-OH iso-$C_{15:0}$) (8.1%). The major hydroxy fatty acid is 2-OH $C_{18:1}$ (6.1%). Ubiquinone is predominantly Q-10. Glycosphingolipids are present. The DNA G+C content of the type strain is 63.3%.

The type strain is TDMA-16^T=NBRC 102120=DSM 18422=CCUG 53607 that was isolated from a freshwater sample collected at Misasa (Tottori, Japan).

2. Materials

2.1. Media and Buffers

1. Nutrient agar: 3 g/L beef extract, 5 g/L peptone, and 15 g/L agar (Difco Laboratories, Detroit, MI, USA).
2. Luria–Bertani (LB) solid medium: 10 g/L tryptone, 5 g/L yeast extract, 5 g/L NaCl, and 15 g/L agar.

2.2. Preparing Media and Cultivation

1. Analytical balance.
2. Flasks for cultivation.
3. Disposable Petri dishes.

4. Micropipettes and sterile tips (0–20, 20–200, and 100–1,000 μL).
5. Autoclave.
6. Incubators (28, 30, 37, 40, and 50°C).
7. Sterile loops, sterile toothpicks, and sterile cotton swabs.
8. Clean bench equipped with UV light.

2.3. Carotenoids' Analysis

1. Butylhydroxytoluene (BHT) (Sigma, St. Louis, MO, USA) (see Note 1).
2. Dim light (see Note 1).
3. Astaxanthin (Sigma, St. Louis, MO, USA).
4. Eppendorf tubes.
5. Racks.
6. Balance.
7. Eppendorf table top vial shaker.
8. Spectrophotometer.
9. 0.45 μm hydrophobic Teflon membrane syringe filter (Advantec, Tokyo, Japan) and syringe (1 mL).
10. 96-well plate.
11. HPLC automated injector with 96 well plate adaptor Class-VP (Shimadzu, Kyoto, Japan) with PDA detector.
12. HPLC/MS system LCMS-2010EV (Shimadzu, Kyoto, Japan).
13. HPLC column CAPCELLPAC C18 (35 mm×4.6 mm; particle size, 5 μm) (Shiseido, Tokyo, Japan).
14. HPLC column Shim-Pack FC-ODS (150 mm×4.6 mm; particle size, 5 μm) (Shimadzu, Kyoto, Japan).

2.4. Identification of Bacterial Isolates: 16S rRNA Gene Sequences

1. Eppendorf tubes, toothpicks, and deionized water (all sterilized).
2. DNA purification kit (Edge BioSystems, Gaithersburg, MD, USA).
3. Primer B8F: 5′-AGAGTTATCCTGGCTCAG-3′.
4. Primer B1492R: 5′-GGTTACCTTGTTACGACTT-3′.
5. Thermocycler.
6. Ex Taq polymerase (Takara-shuzo, Kyoto, Japan).
7. 10× TBE buffer: 108 g Tris base, 55 g boric acid, and 40 mL 0.5 M EDTA (pH 8.0).
8. 1× TBE buffer: 100 mL of 10× TBE buffer and 900 mL of deionized water.
9. 6× blue loading dye: 30% glycerol in distilled water and 0.25% bromophenol blue.

10. Ethidium bromide (see Note 2).
11. Agarose gel electrophoresis system.
12. UV safety goggles and UV box.
13. 0.2-mL thin-walled PCR tubes.
14. BigDye Terminator v3.1 Cycle Sequencing Kit (Applied Biosystems, Foster City, CA, USA).
15. Pipettes and tips (0–10, 20–200, 100–1,000 μL).
16. Gene Clean kit II (Funakoshi, Tokyo, Japan).

2.5. Morphological, Physiological, and Biochemical Characterization

1. Nutrient agar plates and Nutrient broth (5 mL).
2. Incubators (4, 10, 20, 30, 37, 40, and 45°C).
3. BD™ Gram staining kit (BD Biosciences, Franklin Lakes, NJ, USA).
4. Zeiss Axioskop 2 microscope (Carl Zeiss MicroImaging LLC, Thornwood, NY, USA).
5. Scanning electron microscope VE-8800 (Keyence, Tokyo, Japan).
6. GasPak anaerobic system BBL (BD Biosciences, Franklin Lakes, NJ, USA).
7. Cytochrome oxidase test strips (Fluka Biochemika, Sigma-Aldrich GmbH, Taufkirchen, Germany).
8. API 20E, API 20NE, and API 50CH (bioMérieux, Marcy l'Etoile, France).
9. *E. coli* C600 (The Coli Genetic Stock Center CGSC, New Haven, CT, USA).
10. Irradiation at 2.35 kGy (13 Gy/min, ^{137}Cs source).
11. 0.1 M HCl.
12. 0.1 M NaOH.

2.6. Chemotaxonomic Characterization

1. Sherlock Microbial Identification System Version 5.0 (MIDI, Inc. Newark, DE, USA).
2. HPLC automated injector with 96 well plate adaptor Class-VP with PDA detector (Shimadzu, Kyoto, Japan).
3. *Sphingopyxis alaskensis* DSMZ 13593^T (DSMZ culture collection, Braunschweig, Germany).
4. *Sphingomonas asaccharolytica* NBRC 15499^T (NBRC, Chiba, Japan).
5. Thin layer chromatography (TLC) plates size 20 cm × 10 cm (Merck, Whitehouse Station, NJ, USA).
6. Sherlock Microbial Identification System, Version 5·0 (MIDI, Inc. Newark, DE, USA).

3. Methods

3.1. Isolation of the Zeaxanthin-Producing Bacteria

1. Collect water samples from the surface water layer (our samples were collected at Misasa radioactive site in Tottori, Japan).
2. Do serial dilutions with sterile water.
3. Spread 0.1-mL aliquots on Nutrient agar or LB agar.
4. Incubate at 30 and 40°C for up to 3 days.
5. Do screening of colonies based on their color. The colonies of TDMA-5^T and TDMA-16^T were isolated based on their yellow color.
6. For all characteristics, grow TDMA-5^T cells on LB for 3 days at 30°C and TDMA-16^T cells on NA at 40°C.

3.2. Identification of Carotenoids

1. Extract the red pigments by agitating the bacterial cells in methanol on a rotary shaker (100 rpm) at 50°C in the dark until the cells are bleached (see Note 1).
2. Centrifuge at 5,000 × *g* for 5 min.
3. Scan the resultant supernatant for absorbance in the wavelength range (λ = 260–700 nm) at room temperature on a spectrophotometer.
4. Identify each carotenoid on an HPLC system containing a photodiode array detector using two methods (B and C) (see Note 3).
5. Perform HPLC method B using a CAPCELLPAC C18 column (35 × 4.6 mm; particle size, 5 μm; column temperature, 35°C) and elute with methanol:water (9:1, vol/vol) at a flow rate of 1.0 mL/min.
6. Perform HPLC method C on an LC/MS system to determine the molecular weight of carotenoids. Apply each 10 μL sample onto a Shim-Pack FC-ODS (150 × 4.6 mm; particle size, 5 μm; column temperature, 35°C) and elute with 95% methanol at a flow rate 0.8 ml/min. Wash the column regularly (see Note 4).
7. Identify zeaxanthin by retention time, absorption spectrum, and molecular weight, using commercial β-carotene and zeaxanthin as standards. Use an extract of *Sphingomonas paucimobilis* and *S. echinoides* as a reference for nostoxanthin (56, 57).

3.3. Phylogenetic Characterization of the Zeaxanthin-Producing Bacteria

1. Carefully transfer 50 mg wet cells from bacterial isolates into Eppendorf tubes.
2. Extract genomic DNA using the bacterial genomic DNA purification kit.
3. Check the quality of the extracted genomic DNA by agarose gel electrophoresis (steps 4–9).

4. Prepare 1% gel solution by melting 1 g of agarose in 100 mL of 0.5× TBE buffer in a microwave for approximately 2 min. Allow to cool for 5 min, then add 2.5 μL ethidium bromide, and stir to mix (see Note 2).
5. Cast the gel using the supplied tray and comb. Allow the gel to set for a minimum of 30–40 min at room temperature.
6. Load into separate wells: 10 μL of 1 kb ladder, diluted samples (5 μL sample + 5 μL water + 2 μL 6× loading dye).
7. Run the gel for 30 min at 100 V.
8. Expose the gel to UV light and photograph (wear UV goggles).
9. Confirm DNA quality.
10. Set up PCR reaction mixtures in a final volume of 50 μL: 10 μL 10× PCR buffer, 1 μL PCR Nucleotide Mix, 50 U Taq DNA polymerase, 1 μL (10 μM) primer B8F, 1 μL (10 μM) primer B1500R, 2 μL DNA extract, and 33 μL deionized and sterilized water.
11. Perform PCR using a protocol that includes an initial denaturation for a period of 5 min at 94°C; 30 cycles of 94°C for 1 min, 55°C for 1 min, and 72°C for 1 min; 72°C for 3 min; and incubation at 4°C until further processing.
12. Purify PCR amplicons using agarose gel electrophoresis (steps 13–19).
13. Prepare 2% agarose gel, and let it to solidify at room temperature for 30–40 min.
14. Add 10 μL 6× blue loading dye to each PCR product (50 μL).
15. Carefully load all 60 μL into wide gel well.
16. Run the gel for 30 min at 100 V.
17. Wear UV goggles. Remove gel and place on UV box to identify region with PCR products.
18. Cut DNA bands from agarose gel and transfer into sterilized Eppendorf tubes.
19. Purify PCR products from the gel using the Gene Clean kit.
20. Determine the 16S rRNA gene sequences using the BigDye Terminator v3.1 cycle sequencing kit.

3.4. Phylogenetic Analysis

1. Compare the nearly complete 16S rRNA gene (about 1,396–1,450 bp) sequences of isolates to those available in GenBank/EMBL/DDBJ nucleotide sequence database using the BLASTN program (http://www.ncbi.nlm.nih.gov/BLAST/).
2. Using ClustalW, align the obtained sequences with those of closed relatives.

3. Construct a neighbor-joining phylogenetic tree (see Figs. 4 and 5).
4. Use NJ plot to draw and edit the obtained tree.

3.5. Morphological, Physiological, and Biochemical Characterization

1. Cultivate TDMA-5^T and TDMA-16^T cells on LB and Nutrient agar for 48 h at 40°C and 30°C for 2–3 days (see Note 5).
2. Perform Gram staining according to the method described by Smibert and Krieg (66), and observe the bacterial cells under a microscope.
3. Observe the bacterial cells under scanning electron microscope according to the manufacturer's instructions. The specimen for scanning electron microscopy was prepared using the osmium fixation protocol and critical point drying (67).
4. Test the anaerobic growth of both strains by incubation of inoculated nutrient agar plates in a GasPak anaerobic system.
5. Perform the following tests as described in each reference: casein hydrolysis (68); catalase activity, spore formation, and hydrolysis of Tween 80, DNA, and starch (66); and hydrolysis of chitin, carboxymethyl cellulose (high viscosity), and agar (69).
6. Test oxidase activity by using commercial cytochrome oxidase test strips, and according to its manual.
7. Examine other enzyme activities, growth on carbohydrates, acid production from carbohydrates, nitrate reduction, and production of H_2S, indole, and acetoin using the commercial systems API 20E, API 20NE, and API 50CH, according to the manufacturer's instructions.
8. Determine the optimal growth temperature: cultivate the strain on NA at 4, 10, 20, 30, 37, 40, and 45°C.
9. Test the NaCl tolerance using Nutrient broth supplemented with different concentrations of NaCl.
10. Evaluate the growth at various pH values (pH 3.0–10.0) in NB adjusted with HCl or NaOH.
11. Unless otherwise specified, all the characteristics described hereafter are those of the TDMA-5^T cells grown on LB for 3 days at 30°C and of the TDMA-16^T cells grown on NA 40°C.
12. Determine the survival rate after exposure to gamma radiation.
    (a) Grow cells of TDMA-5^T and TDMA-16^T according to step 11.
    (b) Suspend the grown cells in 5 mL of LB (TDMA-5^T) or NB (TDMA-16^T).
    (c) Prepare aliquots of the cell suspension (1 mL) and irradiate at 2.35 kGy (13 Gy/min, ^{137}Cs source).

(d) Dilute serially and plate onto Nutrient agar or LB agar plates.

(e) Aerobically incubate the plates incubated at 30 or 40°C.

(f) Similarly, treat *E. coli* C600.

(g) Determine survival percentage by comparing with unirradiated cultures.

3.6. Chemotaxonomic Characterization

1. Determine the cellular fatty acids of strains TDMA-5^T and TDMA-16^T.
   (a) Grow the cells aerobically on Nutrient agar at 40 or 30°C (see Note 6).
   (b) Extract and analyze the fatty acid methyl esters of strains TDMA-5^T and TDMA-16^T according to the standard protocol of the Sherlock Microbial Identification System (Version 5·0).
2. Determine the respiratory quinone by an HPLC method (70) using cell extracts of *Sphingopyxis alaskensis* DSMZ 13593^T and *Sphingomonas asaccharolytica* NBRC 15499^T as a reference for Q-10.
3. Analyze glycosphingolipids using TLC as described previously (47, 71). Determine the polar lipids according to 72.
4. Determine the DNA G+C content by an HPLC method (73).

4. Notes

1. Carotenoids are highly unstable and sensitive to light, oxygen, and temperature, which induce their oxidation. To minimize the oxidation of carotenoids, the extraction must be quickly done without exposure to light, oxygen, and high temperatures. Therefore, all the experimental procedures should be carried out in a darkroom. Alternatively, dark tubes or tubes foiled with aluminum foil can be used. Nitrogen gas should be also used for flushing the pigment extracts before fridge storage at –20°C to avoid oxidation. In addition, the use of an antioxidant such as BHT (11) can significantly reduce the oxidation of carotenoids during extraction.
2. Ethidium bromide is mutagenic and moderately toxic. You must wear gloves and handle the chemical with care.
3. For details, see HPLC methods B and C in Chapter 3.
4. Usually, DMSO–MeOH carotenoid extracts contain impurities, such as salts, lipids, fatty compounds, proteins, and other cell compounds. These contaminants usually are adsorbed on the

stationary phase of reversed-phase HPLC column, which may become contaminated by the repeated injection of samples. This causes shifts in retention times (0.5–1 min) and also tailing may occur. In addition, this may cause the column backpressure to build up to intolerably high levels, which overpressures the pump and can cause the column to settle and create a void, depending upon where the blockage occurs. To prevent this problem, we clean the columns on a regular basis throughout our work. Washing the HPLC columns with 100% acetonitrile and then with 100% methanol is sufficient to clean and regenerate the column. We strongly recommend using a pre-column.

5. For more details, see the cited references.
6. For more details, see the cited references Asker et al. 28 and 29.

References

1. Sommerburg O, Keunen JEE, Bird AC, van Kuijk F (1998) Fruits and vegetables that are sources for lutein and zeaxanthin: the macular pigment in human eyes. Br J Ophthalmol 82:907–910
2. Nelis HJ, Deleenheer AP (1991) Microbial sources of carotenoid-pigments used in foods and feeds. J Appl Bacteriol 70:181–191
3. Handelman GJ, Cao G, Walter MF, Nightingale ZD, Paul GL, Prior RL et al (1999) Antioxidant capacity of oat (*Avena sativa* L.) extracts. 1. Inhibition of low-density lipoprotein oxidation and oxygen radical absorbance capacity. J Agric Food Chem 47:4888–4893
4. Humphries JM, Khachik F (2003) Distribution of lutein, zeaxanthin, and related geometrical isomers in fruit, vegetables, wheat, and pasta products. J Agric Food Chem 51:1322–1327
5. Hadden WL, Watkins RH, Levy LW, Regalado E, Rivadeneira DM, van Breemen RB et al (1999) Carotenoid composition of marigold (*Tagetes erecta*) flower extract used as nutritional supplement. J Agric Food Chem 47: 4189–4194
6. Perry A, Rasmussen H, Johnson EJ (2009) Xanthophyll (lutein, zeaxanthin) content in fruits, vegetables and corn and egg products. J Food Comp Anal 22:9–15
7. Asker D, Beppu T, Ueda K (2007) *Zeaxanthinibacter enoshimensis* gen. nov., sp. nov., a novel zeaxanthin-producing marine bacterium of the family *Flavobacteriaceae*, isolated from seawater off Enoshima Island, Japan. Int J Syst Evol Microbiol 57:837–843
8. Asker D, Beppu T, Ueda K (2007) *Mesoflavibacter zeaxanthinifaciens* gen. nov., sp. nov., a novel zeaxanthin-producing marine bacterium of the family *Flavobacteriaceae*. Syst Appl Microbiol 30:291–296
9. Berry A, Janssens D, Humbelin M, Jore JP, Hoste B, Cleenwerck I et al (2003) *Paracoccus zeaxanthinifaciens* sp. nov., a zeaxanthin-producing bacterium. Int J Syst Evol Microbiol 53:231–238
10. Hoshino T, Ojima K, Setoguchi Y (2004) Production of zeaxanthin by recombinant *Phaffia rhodozyma* strain. PCT Int Appl US 20030923
11. Bhosale P, Bernstein PS (2005) Microbial xanthophylls. Appl Microbiol Biotechnol 68: 445–455
12. Dufossé L (ed) (2009) Microbial and microalgal carotenoids as colourants and supplements, vol 5. Birkhäuser Verlag Basel, Basel·Boston·Berlin
13. Snodderly DM (1995) Evidence for protection against age-related macular degeneration by carotenoids and antioxidant vitamins. Am J Clin Nutr 62:S1448–S1461
14. Moeller SM, Jacques PF, Blumberg JB (2000) The potential role of dietary xanthophylls in cataract and age-related macular degeneration. J Am Coll Nutr 19:522s–527s
15. Schweitzer D, Jentsch S, Bohm V, Hammer M, Dawczynski J (2010) Suppelementation with lutein and zeaxanthin – a possible protection against age-related macular degeneration. Spektrum Augenheilkd 24:242–247
16. Gale CR, Hall NF, Phillips DIW, Martyn CN (2003) Lutein and zeaxanthin status and risk of age-related macular degeneration. Invest Ophthalmol Vis Sci 44:2461–2465
17. Mares-Perlman JA, Fisher AI, Klein R, Block G, Millen AE, Wright JD (2001) Lutein and zeaxanthin in the diet and serum and their relation

to age-related maculopathy in the Third National Health and Nutrition Examination Survey. Am J Epidemiol 153:424–432

18. Robman L, Vu H, Hodge A, Tikellis G, Dimitrov P, McCarty C et al (2007) Dietary lutein, zeaxanthin, and fats and the progression of age-related macular degeneration. Can J Ophthalmol 42:720–726
19. Moeller SM, Voland R, Tinker L, Blodi BA, Klein ML, Gehrs KM et al (2008) Associations between age-related nuclear cataract and lutein and zeaxanthin in the diet and serum in the carotenoids in the Age-Related Eye Disease study, an ancillary study of the women's health initiative. Arch Ophthalmol 126:354–364
20. Trumbo PR, Ellwood KC (2006) Lutein and zeaxanthin intakes and risk of age-related macular degeneration and cataracts: an evaluation using the Food and Drug Administration's evidence-based review system for health claims. Am J Clin Nutr 84:971–974
21. Keri L, Carpenter H, Veen C, Hird R, Dennis IF, Ding T et al (1997) The carotenoids β-carotene, canthaxanthin and zeaxanthin inhibit macrophage-mediated LDL oxidation. FEBS Lett 401:262–266
22. Nishino H, Tokuda H, Murakoshi M, Satomi Y, Masuda M, Onozuka M et al (2000) Cancer prevention by natural carotenoids. Biofactors 13:89–94
23. Bosma TL, Dole JM, Maness NO (2003) Optimizing marigold (*Tagetes erecta* L.) petal and pigment yield. Crop Sci 43:2118–2124
24. Stankovic I (2004) Zeaxanthin chemical and technical assessment (CTA) in 63rd JECFA, ed FAO
25. Gierhart DL (1994) Production of zeaxanthin and zeaxanthin-containing compositions. Appl Food Biotechnol, Inc. U.S. 5,308,759
26. Kametani K, Matsumura T (1983) Determination of ^{238}U, ^{234}U, ^{226}Ra and ^{228}Ra in spring waters of sanin district. Radioisotopes 32:18–21
27. Asker D, Beppu T, Ueda K (2007) Unique diversity of carotenoid-producing bacteria isolated from Misasa, a radioactive site in Japan. Appl Microbiol Biotechnol 77:383–392
28. Asker D, Beppu T, Ueda K (2007) *Sphingomonas jaspsi* sp. nov., a novel carotenoid-producing bacterium isolated from Misasa, Tottori, Japan. Int J Syst Evol Microbiol 57:1435–1441
29. Asker D, Beppu T, Ueda K (2008) *Nubsella zeaxanthinifaciens* gen. nov., sp. nov., a zeaxanthin-producing bacterium of the family *Sphingobacteriaceae* isolated from freshwater. Int J Syst Evol Microbiol 58:601–606
30. Steyn PL, Segers P, Vancanneyt M, Sandra P, Kersters K, Joubert JJ (1998) Classification of heparinolytic bacteria into a new genus, *Pedobacter*, comprising four species: *Pedobacter heparinus* comb. nov., *Pedobacter piscium* comb. nov., *Pedobacter africanus* sp. nov. and *Pedobacter saltans* sp. nov. proposal of the family *Sphingobacteriaceae* fam. nov. Int J Syst Bacteriol 48:165–177
31. Ntougias S, Fasseas C, Zervakis GI (2007) *Olivibacter sitiensis* gen. nov., sp. nov., isolated from alkaline olive-oil mill wastes in the region of Sitia, Crete. Int J Syst Evol Microbiol 57: 398–404
32. Kim MK, Na JR, Cho DH, Soung NK, Yang DC (2007) *Parapedobacter koreensis* gen. nov., sp. nov. Int J Syst Evol Microbiol 57: 1336–1341
33. Yabuuchi E, Kaneko T, Yano I, Moss CW, Miyoshi N (1983) *Sphingobacterium* gen. nov., *Sphingobacterium spiritivorum* comb. nov., *Sphingobacterium multivorum* comb. nov., *Sphingobacterium mizutae* sp. nov., and *Flavobacterium indologenes* sp. nov. glucose-nonfermenting gram-negative rods in CDC groups IIk-2 and IIb. Int J Syst Bacteriol 33:580–598
34. Shivaji S, Ray MK, Rao NS, Saisree L, Jagannadham MV, Kumar GS et al (1992) *Sphingobacterium antarcticus* sp. nov., a psychrotrophic bacterium from the soils of Schirmacher Oasis, Antarctica. Int J Syst Bacteriol 42:102–106
35. Takeuchi M, Yokota A (1992) Proposals of *Sphingobacterium faecium* sp. nov., *Sphingobacterium piscium* sp. nov., *Sphingobacterium heparinum* comb. nov., *Sphingobacterium thalpophilum* comb. nov. and two genospecies of the genus *Sphingobacterium*, and synonymy of *Flavobacterium yabuuchiae* and *Sphingobacterium spiritivorum*. J Gen Appl Microbiol 38:465–482
36. Kim KH, Ten LN, Liu QM, Im WT, Lee ST (2006) *Sphingobacterium daejeonense* sp. nov., isolated from a compost sample. Int J Syst Evol Microbiol 56:2031–2036
37. Margesin R, Sproer C, Schumann P, Schinner F (2003) *Pedobacter cryoconitis* sp. nov., a facultative psychrophile from alpine glacier cryoconite. Int J Syst Evol Microbiol 53:1291–1296
38. Shivaji S, Chaturvedi P, Reddy GS, Suresh K (2005) *Pedobacter himalayensis* sp. nov., from the Hamta glacier located in the Himalayan mountain ranges of India. Int J Syst Evol Microbiol 55:1083–1088
39. Vanparys B, Heylen K, Lebbe L, De Vos P (2005) *Pedobacter caeni* sp. nov., a novel

species isolated from a nitrifying inoculum. Int J Syst Evol Microbiol 55:1315–1318
40. Gallego V, Garcia MT, Ventosa A (2006) *Pedobacter aquatilis* sp. nov., isolated from drinking water, and emended description of the genus *Pedobacter*. Int J Syst Evol Microbiol 56:1853–1858
41. Hwang CY, Choi DH, Cho BC (2006) *Pedobacter roseus* sp. nov., isolated from a hypertrophic pond, and emended description of the genus *Pedobacter*. Int J Syst Evol Microbiol 56:1831–1836
42. Ten LN, Liu QM, Im WT, Lee M, Yang DC, Lee ST (2006) *Pedobacter ginsengisoli* sp. nov., a DNase-producing bacterium isolated from soil of a ginseng field in South Korea. Int J Syst Evol Microbiol 56:2565–2570
43. Yoon JH, Lee MH, Kang SJ, Park SY, Oh TK (2006) *Pedobacter sandarakinus* sp. nov., isolated from soil. Int J Syst Evol Microbiol 56:1273–1277
44. Kwon SW, Kim BY, Lee KH, Jang KY, Seok SJ, Kwon JS et al (2007) *Pedobacter suwonensis* sp. nov., isolated from the rhizosphere of Chinese cabbage (*Brassica campestris*). Int J Syst Evol Microbiol 57:480–484
45. Yoon MH, Ten LN, Im WT, Lee ST (2007) *Pedobacter panaciterrae* sp. nov., isolated from soil in South Korea. Int J Syst Evol Microbiol 57:381–386
46. Yabuuchi E, Kosako Y (2005) Family *Sphingomonadaceae* Kosako, Yabuuchi, Naka, Fijiwara and Kobayashi 2000b, 1953VP. In: Brenner DJ, Kreig NP, Staley JT, Garrity GM (eds) Bergey's manual of systematic bacteriology, 2nd edn, vol 2, the Proteobacteria, part C, The Betaproteobacteria, the Deltaproteobacteria and the Epsilonproteobacteria. Springer, New York
47. Takeuchi M, Hamana K, Hiraishi A (2001) Proposal of the genus *Sphingomonas sensu stricto* and three new genera, *Sphingobium*, *Novosphingobium* and *Sphingopyxis*, on the basis of phylogenetic and chemotaxonomic analyses. Int J Syst Evol Microbiol 51:1405–1417
48. Yabuuchi E, Kosako Y, Fujiwara N, Naka T, Matsunaga I, Ogura H et al (2002) Emendation of the genus *Sphingomonas* Yabuuchi et al. 1990 and junior objective synonymy of the species of three genera, *Sphingobium*, *Novosphingobium* and *Sphingopyxis*, in conjunction with *Blastomonas ursincola*. Int J Syst Evol Microbiol 52:1485–1496
49. Yabuuchi E, Yano I, Oyaizu H, Hashimoto Y, Ezaki T, Yamamoto H (1990) Proposals of *Sphingomonas paucimobilis* gen. nov. and comb. nov., *Sphingomonas parapaucimobilis* sp. nov., *Sphingomonas yanoikuyae* sp. nov., *Sphingomonas adhaesiva* sp. nov., *Sphingomonas capsulata* comb. nov., and two genospecies of the genus *Sphingomonas*. Microbiol Immunol 34:99–119
50. Ederer MM, Crawford RL, Herwig RP, Orser CS (1997) PCP degradation is mediated by closely related strains of the genus *Sphingomonas*. Mol Ecol 6:39–49
51. Zablotowicz RM, Leung KT, Alber T, Cassidy MB, Trevors JT, Lee H et al (1999) Degradation of 2,4-dinitrophenol and selected nitroaromatic compounds by *Sphingomonas* sp. UG30. Can J Microbiol 45:840–848
52. Zipper C, Nickel K, Angst W, Kohler HP (1996) Complete microbial degradation of both enantiomers of the chiral herbicide mecoprop [(RS)-2-(4-chloro-2-methylphenoxy) propionic acid] in an enantioselective manner by *Sphingomonas herbicidovorans* sp. nov. Appl Environ Microbiol 62:4318–4322
53. White DC, Sutton SD, Ringelberg DB (1996) The genus *Sphingomonas*: physiology and ecology. Curr Opin Biotechnol 7:301–306
54. Silva C, Cabral JM, van Keulen F (2004) Isolation of a beta-carotene over-producing soil bacterium, *Sphingomonas* sp. Biotechnol Lett 26:257–262
55. Sajilata MG, Singhal RS, Kamat MY (2008) The carotenoid pigment zeaxanthin – a review. Compr Rev Food Sci Food Saf 7:29–49
56. Jenkins CL, Andrewes AG, McQuade TJ, Starr MP (1979) The pigment of *Pseudomonas paucimobilis* is a carotenoid (nostoxanthin), rather than a brominated aryl-polyene (xanthomonadin). Curr Microbiol 3:1–4
57. Rowe NJ, Tunstall J, Galbraith L, Wilkinson SG (2000) Lipid composition and taxonomy of [*Pseudomonas*] *echinoides*: transfer to the genus *Sphingomonas*. Microbiology 146: 3007–3012
58. Brosius J, Palmer ML, Kennedy PJ, Noller HF (1978) Complete nucleotide sequence of a 16S ribosomal RNA gene from *Escherichia coli*. Proc Natl Acad Sci USA 75:4801–4805
59. Busse HJ, Denner EB, Buczolits S, Salkinoja-Salonen M, Bennasar A, Kämpfer P (2003) *Sphingomonas aurantiaca* sp. nov., *Sphingomonas aerolata* sp. nov. and *Sphingomonas faeni* sp. nov., air- and dustborne and Antarctic, orange-pigmented, psychrotolerant bacteria, and emended description of the genus *Sphingomonas*. Int J Syst Evol Microbiol 53:1253–1260
60. Busse HJ, Hauser E, Kämpfer P (2005) Description of two novel species, *Sphingomonas abaci* sp. nov. and *Sphingomonas panni* sp. nov. Int J Syst Evol Microbiol 55:2565–2569

61. Busse HJ, Kämpfer P, Denner EB (1999) Chemotaxonomic characterisation of *Sphingomonas*. J Ind Microbiol Biotechnol 23:242–251
62. Yang DC, Im WT, Kim MK, Ohta H, Lee ST (2006) *Sphingomonas soli* sp. nov., a beta-glucosidase-producing bacterium in the family *Sphingomonadaceae* in the alpha-4 subgroup of the *Proteobacteria*. Int J Syst Evol Microbiol 56:703–707
63. Takeuchi M, Hiraishi A (2001) Taxonomic significance of 2-hydroxy fatty acid profiles of the species in the genus *Sphingomonas* and related taxa. IFO Res Commun 20:72–82
64. Collins MD, Jones D (1981) Distribution of isoprenoid quinone structural types in bacteria and their taxonomic implication. Microbiol Rev 45:316–354
65. Kämpfer P, Denner EB, Meyer S, Moore ER, Busse HJ (1997) Classification of "*Pseudomonas azotocolligans*" Anderson 1955, 132, in the genus *Sphingomonas* as *Sphingomonas trueperi* sp. nov. Int J Syst Bacteriol 47:577–583
66. Smibert RM, Krieg NR (1994) Phenotypic characterization. In: Gerhardt P, Murray RGE, Wood WA, Krieg NR (eds) Methods for general and molecular bacteriology. American Society for Microbiology, Washington, DC, pp 607–654
67. Gerhardt P, Murray RGE, Costilow RN, Nester EW, Wood WA, Krieg NR et al (eds) (1981) American Society for Microbiology, Washington, DC
68. Norris JR, Ribbons DW, Varma AK (1985) Methods in microbiology. Academic, London
69. Cowan ST, Steel KJ (1993) Manual for the identification of medical bacteria, 3rd edn. Cambridge University Press, London
70. Collins MD (1994) Isoprenoid quinones. In: O'Donnell MGAG (ed) Chemical methods in prokaryotic systematics. Wiley, Chichester, pp 265–310
71. Kawahara K, Seydel U, Matsuura M, Danbara H, Rietschel ET, Zahringer U (1991) Chemical structure of glycosphingolipids isolated from *Sphingomonas paucimobilis*. FEBS Lett 292:107–110
72. Tindall BJ (1990) Lipid composition of *Halobacterium lacusprofundi*. FEMS Microbiol Lett 66:199–202
73. Mesbah M, Whitman WB (1989) Measurement of deoxyguanosine/thymidine ratios in complex mixtures by high-performance liquid chromatography for determination of the mole percentage guanine + cytosine of DNA. J Chromatogr 479:297–306
74. Chen F, Li HB, Wong RN, Ji B, Jiang Y (2005) Isolation and purification of the bioactive carotenoid zeaxanthin from the microalga *Microcystis aeruginosa* by high-speed counter-current chromatography. J Chromatogr A 1064:183–186
75. Liao WL, Nur-E-Borhan SA, Okada S, Matsui T, Yamaguchi K (1993) Pigmentation of cultured black tiger prawn by feeding with a *Spirulina* supplemented diet. Nippon Suisan Gakkaishi/Bull Japan Soc Sci Fisher 59:165–169
76. Murata N, Sato N, Omata T, Kuwabara T (1981) Separation and characterization of thylakoid and cell envelope of the blue-green alga (cyanobacterium) *Anacystis nidulans*. Plant Cell Physiol 22:855–866
77. Withers NW, Alberte RS, Lewin RA, Thornber JP, Britton G, Goodwin TW (1978) Photosynthetic unit size, carotenoids, and chlorophyll-protein composition of *Prochloron* sp., a prokaryotic green alga. Proc Natl Acad Sci USA 75:2301–2305
78. Liao HH, Medwid RD, Heefner DL, Sniff KS, Hassler RA, Yarus MJ (1995) Carotenoid producing culture using *Nespongiococcum excentricum*. U.S. patent 5,437,997
79. Jin ES, Feth B, Melis A (2003) A mutant of the green alga *Dunaliella salina* constitutively accumulates zeaxanthin under all growth conditions. Biotechnol Bioeng 81:115–124
80. Andrew Y, Britton G (1990) Photobleaching in the unicellular green alga *Dunaliella parva* 19/9. Photosynth Res 25:129–136
81. Ben-Amotz A, Katz A, Avron M (1982) Accumulation of β-carotene in halotolerant algae: purification and characterization of β-carotene rich globules from *Dunaliella bardawil* (*chlorophyceae*). J Phycol 18:529–537
82. Goodwin TW (1976) Distribution of carotenoids. In: Goodwin TW (ed) Chemistry and biochemistry of plant pigments. Academic, New York
83. Palermo JA, Gros EG, Seldes AM (1991) Carotenoids from three red algae of the *Corallinaceae*. Phytochemistry 30:2983–2986
84. Shepherd D, Dasek J, Suzanne M, Carels C (1976) Production of zeaxanthin. U.S. patent 3,951,743
85. Bhosale P, Larson AJ, Frederick JM, Southwick K, Thulin CD, Bernstein PS (2004) Identification and characterization of a Pi isoform of glutathione S-transferase (GSTP1) as a zeaxanthin-binding protein in the macula of the human eye. J Biol Chem 279: 49447–49454

86. Lagarde D, Beuf L, Vermaas W (2000) Increased production of zeaxanthin and other pigments by application of genetic engineering techniques to *Synechocystis* sp. strain PCC 6803. Appl Environ Microbiol 66:64–72
87. Hundle BS, O'Brien DA, Beyer P, Kleinig H, Hearst JE (1993) In vitro expression and activity of lycopene cyclase and beta-carotene hydroxylase from *Erwinia herbicola*. FEBS Lett 315:329–334
88. McDermott JC, Britton G, Goodwin TW (1973) Carotenoid biosynthesis in a *Flavobacterium* sp.: stereochemistry of hydrogen elimination in the desaturation of phytoene to lycopene, rubixanthin and zeaxanthin. Biochem J 134:1115–1117
89. Albrecht M, Misawa N, Sandmann G (1999) Metabolic engineering of the terpenoid biosynthetic pathway of *Escherichia coli* for production of the carotenoids beta-carotene and zeaxanthin. Biotechnol Lett 21:791–795
90. Denner EB, Kämpfer P, Busse HJ, Moore ER (1999) Reclassification of *Pseudomonas echinoides* Heumann 1962, 343AL, in the genus *Sphingomonas* as *Sphingomonas echinoides* comb. nov. Int J Syst Bacteriol 49: 1103–1109
91. Yoon JH, Lee MH, Kang SJ, Lee SY, Oh TK (2006) Sphingomonas dokdonensis sp. nov., isolated from soil. Int J Syst Evol Microbiol 56:2165–2169
92. Lee JS, Shin YK, Yoon JH, Takeuchi M, Pyun YR, Park YH (2001) Sphingomonas aquatilis sp. nov., Sphingomonas koreensis sp. nov., and Sphingomonas taejonensis sp. nov., yellow-pigmented bacteria isolated from natural mineral water. Int J Syst Evol Microbiol 51:1491–1498
93. Buonaurio R, Stravato VM, Kosako Y, Fujiwara N, Naka T, Kobayashi K et al (2002) Sphingomonas melonis sp. nov., a novel pathogen that causes brown spots on yellow Spanish melon fruits. Int J Syst Evol Microbiol 52:2081–2087
94. Saitou N, Nei M (1987) The neighbor-joining method: a new method for reconstructing phylogenetic trees. Mol Biol Evol 4:406–425

Chapter 6

Novel Approach in the Biosynthesis of Functional Carotenoids in *Escherichia coli*

Hisashi Harada and Norihiko Misawa

Abstract

Many carotenoid pigments are present in a small quantity in nature or low yielding from their natural sources, despite these vivid colorations. Thus, the synthesis of useful carotenoids with metabolic pathway-engineered microorganisms should offer an alternative and promising approach for their efficient production. Here, we describe a novel method for an efficient production of such carotenoids, using *E. coli* cells that carry heterologous mevalonate pathway-based genes. This method also enables relevant researchers to efficiently identify the function of an isolated carotenogenic gene candidate. For example, the recombinant *E. coli* cells, which harbor a lycopene-producing plasmid, can synthesize 12.5 mg/g dry cell weight of lycopene with the addition of lithium acetoacetate to the medium. This level corresponded to an 11.8-fold increase of that of *E. coli* cells carrying only the lycopene-producing plasmid.

Key words: Pathway engineering, Mevalonate pathway, Lycopene, Astaxanthin

1. Introduction

Carotenoids are naturally occurring pigments in some yeasts, fungi and nonphotosynthetic bacteria as well as all photosynthetic organisms, including higher plants, algae, cyanobacteria, and photosynthetic bacteria. Although more than 750 carotenoids have been isolated from natural sources that usually possess vivid colorations (1), many of them are present in a small quantity or low yielding from their sources. Several carotenoids, including lycopene and astaxanthin, are expected to protect animals from oxidative damages, which are considered as one of the main causes of chronic diseases, such as cancer and cardiovascular diseases (2). The synthesis of such functional carotenoids with metabolic pathway-engineered microorganisms should offer an alternative and promising approach for their efficient production. *Escherichia coli*, which is a non-carotenogenic bacterium, is one of the most

José-Luis Barredo (ed.), *Microbial Carotenoids from Bacteria and Microalgae: Methods and Protocols*, Methods in Molecular Biology, vol. 892, DOI 10.1007/978-1-61779-879-5_6, © Springer Science+Business Media, LLC 2012

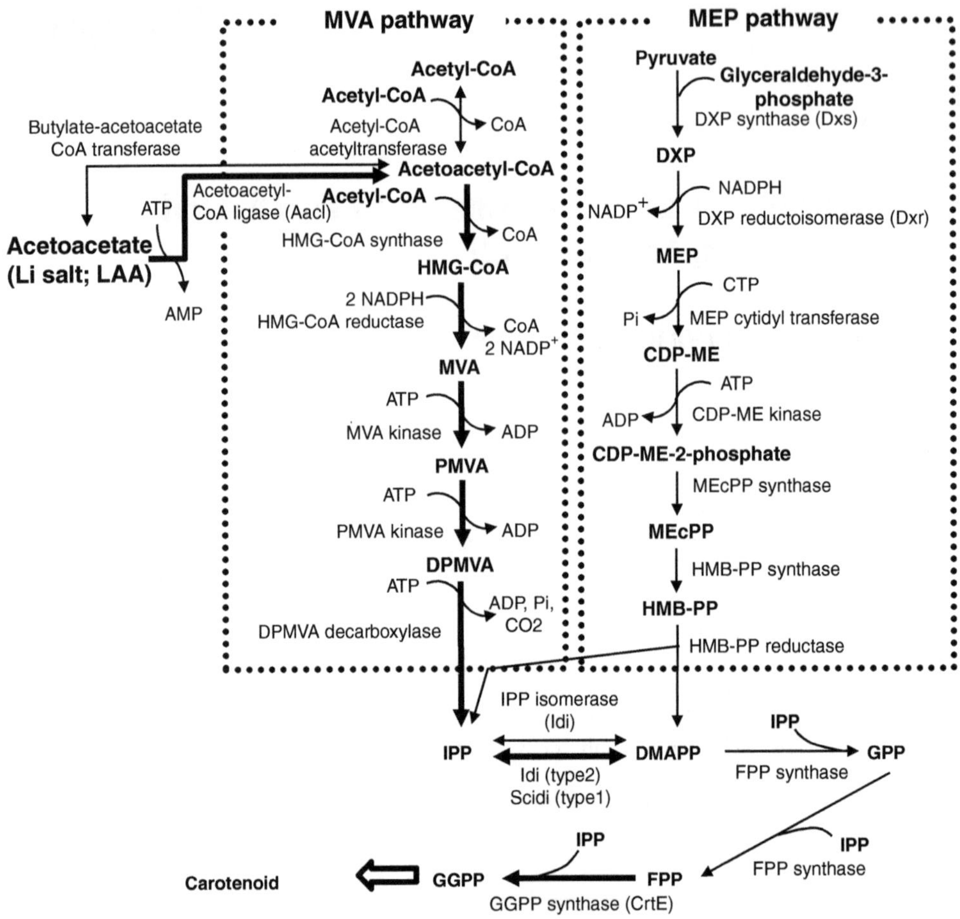

Fig. 1. Biosynthesis of isoprenoid basic structures and isoprenoid pathway engineering performed in *E. coli*. Two distinct biosynthetic routes to IPP, the mevalonate (MVA) pathway and the non-mevalonate (MEP) pathway, are known. Genes mediating steps represented with *bold arrows* are contained in plasmid pAC-Mev/Scidi/Aacl.

appropriate hosts for microbial production because of its full genetic manipulation system.

In *E. coli*, isopentenyl diphosphate (IPP), which is the initial precursor of isoprenoids, including carotenoids, is biosynthesized through non-mevalonate [2-*C*-methyl-D-erythritol 4-phosphate (MEP)] pathway (3) (Fig. 1). IPP is isomerized to dimethylallyl diphosphate (DMAPP) by IPP isomerase (Idi), and then this C_5 isomer sequentially combine with IPP to yield geranyl diphosphate (GPP, C_{10}) and farnesyl diphosphate (FPP, C_{15}) (Fig. 1). The addition of further IPP to FPP results in the formation of geranylgeranyl diphosphate (GGPP, C_{20}), which is the precursor of carotenoids (Fig. 1). In order to produce carotenoids in *E. coli*, it is needed to introduce a series of biosynthesis genes from FPP to a carotenoid of interest. However, resultant recombinant *E. coli* cells synthesize small amounts of the carotenoid, which are far from their practical use, since *E. coli* synthesizes small amounts of FPP originally. Thus, pathway-engineering approaches with this bacterium have extensively been performed to elevate carotenoid levels (reviewed in refs. 4, 5). The introduction of heterologous mevalonate (MVA) pathway (Fig. 1) genes into *E. coli* has been shown to improve the productivity of carotenoids (6–11). Some of these studies also showed that production levels of generated carotenoids were significantly increased with the addition of exogenous MVA, such as D-mevalonolactone (D-mevalonic acid lactone; D-MVL) in the culture medium. Here, we describe methods to efficiently synthesize desirable carotenoids, such as lycopene and astaxanthin, from a cheaper substrate, acetoacetate salt, as well as D-MVL.

2. Materials

2.1. Carotenoid Production with Isoprenoid Pathway-Engineered E. coli

1. *E. coli* JM109 competent cells (Toyobo, Osaka, Japan).
2. pAC-Mev (11) (see Fig. 2 and Note 1).
3. pAC-Mev/Scidi (11) (see Fig. 2 and Note 1).
4. pAC-Mev/Scidi/Aacl (11) (see Fig. 2 and Note 1).
5. pCRT-EIB (11) (see Note 2).

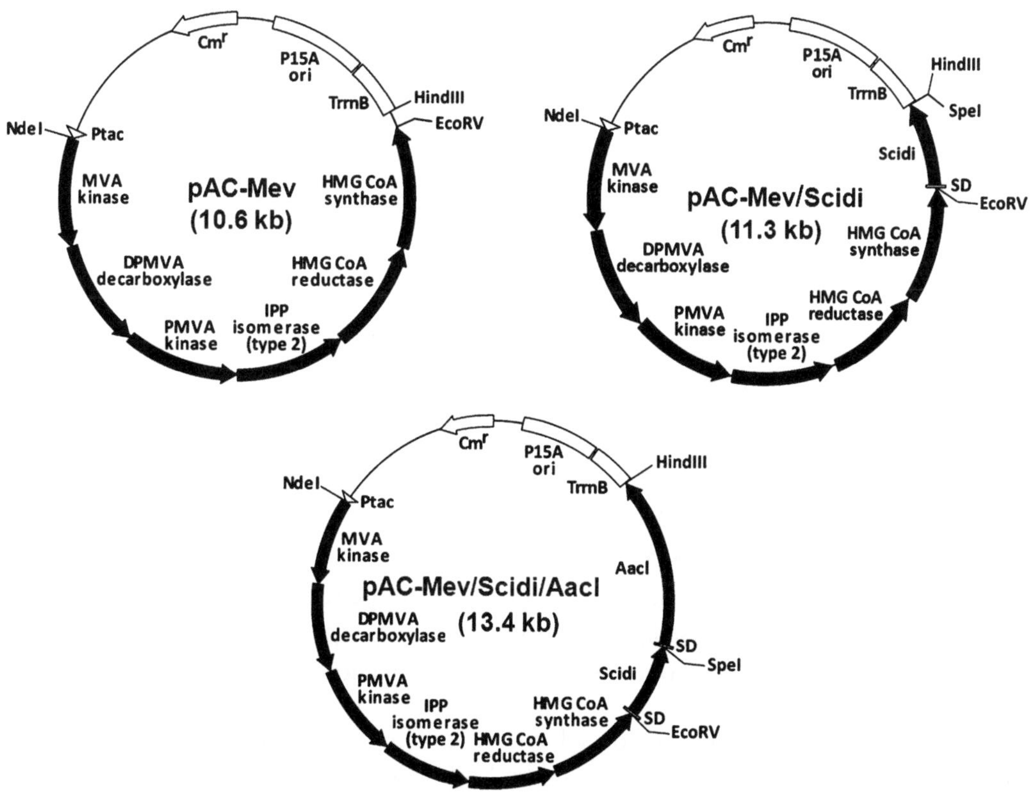

Fig. 2. Structures of three plasmids carrying heterologous mevalonate pathway-based genes.

6. pUC-Asta (11) (see Note 2).
7. LB: 10 g/L of tryptone, 5 g/L of yeast extract, and 10 g/L of NaCl, adjust pH to 7.0 with 5 N NaOH, and sterilize by autoclaving (121°C for 15 min).
8. LB solid: LB and 15 g/L of agar.
9. LB (Ap): Add 1 mL of 100 g/L of Ap to 1 L LB before use. Store at 4°C.
10. LB (Cm): Add 1 mL of 30 g/L of Cm to 1 L LB before use. Store at 4°C.
11. 2× YT: 16 g/L of tryptone, 10 g/L of yeast extract, and 5 g/L of NaCl, adjust pH to 7.0 with 5 N NaOH, and sterilize by autoclaving (121°C for 15 min).
12. 2× YT (Ap): Add 1 mL of 100 g/L of Ap to 1 L 2× YT before use. Store at 4°C.
13. 2× YT (Cm): Add 1 mL of 30 g/L of Cm to 1 L 2× YT before use. Store at 4°C.
14. SOC: 20 g/L of tryptone, 5 g/L of yeast extract, 0.5 g/L of NaCl, and 0.186 g/L of KCl, adjust pH to 7.0 with 5 N NaOH, and sterilize by autoclaving (121°C for 15 min). Add 5 mL of a sterile solution containing 2 M of $MgCl_2$ and 20 mL of a sterile solution containing 1 M of glucose before use. Store at 4°C.
15. Isopropyl β-D-thiogalactopyranoside (IPTG) solution 1 M (1,000×): Dissolve 0.238 g of IPTG in 1 mL of distilled H_2O and sterilize by filtration. Store at −20°C.
16. D-mevalonolactone (D-MVL) (Tokyo Kasei, Tokyo, Japan).
17. D-mevalonolactone (D-MVL) solution 0.5 g/mL (1,000×): Dilute 1 g of D-MVL in 1 mL of distilled H_2O and sterilize by filtration. Store at 4°C.
18. Lithium acetoacetate (LAA) solution 1 g/mL (1,000×): Dissolve 1 g of LAA in 1 mL of distilled H_2O and sterilize by filtration. Store at 4°C.
19. Water bath.
20. Heating block.
21. Incubator.
22. Shaking incubator.
23. Spectrophotometer.
24. Sterile polypropylene centrifuge tubes (15 mL).
25. Baffled flasks (200 mL) with an air-permeable plug (cotton or silicone sponge) sterilized by autoclaving (121°C for 15 min).

2.2. HPLC Analysis of Generated Carotenoids

1. Extraction solvent: Chloroform/methanol (50/50, *v*/*v*). Both solvents should be HPLC grade.
2. Separation solvent (mobile phase) A: Methanol/H_2O (95/5, *v*/*v*). Methanol should be HPLC grade.

3. Separation solvent (mobile phase) B: Methanol/tetrahydrofuran (70/30, *v/v*). Both reagents should be HPLC grade.
4. Vortex mixer.
5. HPLC apparatus equipped with photodiode array (PDA) detector.
6. TSKgel ODS-80Ts C18 reverse phase column (4.6 × 150 mm, 60 Å pore size, and 5 μm particle size) attached to the separation module (Tosoh, Tokyo, Japan).
7. Centrifugal evaporator.
8. Syringe filters: 0.45 μm pore size, and resistance against the organic HPLC solvents.
9. PP screw neck vial (300 μL; 12 mm × 32 mm) with LectraBond cap and PTFE/Silicone septa (Waters, Milford, MA, USA).
10. PP centrifuge tubes (15 mL).

3. Methods

3.1. Transformation and Screening of Carotenoid-Producing *E. coli*

1. Add 25–50 ng of each combination of pACYC-based plasmid (pAC-Mev, pAC-Mev/Scidi or pAC-Mev/Scidi/Aacl; Cm^r) and pUC-based plasmid (pCRT-EIB or pUC-Asta; Ap^r) to 50 μL of *E. coli* JM109 competent cells. Swirl the tube to mix and immediately store the tube on ice for 30 min.
2. Heat the tube for 45 s exactly in a water bath set at 42°C, and transfer the tube rapidly on ice for a few minutes.
3. Add 450 μL of SOC medium. Transfer the tube to a shaking incubator set at 30°C, and then incubate for 45 min.
4. Transfer the 50–100 μL of transformed competent cells onto LB (Ap, Cm) plate. Store the plate overnight (by 1 day) at 30°C. Red colored colonies should be generated.

3.2. Production of Carotenoids in Isoprenoid Pathway-Engineered *E. coli*

1. Inoculate one to three colonies of carotenoid-synthesizing *E. coli* into 3 mL of LB (Ap, Cm) medium in a 15 mL sterilized PP centrifuge tube. Incubate the culture overnight at 30°C in a shaking incubator.
2. Inoculate 300 μL (1% *v/v*) of the overnight culture into 30 mL of 2× YT (Ap, Cm) medium in a 200 mL baffled flask (see Note 3). Incubate the culture for 3–5 h at 30°C in a shaking incubator until OD_{600} of the culture reached 0.5.
3. Cool the culture to room temperature. Add 30 μL of 1,000× IPTG solution, and 30 μL of 1,000× D-MVL solution or 1,000× LAA solution (see Note 4).
4. Incubate the culture with shaking at 20°C for 2 days (see Note 5). After the cultivation, divide 30 mL of the culture into two

15 mL clear PP centrifuge tube, and then collect the cells by centrifugation at 5,000 × *g* for 15 min at 4°C. Discard supernatant and store cell pellet at −70°C before HPLC analysis. One of two cell pellets should be used for measurement of dry cell weight (DCW).

3.3. Extraction of Carotenoids from Isoprenoid Pathway-Engineered E. coli

Whenever possible, all procedure should be carried out on ice under weak light.

1. Add 1 mL of chloroform/methanol extraction solvent to the harvested cells, and vortex the tube vigorously for 30 min.
2. Centrifuge the tube at 5,000 × *g* for 15 min at 4°C.
3. Collect the chloroform/methanol phase into a new 15 mL tube. If colored cell pellet remains, add another 1 mL of chloroform/methanol extraction solvent to the cell pellet and extract again.
4. Dry up the pooled chloroform/methanol extract by centrifugal evaporation (or under a stream of nitrogen). The dried extract can be stored under an atmosphere of nitrogen at −20°C prior to HPLC.
5. Dissolve the dried extract with appropriate volume of ethyl acetate. Filtrate by syringe filter and transfer it into a 300 μL vial for HPLC.

3.4. HPLC Analysis of Carotenoids

HPLC analysis of carotenoids described by Yokoyama and Miki (12), which takes less than 20 min for each run, is specifically suitable for separation of xanthophylls (carotenoids including oxygen atom(s)).

1. Set the vial on a rack and inject 10–50 μL of the sample for HPLC analysis.
2. Elute the carotenoid extract at a rate of 1 mL/min with solvent A for 5 min, followed by a linear gradient from solvent A to B for 5 min, solvent B alone for 8 min, and then back to solvent A.
3. Identify and quantify carotenoids by comparing retention time, spectrum, and peak area with those of authentic samples. Figure 3 shows results of quantification analysis on *E. coli* carrying pACYC-based plasmid (pAC-Mev, pAC-Mev/Scidi or pAC-Mev/Scidi/Aacl) and pUC-based plasmid (pCRT-EIB or pUC-Asta).

In our hands, *E. coli* cells harboring pCRT-EIB and pAC-Mev/Scidi/Aacl produce 12.5 mg/g DCW of lycopene in the presence of LAA, which is an 11.8-fold increase of level of those carrying pCRT-EIB alone. When *E. coli* cells harboring pUC-Asta and pAC-Mev/Scidi/Aacl are grown in the presence of LAA, total amounts of carotenoids are significantly increased to 7.5-fold compared

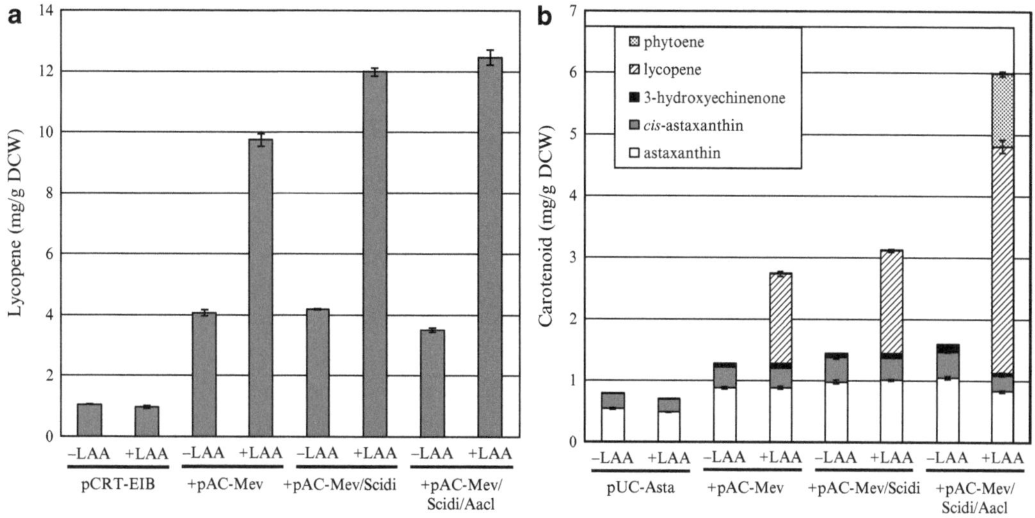

Fig. 3. Carotenoid production by *E. coli* JM109 harboring respective plasmids with or without addition of LAA to the culture medium. (a) Lycopene production with respective recombinant *E. coli* cells. (b) Production of astaxanthin and other carotenoids with respective recombinant *E. coli* cells.

with the control strain carrying only pUC-Asta. Whereas there is a little or no effect on the enhancement of astaxanthin content in each engineered *E. coli*, since large amounts of carotenoid intermediates from lycopene or phytoene to astaxanthin accumulate (11).

The three MVA pathway-engineered plasmids (Fig. 2) increase FPP content in *E. coli*. Thus, it is feasible to synthesize other isoprenoids, such as sesquiterpenes instead of carotenoids, if their biosynthesis genes from FPP are properly co-expressed using a compatible vector like pUC. For example, several sesquiterpenes are produced using such a method (4, 11, 13).

4. Notes

1. pAC-Mev, pAC-Mev/Scidi, and pAC-Mev/Scidi/Aacl carry the chloramphenicol resistance (Cm^r) gene and heterologous MVA pathway-based genes. We constructed plasmid pAC-Mev (11), which contains the gene cluster derived from *Streptomyces* sp. strain CL190, which encodes five enzymes in the MVA pathway [3-hydroxy-3-methylglutaryl CoA (HMG-CoA) synthase, HMG-CoA reductase, MVA kinase, phosphomevalonate (PMVA) kinase, and diphosphomevalonate (DPMVA) decarboxylase] and type 2 IPP isomerase (*idi*) (14, 15), using pACYC184 vector (see Figs. 1 and 2). To construct plasmids pAC-Mev/Scidi and pAC-Mev/Scidi/Aacl (11), the type 1 *idi*

gene from *Saccharomyces cerevisiae* (*Scidi*) and the acetoacetate-CoA ligase (EC 6.2.1.16) gene from *Rattus norvegicus* (*Aacl*) as well as *Scidi* were respectively ligated into the downstream of the six MVA pathway-related genes along with the addition of artificial Shine-Dalgarno sequence. These plasmids are compatible with most of other *E. coli* plasmids containing pMB1/*colE1* origin, such as pET, pUC, and pBluescript.

2. pCRT-EIB and pUC-Asta carry ampicillin resistance (Ap^r) gene, and genes for the production of lycopene and astaxanthin, respectively. A lycopene-synthesizing plasmid pCRT-EIB, in which the GGPP synthase gene (*crtE*), the phytoene synthase gene (*crtB*), and the phytoene desaturase gene (*crtI*) from soil bacterium *Pantoea ananatis* are inserted in pUC19 to have transcriptional read-through from the *lac* promoter, was described as pCAR-ADE (16). A plasmid for astaxanthin production (pUC-Asta) was described (11), which carries the six carotenoid biosynthesis genes, *crtE*, *crtB*, *crtI*, the lycopene cyclase gene (*crtY*) and the β-carotene hydroxylase gene (*crtZ*) from *P. ananatis*, and the β-carotene ketolase gene (*crtW*) from marine bacterium *Brevundimonas* sp. strain SD212 (17, 18) on vector pUC18.
3. If we want to perform structural determination of generated carotenoids by NMR (and other spectrometric) analysis, it is needed to scale up the following cultivation (often more than ten times). A novel carotenoid can be identified using this method, as described (19).
4. All the three MVA pathway-engineered plasmids (pAC-Mev, pAC-Mev/Scidi, and pAC-Mev/Scidi/Aacl) can utilize D-MVL as a substrate. The addition of LAA instead of D-MVL to these culture media can also improve the productivity of carotenoids, while the maximum productivity is achieved when using plasmid pAC-Mev/Scidi/Aacl.
5. Inoculation temperature after the addition of IPTG significantly influences the yield of generated carotenoids. Temperature should be maintained around 20°C until the end of cultivation.

Acknowledgments

This work has been supported as a part of the Research and Development Program for New Bio-Industry Initiatives (2006–2010) of Bio-Oriented Technology Research Advancement Institution (BRAIN).

References

1. Britton G, Liaaen-Jensen S, Pfander H (2001) Carotenoids handbook. Birkhauser, Boston
2. Donaldon MS (2004) Nutrition and cancer: a review of the evidence for an anticancer diet. Nutr J 3:19
3. Rohmer M, Knani M, Simonin P, Sutter B, Sahm H (1993) Isoprenoid biosynthesis in bacteria: a novel pathway for early steps leading to isopentenyl diphosphate. Biochem J 295: 517–524
4. Harada H, Misawa N (2009) Novel approaches and achievements in biosynthesis of functional isoprenoids in *Escherichia coli*. Appl Microbiol Biotechnol 84:1021–1031
5. Misawa N (2010) Carotenoids. In: Mander L, Lui H-W (eds) Comprehensive natural products II chemistry and biology, vol 1. Elsevier, Oxford, pp 733–753
6. Kakinuma K, Dekishima Y, Matsushima Y, Eguchi T, Misawa N, Takagi M, Kuzuyama T, Seto H (2001) New approach to multiply deuterated isoprenoids using triply engineered *Escherichia coli* and its potential as a tool for mechanistic enzymology. J Am Chem Soc 123: 1238–1239
7. Campos N, Rodriguez-Concepcion M, Sauret-Gueto S, Gallego F, Lois LM, Boronat A (2001) *Escherichia coli* engineered to synthesize isopentenyl diphosphate and dimethylallyl diphosphate from mevalonate: a novel system for the genetic analysis of the 2-*C*-methyl-D-erythritol 4-phosphate pathway for isoprenoid biosynthesis. Biochem J 353:59–67
8. Vadali RV, Fu Y, Bennett GN, San KY (2005) Enhanced lycopene productivity by manipulation of carbon flow to isopentenyl diphosphate in *Escherichia coli*. Biotechnol Prog 21: 1558–1561
9. Yoon SH, Park HM, Kim JE, Lee SH, Choi MS, Kim JY, Oh DK, Keasling JD, Kim SW (2007) Increased β-carotene production in recombinant *Escherichia coli* harboring an engineered isoprenoid precursor pathway with mevalonate addition. Biotechnol Prog 23:599–605
10. Rodriguez-Villalon A, Perez-Gil J, Rodriguez-Concepcion M (2008) Carotenoid accumulation in bacteria with enhanced supply of isoprenoid precursors by upregulation of exogenous or endogenous pathways. J Biotechnol 135:78–84
11. Harada H, Yu F, Okamoto S, Kuzuyama T, Utsumi R, Misawa N (2009) Efficient synthesis of functional isoprenoids from acetoacetate through metabolic pathway-engineered *Escherichia coli*. Appl Microbiol Biotechnol 81: 915–925
12. Yokoyama A, Miki W (1995) Composition and presumed biosynthetic pathway of carotenoids in the astaxanthin-producing bacterium *Agrobacterium aurantiacum*. FEMS Microbiol Lett 128:139–144
13. Fujisawa M, Harada H, Kenmoku H, Mizutani S, Misawa N (2010) Cloning and characterization of a novel gene that encodes (*S*)-β-bisabolene synthase from ginger, *Zingiber officinale*. Planta 232:121–130, Erratum, 232, 131
14. Takagi M, Kuzuyama T, Takahashi S, Seto H (2000) A gene cluster for the mevalonate pathway from *Streptomyces* sp. strain CL190. J Bacteriol 182:4153–4157
15. Kaneda K, Kuzuyama T, Takagi M, Hayakawa Y, Seto H (2001) An unusual isopentenyl diphosphate isomerase found in the mevalonate pathway gene cluster from *Streptomyces* sp. strain CL190. Proc Natl Acad Sci USA 98: 932–937
16. Misawa N, Nakagawa M, Kobayashi K, Yamano S, Izawa Y, Nakamura K, Harashima K (1990) Elucidation of the *Erwinia uredovora* carotenoid biosynthetic pathway by functional analysis of gene products expressed in *Escherichia coli*. J Bacteriol 172:6704–6712
17. Nishida Y, Adachi K, Kasai H, Shizuri Y, Shindo K, Sawabe A, Komemushi S, Miki W, Misawa N (2005) Elucidation of a carotenoid biosynthesis gene cluster encoding a novel enzyme, 2,2'-β-hydroxylase, from *Brevundimonas* sp. strain SD212 and combinatorial biosynthesis of new or rare xanthophylls. Appl Environ Microbiol 71:4286–4296
18. Choi SK, Nishida Y, Matsuda S, Adachi K, Kasai H, Peng X, Komemushi S, Miki W, Misawa N (2005) Characterization of β-carotene ketolases, CrtW, from marine bacteria by complementation analysis in *Escherichia coli*. Mar Biotechnol (NY) 7:515–522
19. Osawa A, Harada H, Choi SK, Misawa N, Shindo K (2011) Production of caloxanthin 3'-β-D-glucoside, zeaxanthin β-D-diglucoside, and nostoxanthin in a recombinant *Escherichia coli* expressing system harboring seven carotenoid biosynthesis genes, including *crtX* and *crtG*. Phytochemistry 72:711–716

References

Chapter 7

Engineering *Escherichia coli* for Canthaxanthin and Astaxanthin Biosynthesis

Qiong Cheng and Luan Tao

Abstract

Escherichia coli is a non-carotenogenic bacterium that could synthesize farnesyl pyrophosphate precursor through the isoprenoid pathway. Carotenoid production in *E. coli* requires heterologous expression of carotenoid synthesis genes. The carotenoid synthesis operons are assembled from genes isolated from carotenogenic bacterial sources. Expression of the different operons yields different carotenoid titers. The operons containing the *idi* gene give more than fivefold higher carotenoid titers than the operons lacking the *idi* gene. The carotenoid modification genes encoding ketolases and hydroxylases are incorporated into the operons for canthaxanthin and astaxanthin production. The ketolases and hydroxylases from different bacterial sources produce astaxanthin of different purity relative to the total carotenoids. Expression of the ketolases and hydroxylases closer to the promoter appears to give higher astaxanthin purity than expression farther from the promoter at the end of the operons. Balanced expression of ketolases and hydroxylases is critical to achieve high astaxanthin purity. Here, we describe methods to assemble carotenoid biosynthesis operons from carotenogenic gene clusters isolated from different bacterial sources and evaluate canthaxanthin or astaxanthin production in *E. coli*.

Key words: Carotenoid biosynthesis, Overlapping PCR, Gene cluster, Astaxanthin, Canthaxanthin

1. Introduction

Astaxanthin and canthaxanthin are high value carotenoids that are widely used in aquaculture and poultry industries. They have also recently been introduced into nutraceutical and personal care markets. Study synthesis of astaxanthin and canthaxanthin in *E. coli* provides valuable insights for biological production of these carotenoids in microbes.

Carotenoid biosynthesis starts with the common isoprenoid pathway (1) to generate farnesyl pyrophosphate precursor (Fig. 1). The *idi* gene encoding isopentenyl pyrophosphate isomerase is one of the genes in the isoprenoid pathway that is responsible for

José-Luis Barredo (ed.), *Microbial Carotenoids from Bacteria and Microalgae: Methods and Protocols*, Methods in Molecular Biology, vol. 892, DOI 10.1007/978-1-61779-879-5_7, © Springer Science+Business Media, LLC 2012

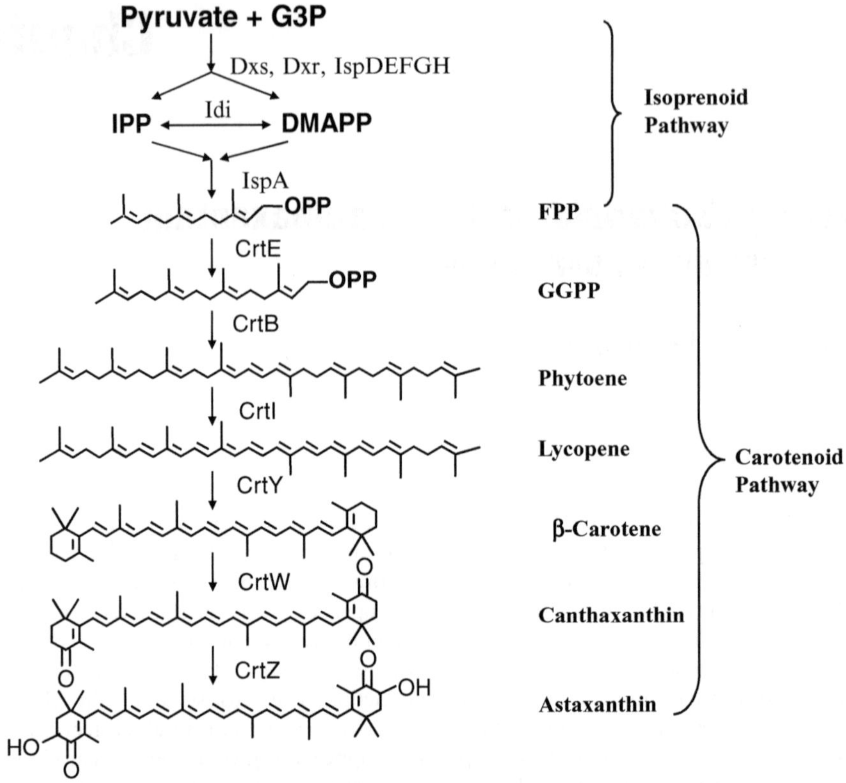

Fig. 1. Pathway for synthesis of canthaxanthin and astaxathin. Multiple steps in the non-mevalonate isoprenoid pathway catalyzed by Dxs, Dxr, IspD, IspE, IspF, IspG, and IspH are not shown in detail. Idi: isopentenyl pyrophosphate isomerase; IspA: fanesyl pyrophosphate synthase; CrtE: geranylgeranyl pyrophosphate synthase; CrtB: phytoene synthase; CrtI: phytoene desaturase; CrtY: lycopene β-cyclase; CrtW: 4, (4′)-β-ionone ring ketolase; CrtZ: 3, (3′)-β-ionone ring hydroxylase. The abbreviations for the compounds are as follows. *G3P* glyceraldehyde 3-phosphate, *IPP* isopentenyl pyrophosphate, *DMAPP* dimethylallyl pyrophophate, *FPP* farnesyl pyrophosphate, *GGPP* geranylgeranyl pyrophosphate.

the intramolecular conversion of isopentenyl pyrophosphate to dimethylallyl pyrophosphate, both of which are the substrates for synthesis of farnesyl pyrophosphate. The carotenoid pathway (2) enzymes (CrtEBIY) present in common carotenogenic organisms are involved in subsequent synthesizing C_{40} carotenes, such as β-carotene.

Overexpression of endogenous or exogenous *idi* gene not linked to the carotenogenic gene cluster was reported to increase carotenoid titer (3–5). The ketolase and hydroxylase (CrtWZ) modifies the β-carotene ring to create oxygenated derivatives, such as canthaxanthin and astaxanthin. The oxygenation is the rate-limiting step for astaxanthin synthesis which could generate multiple intermediates that decrease astaxanthin purity as measured by the percentage of astaxanthin produced relative to the total carotenoid content. Previous in vitro (6) and in vivo (7, 8) studies showed that the CrtW and CrtZ enzymes from different bacterial sources had different activities and substrate specificities.

Here, we show the method to assemble β-carotene synthesis operons from carotenogenic gene clusters isolated from different bacterial sources. We also describe how to incorporate different *crtWZ* gene combinations into the operon for canthaxanthin and astaxanthin synthesis. Balanced high expression of ketolases and hydroxylases is critical to achieve high astaxanthin purity.

2. Materials

2.1. Sources of β-Carotene Synthesis Gene Clusters

1. *crtEXYIBZ* gene cluster from *Enterobacteriaceae* bacterium DC260 (GenBank accession number DQ090833).
2. *crtEidiYIBZ* gene cluster from *Enterobacteriaceae* bacterium DC404 (GenBank accession number DQ090834).
3. *crtEXYIBZ* gene cluster from *Enterobacteriaceae* bacterium DC416 (GenBank accession number DQ090835).
4. *crtEidiXYIBZ* gene cluster from *Enterobacteriaceae* bacterium DC413 (GenBank accession number DQ090836).

2.2. Sources of Carotenoid Modification Genes

1. *crtW* from *Sphingomonas* sp. DC18 (GenBank accession number DQ400932).
2. *crtW* from *Brevundimonas vesicularis* DC263 (GenBank accession number DQ309446).
3. *crtW* synthetic gene from *Paracoccus* sp. MBIC1143 (formerly classified as *Agrobacterium aurantiacum*) (GenBank accession number D58420).
4. *crtW* from *Algoriphagus* sp. KK10202C (GenBank accession number DQ286432).
5. *crtZ* from *Brevundimonas vesicularis* DC263 (GenBank accession number DQ309446).
6. *crtZ* from *Novosphingobium aromaticivorans* ATCC700278 (Accession number AAAV01000142.1).
7. *crtZ* synthetic gene from *Paracoccus* sp. MBIC1143 (formerly classified as *Agrobacterium aurantiacum*) (GenBank accession number D58420).
8. *crtZ* from *Enterobacteriaceae* strain DC260 (GenBank accession number DQ090833).
9. *crtZ* from *Enterobacteriaceae* strain DC404 (GenBank accession number DQ090834).

2.3. DNA Oligonucleotide Primers

1. Forward primer 1: 5′-tata<u>GAATTC</u>**CCTAGG**TGCAAGTAAGGACTGCCATTATGAC-3′.
2. Reverse primer 1: 5′-GATGTCATAGTACTTCTCCGCGCGCTGTACATCATAGCGTTTCTCCGGCGCGAT-3′.

3. Forward primer 2: 5′-ATCGCGCCGGAGAAACGCTATGA TGTACAGCGCGCGGAGAAGTACTATGACATC-3′.
4. Reverse primer 2: 5′-tataGAATTCACTAGTTAACGCGGA CGCTGCCAGAG-3′.
5. Sequencing primer 1: 5′-AACCAGACCGTTCAGCTGGA-3′.
6. Sequencing primer 2: 5′-GGTGTAACAAGGGTGAACAC-3′.
7. crtW forward primer: 5′-tataTCTAGA**AAGGAGGAAT AAACC**ATGCGGCAAGCGAACAGGATG-3′.
8. crtW reverse primer: 5′-tataCCTAGGCTAGCTGAACAAA CTCCACCAG -3′.
9. crtZ forward primer: 5′-tataTCTAGA**AAGGAGGAAT AAACC**ATGTCCTGGCCGACGATGATC-3′.
10. crtZ reverse primer: 5′-tataACTAGTCAGGCGCC GTTGCTGGATGA-3′.

2.4. Reagents and Equipment

1. *PfuUltra*® II Fusion HS DNA Polymerase (Stratagene, Santa Clara, CA, USA).
2. GeneAmp® PCR System 9700 (Applied Biosystems, Carlsbad, CA, USA).
3. DNA Clean & Concentrator™-5 Kit (Zymo Research, Orange, CA, USA).
4. Zymoclean™ Gel DNA Recovery Kit (Zymo Research, Orange, CA, USA).
5. Nanodrop 1000 Spectrophotometer (Thermo Fisher Scientific, Waltham, MA, USA).
6. Reliant® Gel System 1.0% Seakem® Gold agarose gel (Cambrex BioSciences, Rockland, ME, USA).
7. High DNA Mass Ladder (Invitrogen, Carlsbad, CA, USA).
8. 10× BlueJuice™ Gel Loading Buffer (Invitrogen, Carlsbad, CA, USA).
9. 10× TBE buffer (10× electrophoresis TBE buffer stock solution pH 8.3): Weight 86.4 g of Tris, 44 g boric acid and 7.36 g of EDTA. Add water and dissolve. Bring to 800 mL with water. For 1× working solution, dilute stock solution 1:10 with sterilized water.
10. pBHR1 plasmid (Mobitec, Marco Island, FL, USA).
11. Restriction enzymes: *EcoR*I, *Spe*I, *Xba*I, *Avr*II.
12. T4 DNA ligase.
13. *E. cloni* 10G electrocompetent cells (Lucigen, Middletown, WI, USA).
14. Gene Pulser® (Bio-Rad, Hercules, CA, USA).
15. Gene Pulser® Cuvette, 1 mm electrode gap (Bio-Rad, Hercules, CA, USA).

16. Recovery medium (Lucigen, Middletown, WI, USA).
17. LB: 5 g/L yeast extract, 10 g/L tryptone, and 5 g/L NaCl. Adjust pH to 7.5 with 1 N NaOH. Autoclave on liquid cycle at 15 psi and 121°C for 20 min. Store at 4°C.
18. LB agar: LB and 15 g/L agar before autoclaving. Autoclave on liquid cycle at 15 psi and 121°C for 20 min. Store at 4°C.
19. Kanamycin (Sigma, St. Louis, MO, USA).
20. QIAprep Spin Miniprep Kit (Qiagen, Valencia, CA, USA).
21. Agilent Series 1100 LC/MSD SI (Agilent Technologies, Santa Clara, CA, USA).
22. 150 mm×4.6 mm ZORBAX C18 (3.5 μm particles) HPLC column (Agilent Technologies, Santa Clara, CA, USA).
23. HPLC Buffer A: 95% acetonitrile and 5% dH_2O; Buffer B: 100% tetradehydrofuran.
24. β-carotene (Sigma, St. Louis, MO, USA).
25. Canthaxanthin (CaroteNature, Lupsingen, Swizerland).
26. Zeaxanthin (CaroteNature, Lupsingen, Swizerland).
27. Astaxanthin (CaroteNature, Lupsingen, Swizerland).

3. Methods

3.1. Overlapping PCR Amplification of crtEidiYIB Genes from Strain DC413

The native carotenoid synthesis gene clusters isolated from *Enterobacteriaceae* strains are usually organized as *crtEXYIBZ*; whereas the *crtEXYIB* genes are transcribed in one direction as an operon, the downstream *crtZ* gene is transcribed in the opposite direction (9). The genes required for β-carotene synthesis are assembled as the *crtE crtY crtI crtB* operon by overlapping PCR. We remove the *crtX* gene encoding zeaxanthin glucosyl transferase, which converts zeaxanthin to zeaxanthin-β-diglucoside. The *idi* gene that is naturally present in several clusters is kept intact and in these cases the genes in the operon are *crtE idi crtY crtI crtB*. Following is an example on how to perform overlapping PCR to assemble the *crtEidiYIB* genes together as an operon and to clone the operon into pBHR1 vector to be expressed under the control of the chloramphenicol resistance gene promoter (see Note 1), resulting in construct pDCQ380.

3.1.1. Set up First Round PCR Reactions to Amplify crtEidi and crtYIB, Respectively

1. Set up PCR mixture for each PCR reaction as following: 40.6 μL of distilled water (dH_2O), 5.0 μL of 10× PfuUltra® II Fusion HS DNA polymerase buffer, 0.4 μL of dNTPs (25 mM each dNTP), 1.0 μL of DNA template (~10 ng/μL), 1.0 μL of forward primer (100 ng/μL), 1.0 μL of reverse primer

(100 ng/μL), and 1.0 μL of PfuUltra® II Fusion HS DNA polymerase (2.5 U/μL) in a total reaction volume of 50.0 μL (see Note 2).

2. Add the components in order while mixing gently into a thin-wall PCR tube.
3. Cover the PCR tube with a PCR tube cap or an aluminum foil.
4. Put the tube into a PCR cycler with a preheated lid, and close the lid.
5. Perform the PCR using the following cycling conditions: (I) 95°C for 2 min. (II) 30 cycles of 95°C for 30 s, 58°C for 30 s, 72°C for 1 min (for <3 kb PCR product) or 3 min (for ≥3 kb PCR product). (III) 72°C for 5 min.
6. To analyze PCR amplified crtEidi and crtYIB products on 1.0% Seakem® Gold agarose gel, mix 9 μL of PCR amplification product with 1 μL 10× BlueJuice™ gel loading buffer. Additionally, mix 4.5 μL High DNA Mass Ladder with 0.5 μL 10× BlueJuice™ gel loading buffer.
7. Load the sample and DNA marker into different wells of a 1.0% Seakem® Gold agarose gel.
8. Run the gel in 1× TBE buffer at 90 V for about 30 min.
9. Transfer the gel onto an UV transilluminator. Turn on the power and check the size and purity of the amplified PCR product (see Note 3).

3.1.2. Purification of PCR Amplified crtEidi and crtYIB Products Using DNA Clean & Concentrator™-5 Kit

1. Add DNA-binding buffer to the tube containing the DNA sample. The amount of buffer added is two times the volume of DNA sample to be purified.
2. Mix by vortex briefly.
3. Load the solution into a Zymo-Spin™ Column in a Collection Tube.
4. Centrifuge at maximal speed in the microcentrifuge for 30 s. Discard the flow-through.
5. Add 200 μL of DNA Wash buffer to the column and centrifuge for 30 s. Discard the flow-through. Repeat the wash step once.
6. Add 10 μL of distilled water to the column matrix. Place column into a new 1.5 mL microcentrifuge tube and centrifuge for 60 s to elute DNA.
7. Measure DNA concentration using Nanodrop 1000 spectrophotometer (see Subheading 3.1.4).

3.1.3. Purification of PCR Amplified crtEidi and crtYIB Products from Agarose Gels Using Zymoclean™ Gel DNA Recovery Kit

1. Put the agarose gel on a UV transilluminator, turn on the UV light to see the DNA bands.
2. Locate the position of the DNA band with correct size.
3. Cut the correct gel band containing the needed DNA using a razor blade and put it in a 1.5 mL microcentrifuge tube.

4. Add agarose gel dissolving buffer to the tube. The amount of buffer added is three times the volume of the gel slice in the tube.
5. Incubate at 55°C for 10 min until gel slice is completely dissolved.
6. Load the solution into a Zymo-Spin™ Column in a collection tube.
7. Centrifuge at maximal speed in the microcentrifuge for 30 s. Discard the flow-through.
8. Add 200 μL of DNA Wash buffer to the column and centrifuge for 30 s. Discard the flow-through. Repeat the wash step once.
9. Add 10 μL of distilled water to the column matrix. Place column into a new 1.5 mL microcentrifuge tube and centrifuge for 60 s to elute DNA.
10. Measure DNA concentration using Nanodrop 1000 spectrophotometer (see Subheading 3.1.4).

3.1.4. Measuring DNA Concentrations of Purified PCR Amplified crtEidi and crtYIB Products Using Nanodrop 1000 Spectrophotometer

1. Turn on the Nanodrop 1000 spectrophotometer and its computer software.
2. Choose "DNA-50" for the sample type selection column on the computer screen.
3. Open the arm of the spectrophotometer.
4. Pipet 1 μL distilled water onto the sample pedestal. Close the arm.
5. Press "blank" button on the screen. Wait until the calibration completes.
6. Open the arm and dry the surface of the pedestal with a lint-free lab tissue wiper.
7. Pipet 1 μL DNA sample onto the sample pedestal. Close the arm.
8. Press "measure" button on the screen. Wait until the measurement completes. Record the concentration reading of the DNA sample.

3.1.5. Set up Second Round Overlapping PCR to Amplify crtEidiYIBgene Cluster

1. The PCR mixture (see Note 4) and running condition is the same as described in Subheading 3.1.1. The DNA template used here is 5 ng each of the purified PCR products of *crtEidi* and *crtYIB* from Subheading 3.1.4.
2. The PCR amplified *crtEidiYIB* product is purified and quantified using the same protocols described from Subheadings 3.1.1 to 3.1.4.

3.2. Cloning of the Overlapping PCR Product of the Assembled crtEidiYIB Genes into the EcoRI Site in pBHR1 Vector to Create Construct pDCQ380

3.2.1. Restriction Enzyme EcoRI Digestion of pBHR1 or Purified PCR Product of crtEidiYIB

1. Set up *Eco*RI digestion mixture as following: 7 μL of distilled water (dH_2O), 2 μL of 10× NEBuffer 4, 10 μL of pBHR1 plasmid DNA (vector) or purified crtEidiYIB PCR product (50–100 ng/μL) from Subheading 3.1.5, and 1 μL of *Eco*RI (20 U/μL) in a total reaction volume of 20 μL.
2. Add the components in order while mixing gently into a 1.5 mL microcentrifuge tube.
3. Incubate at 37°C for 1 h.
4. Purify the *Eco*RI-digested *pBHR1* and the *Eco*RI-digested PCR products of crtEidiYIB with DNA Clean & Concentrator™-5 Kit (see Subheading 3.1.2).
5. Measure DNA concentration using Nanodrop 1000 spectrophotometer (see Subheading 3.1.4).

3.2.2. Ligation of Purified EcoRI-Digested pBHR1 and Purified EcoRI-Digested PCR Product of crtEidiYIB

1. Setting up ligation mixture as following: 1 μL of distilled water (dH_2O), 2 μL of 5×T4 DNA ligase buffer, 1 μL of *Eco*RI-digested pBHR1 (diluted to 10 ng/μL with dH_2O) from Subheading 3.2.1, 5 μL of *Eco*RI-digested *crtEidiYIB* PCR product (diluted to 10 ng/μL with dH_2O) from Subheading 3.2.1, and 1 μL of T4 DNA ligase (1 U/μL) in a total reaction volume of 10 μL.
2. Add the components in order while mixing gently into a 1.5 mL microcentrifuge tube.
3. Incubate at room temperature for 1 h. The DNA mixture is ready for transformation.

3.2.3. Transformation of Ligation Reaction Mixture into E. cloni 10G Electrocompetent Cells

1. Thaw one tube of competent cells on ice.
2. Put a Gene Pulser® cuvette (1 mm) and a 1.5 mL microcentrifuge tube on ice for 10 min.
3. Add 0.5 μL DNA ligation reaction mixture from Subheading 3.2.2 into the tube.
4. Aliquot 20 μL competent cells into the tube. Mix gently. Make sure there are no bubbles.
5. Transfer mixture of DNA and cells into the ice cold cuvette.
6. Set the electroporator parameters: voltage = 1.8 kV, resistance = 200 Ω, and capacitance = 25 microFaradays (μF).
7. Electroporate the cells.
8. Immediately after electroporation, add 1 mL recovery medium into the electroporation cuvette. Mix by pipetting.
9. Transfer cells into a Falcon 2059 tube, incubate at 250 rpm, 37°C for 45 min.
10. Spread 1 μL to 100 μL of the cells onto LB plates with 50 μg/mL kanamycin.
11. Incubate the plates at 30°C for 2 days or longer until some colonies turn yellow.

3.2.4. Preparation and Confirmation of pDCQ380 Plasmid DNA

1. Pick several yellow colored clones from the LB plate with 50 μg/mL kanamycin in Subheading 3.2.3 and inoculate each in 3 mL LB medium with 50 μg/mL kanamycin in a Falcon 2059 tube. Incubate at 250 rpm, 37°C for 16 h.
2. Transfer 1.5 mL cell culture into a 1.5 mL microcentrifuge tube, centrifuge in the microcentrifuge at maximal speed for 1 min. Discard the supernatant completely.
3. Add 0.25 mL ice-cold buffer P1 in QIAprep Spin Miniprep Kit. Resuspend the cell pellet completely.
4. Add 0.25 mL buffer P2, invert the tube five times, let stay for 4 min at room temperature to lyse the cells.
5. Add 0.35 mL buffer N3, vortex for 10 s. Centrifuge at maximal speed for 10 min.
6. Transfer the supernatant to QIAprep Spin column. Centrifuge at maximal speed for 30 s. Discard the flow-through.
7. Add 0.8 mL buffer PE to the column. Centrifuge at maximal speed for 30 s. Discard the flow-through. Centrifuge at maximal speed for another 30 s to dry the column.
8. Add 50 μL buffer EB onto the matrix of the column. Centrifuge at maximal speed for 30 s to elute plasmid DNA.
9. The pDCQ380 plasmid DNA concentration is measured using Nanodrop 1000 spectrophotometer (see Subheading 3.1.4).
10. Plasmid is confirmed by sequencing (see Note 5).

3.3. Assembly of Gene Operons for Canthaxanthin Synthesis

The β-carotene ketolase gene (*crtW*) was added to the β-carotene synthesis operon for canthaxanthin synthesis. Typical cathaxanthin synthesis gene operons (Fig. 2) we assembled contained the *crtW* gene added as the first gene in the operon. Many *crtW* genes can fulfill this reaction and achieve almost complete (>90%) conversion of β-carotene to canthaxanthin. Materials section (see Subheading 2) listed several sources of *crtW* genes we used. More can be found in the databases from different bacterial sources: *Paracoccus marcusii* (accession number Y15112), *Paracoccus haeundaensis* (accession number AY957386), *Alcaligenes* sp. (accession number D58422), *Bradyrhizobium* sp. (accession number AF218415), *Brevundimonas aurantiaca* (accession number AY166610), *Brevundimonas* sp. SD212 (accession number AB181388). Following is an example of how to assemble the *crtWEidiYIB* cluster on pBHR1 vector, resulting in canthaxanthin expression construct pDCQ392W.

1. Set up PCR reaction as described in Subheading 3.1.1 to amplify a *crtW* gene (see Note 6).
2. Purify the PCR product as described in Subheading 3.1.2. Measure the DNA concentration as described in Subheading 3.1.4.
3. Double digest the purified *crtW* PCR product with *Xba*I *and* *Avr*II as following: 22.5 μL of distilled water (dH_2O), 4.0 μL

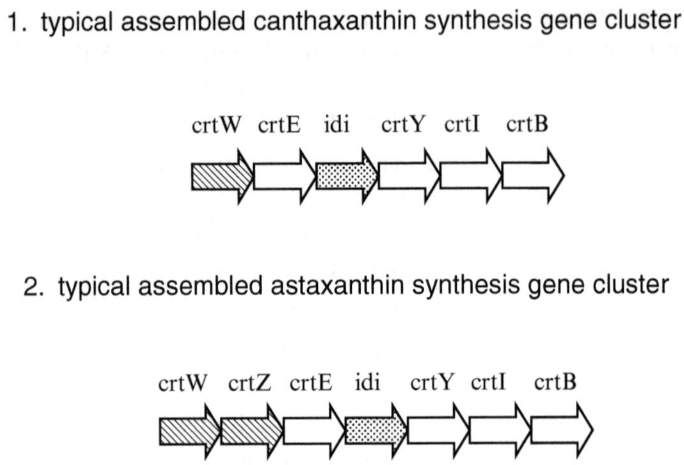

Fig. 2. Assembled typical gene clusters for canthaxanthin and astaxanthin synthesis. The genes required for β-carotene synthesis (*crtEYIB*) were shown as *open arrows*. The genes required for β-carotene modification (*crtWZ*) were shown as *striped arrows*. The *idi* gene in the isoprenoid pathway is shown as the *dotted arrow* and increases carotenoid titers.

of 10× NEBuffer 4, 10.0 μL of purified *crtW* PCR product (50–100 ng/μL), 1.0 μL of *Xba*I (20 U/μL), and 2.5 μL of *Avr*II (4 U/μL) in a total reaction volume of 40.0 μL; incubate at 37°C for 1 h.

4. Digest pDCQ380 plasmid DNA with *Avr*II as following: 10.0 μL of distilled water (dH_2O), 2.5 μL of 10× NEBuffer 4, 10.0 of μL purified *pDCQ380* plasmid DNA (50–100 ng/μL) from Subheading 3.2.4, and 2.5 μL of *Avr*II (4 U/μL) in a total reaction volume of 25.0 μL; incubate at 37°C for 1 h.
5. Purify the digested DNA products as described in Subheading 3.1.2. Measure the DNA concentration as described in Subheading 3.1.4.
6. Setting up ligation mixture as following: 3 μL of distilled water (dH_2O), 2 μL of 5×T4 DNA ligase buffer, 2 μL of *Avr*II digested pDCQ380 (diluted to 10 ng/μL with dH_2O), 2 μL of *Xba*I/*Avr*II double-digested *crtW* PCR product (diluted to 10 ng/μL with dH_2O), and 1 μL of T4 DNA ligase (1 U/μL) in a total reaction volume of 10 μL; incubate at room temperature for 1 h.
7. Transform the ligation mixture into *E. cloni* 10G electrocompetent cells following the same procedure in Subheading 3.2.3. Incubate the plates at 30°C for 2 days or longer until some colonies turn orange.
8. Pick several orange colored clones and perform plasmid DNA preparation of pDCQ392W following the same procedure as described in Subheading 3.2.4.
9. Confirm plasmid pDCQ392W by sequencing (see Note 7).

3.4. Assembly of Gene Operons for Astaxanthin Synthesis

Addition of *crtZ* gene to the canthaxanthin gene cluster could result in astaxanthin synthesis. Typical astaxanthin synthesis gene clusters we assembled (Fig. 2) contain the *crtWZ* genes upstream of the *crtEYIB* genes in the same operon (see Note 8). Several sources of *crtZ* genes we used are listed in Materials (see Subheading 2). More can be found in the databases from different bacterial sources: *Paracoccus marcusii* (accession number Y15112), *Paracoccus haeundaensis* (accession number AY957386), *Alcaligenes* sp. (accession number D58422), *Brevundimonas* sp. SD212 (accession number AB181388), *Erwinia uredovora* (reclassified as *Pantoea ananatis*, accession number D90087), *Pantoea agglomerans* (accession number M87280), *Pantoea stewartii* (AY166713), *Enterobacteriaceae* strains DC416 (accession number DQ090835) and DC413 (accession number DQ090836). Different combinations of *crtW* and *crtZ* genes from different bacterial sources (see Note 9) are used to improve astaxanthin purity (Fig. 3). Following is an example of how to assemble the *crtWZEidiYIB* cluster on pBHR1 vector, resulting in astaxanthin expression construct pDCQ392.

1. Set up PCR reaction as described in Subheading 3.1.1 to amplify a *crtZ* gene (see Note 10).
2. Purify the PCR product as described in Subheading 3.1.2. Measure the DNA concentration as described in Subheading 3.1.4.
3. Double digest the purified *crtZ* PCR product with *Xba*I *and Spe*I (see Note 11) as following: 6 μL of distilled water (dH_2O), 2 μL of 10× NEBuffer 4, 10 μL of purified *crtZ* PCR product (50–100 ng/μL), 1 μL of *Xba*I (20 U/μL), and 1 μL of *Spe*I (10 U/μL) in a total reaction volume of 20 μL; incubate at 37°C for 1 h.

Plasmid	Ast%	Genes expressed on plasmid
pDCQ334:	75%	W_{agro} Z_{agro} $EiYIB_{404}$
pDCQ337:	29%	Z_{404} W_{agro} Z_{agro} $EiYIB_{404}$
pDCQ339:	81%	W_{202C} W_{agro} Z_{agro} $EiYIB_{404}$
pDCQ338:	85%	W_{202C} Z_{404} W_{agro} Z_{agro} $EiYIB_{404}$
pDCQ343:	64%	W_{18} Z_{263} E YIB_{260}
pDCQ378:	45%	W_{18} Z_{260} E YIB_{260}
pDCQ344:	86%	Z_{263} W_{263} E YIB_{260}
pDCQ377:	85%	W_{263} Z_{263} $EiYIB_{404}$
pDCQ381:	81%	$EiYIB_{413}$ W_{263} Z_{263}
pDCQ392:	90%	W_{263} Z_{263} $EiYIB_{413}$

Fig. 3. Astaxanthin purity from *E. coli* containing different astaxanthin synthesis plasmids. The *subscript letters* refer to the bacterial sources where the genes were isolated.

4. Digest pDCQ392W plasmid DNA with *Avr*II as following: 10.0 μL of distilled water (dH_2O), 2.5 μL of 10× NEBuffer 4, 10.0 μL of purified pDCQ392W plasmid DNA (50–100 ng/μL) from Subheading 3.3, and 2.5 μL of *Avr*II (4 U/μL) in a total reaction volume of 25.0 μL; incubate at 37°C for 1 h.
5. Purify the digested DNA products as described in Subheading 3.1.2.
6. Setting up ligation mixture as following: 3 μL of distilled water (dH_2O), 2 μL of 5×T4 DNA ligase buffer, 2 μL of *Avr*II digested pDCQ392W (diluted to 10 ng/μL with dH_2O), 2 μL of *Xba*I/*Spe*I double-digested *crtZ* PCR product (diluted to 10 ng/μL with dH_2O), and 1 μL of T4 DNA ligase (1 U/μL) in a total reaction volume of 10 μL; incubate at room temperature for 1 h.
7. Transform the ligation mixture into *E. cloni* 10G electrocompetent cells following the same procedure in Subheading 3.2.3. Incubate the plates at 30°C for 2 days or longer until some colonies turn red.
8. Pick several red colored clones and perform plasmid DNA preparation of pDCQ392 following the same procedure as described in Subheading 3.2.4.
9. Confirm plasmid pDCQ392 by sequencing (see Note 12).

3.5. Analysis of Canthaxanthin and Astaxanthin in E. coli

3.5.1. Growing of the Carotenoids Producing E. coli Cultures in Shake Flasks

1. Transfer cells from streaked single colony on the kanamycin plate to 100 mL LB medium in a 500 mL flask (see Note 13) and culture at 30°C and 220 rpm for 2 days.
2. Aliquot 50 mL cell cultures each into two 50 mL conical polypropylene tubes for carotenoid extraction and dry cell weight measurement.

3.5.2. Extraction of Carotenoids from Cells

1. Pellet cells by centrifugation at 4,000 × *g* for 15 min. Remove supernatant completely.
2. Add 5 mL of an acetone–methanol mixture (50:50, v/v) to the cell pellet, vortex for 10 min to extract carotenoids.
3. Centrifuge at 4,000 × *g* for 15 min. Transfer supernatant to a new 50 mL conical tube.
4. Repeat steps 2 and 3 once. Combine the extracted solution.
5. Dry the extracted solution under nitrogen completely.
6. Add 0.5 mL of acetone–methanol mixture to redissolve the carotenoids.
7. Filter the solution into an autosampler vial with a 0.2 μm pore size Teflon filter and a 1 mL syringe.

3.5.3. Analysis of Carotenoids by HPLC

1. The high performance liquid chromatography (HPLC) system used has an autosampler and a DAD detector. A 150 mm × 4.6 mm ZORBAX C18 (3.5 μm particles) column is used for the separation of carotenoids.
2. Load the sample vials into the autosampler of the HPLC system.
3. Program the HPLC system as following: Keep the column temperature at 40°C. The solvent flow rate is 1 mL/min. The solvent running program used is: 0–2 min: 95% Buffer A and 5% Buffer B; 2–10 min: linear gradient from 95% Buffer A and 5% Buffer B to 60% Buffer A and 40% Buffer B; 10–12 min: linear gradient from 60% Buffer A and 40% Buffer B to 50% Buffer A and 50% Buffer B; 12–18 min: 50% Buffer A and 50% Buffer B; and, 18–20 min: 95% Buffer A and 5% Buffer B.
4. Pre-run the HPLC with the starting solvent (95% Buffer A and 5% Buffer B) at 1 mL/min until the DAD detector shows stable baseline.
5. Start the HPLC and inject 10 μL sample for each run.
6. After the run, analyze data by comparing the peak area value of the sample with that of the standard on the standard curve (see Note 14).

3.5.4. Carotenoids Standard Preparation

The elution time of the standard under the above condition is: β-carotene, 13.2 min; zeaxanthin, 5.9 min; canthaxanthin, 7.5 min; astaxanthin, 4.8 min. The spectra data and molecular weight of the carotenoids are: β-carotene, λ_{max}: 456, 484 nm; MH^+: 537; zeaxanthin, λ_{max}: 456, 480 nm; MH^+: 569; canthaxanthin, λ_{max}: 475 nm; MH^+: 565; astaxanthin, λ_{max}: 480 nm; MH^+: 597.

1. Dissolve 2 mg carotenoid standard in 5 mL acetone first, then add 5 mL of methanol to the mixture.
2. Filter this 200 μg/mL carotenoid stock solution into a 15 mL conical tube with a 0.2 μm pore size Teflon filter and a 10 mL syringe.
3. Transfer 1 mL of the 200 μg/mL carotenoid stock solution into 9 mL acetone–methanol mixture (50:50, v/v) to make 20 μg/mL standard solution.
4. Dilute this 20 μg/mL standard solution with acetone–methanol mixture (50:50, v/v) to make 15, 10, 5 μg/mL standard solutions. Transfer 1 mL of each standard solution into an autosampler vial.
5. Run 20 μL of each of the 5, 10, 15, 20 μg/mL standard solutions on LC/MS.
6. Identify the carotenoid peak position and record the peak area value for each concentration of standard.
7. A standard curve can be plotted using the peak area value against the concentration of each standard.

3.5.5. Determination of Carotenoid Titer

The titer of canthaxanthin or astaxanthin produced could be calculated as μg of canthaxanthin or astaxanthin per g of dry cell weight (ppm). Dry cell weight of the cultures could be determined as following.

1. Pre-weight an aluminum weight boat and 47-mm 0.2 μm pore size polycarbonate Whatman Nuclepore membrane filter on an analytic balance using tweezers.
2. Put the filter into vacuum filter unit, load the cell culture and turn on house vacuum.
3. Let medium pass the filter and the retain cells on the filter under vacuum.
4. When filter become dry, remove it from the unit using tweezers and put into the same weight boat.
5. Place weight boat and filter into 105°C oven for 24 h.
6. Calculate the dry cell weight by subtracting the weight of the filter and weight boat alone from the weight of the filter and weight boat plus the cells.

The carotenoid produced from the *idi*-containing gene clusters in the native *E. coli* host (no host modification) is approximately 1,400–1,800 ppm. The carotenoid produced from the *idi*-lacking gene clusters in the native *E. coli* host (no host modification) is approximately 200–300 ppm. With host modifications including increasing isoprenoid flux, it is possible to increase carotenoid titer in *E. coli* to about 6,000–10,000 ppm (9, 10).

4. Notes

1. It is possible to use other expression vectors of different origins with different promoters to express the carotenoid biosynthesis genes.
2. The DNA template used here is pWEB413 (11), a cosmid derived from strain DC413 that contains the caroteinoid gene cluster. To amplify the *crtEidi* genes (~2 kb), the forward primer 1 and the reverse primer 1 are used. To amplify the *crtYIB* genes (~3.6 kb), the forward primer 2 and the reverse primer 2 are used. The underlined primer sequences are *Eco*RI sites, and the bold primer sequence is an *Avr*II site. These sites are added for cloning purposes.
3. If a single correct sized PCR band is observed, purify the rest of the PCR amplification product with DNA Clean & Concentrator™-5 Kit (see Subheading 3.1.2). If other by product

bands are present, load the rest of the PCR amplification product into a 1.0% agarose gel and purify the correct size DNA band with Zymoclean™ Gel DNA Recovery Kit (see Subheading 3.1.3).

4. To amplify the *crtEidiYIB* gene cluster (~5.6 kb), forward primer 1 and reverse primer 2 are used. The underlined primer sequences are *Eco*RI sites, and the bold primer sequence is an *Avr*II site. These sites are added for cloning purposes.
5. Plasmid is sequenced using sequencing primer 1 and sequencing primer 2. These two primers target the region on the pBHR1 vector flanking the *Eco*RI-cloning site. Sequencing results should confirm the existence and orientation of the *crtEidiYIB* gene cluster in the pDCQ380 plasmid. The yellow color of the colonies is also a good indication of expression of the cloned β-carotene synthesis genes.
6. The DNA template used here is pWEB263, a cosmid derived from strain DC263 that contains the *crtW* gene (12). To amplify the *crtW* gene (~0.8 kb), crtW forward primer and crtW reverse primer are used. The underlined primer sequence is an *Xba*I or *Avr*II site. These sites are added for cloning purposes. The bold primer sequence is an incorporated ribosome-binding site.
7. The pDCQ392W plasmid is sequenced using sequencing primer 1, which binds to the region flanking the *Eco*RI-cloning site on the pBHR1 vector. The sequencing result should confirm the existence and orientation of the *crtW* gene in the pDCQ392W construct. The orange color of the colonies is also a good indication of expression of the cloned β-carotene ketolase.
8. Balanced expression of ketolases and hydroxylases is critical to achieve high astaxanthin purity. Stronger expression of ketolases and hydroxylases closer to the promoter appears to give higher astaxanthin purity than expression farther from the promoter at the end of the operons.
9. The ketolases and hydroxylases from different bacterial sources have different activities and substrate specificities. Some mutant enzymes with altered activity/substrate specificity (13) may be useful for production of astaxanthin at high purity.
10. The DNA template used here is pWEB263 (12), a cosmid derived from strain DC263 that contains the *crtZ* gene. To amplify the *crtZ* gene (~0.5 kb), crtZ forward primer and crtZ reverse primer are used. The underlined primer sequence is an *Xba*I or *Spe*I site. These sites are added for cloning purposes. The bold primer sequence is an incorporated ribosome-binding site.

11. Restriction enzymes *Avr*II/*Spe*I/*Xba*I have compatible ends and can be used for successive cloning of different fragments into one construct.
12. The pDCQ392 plasmid is sequenced using the *crtZ* reverse primer. The sequencing result should confirm the existence and orientation of the *crtZ* gene in the pDCQ392 construct. The red color of the colonies is also a good indication of astaxanthin biosynthesis.
13. Astaxanthin purity could also be affected by growth conditions, especially oxygen concentration in the medium. Make sure there is plenty of aeration in shake flasks and maintain dissolved oxygen >20% in fermentors.
14. Canthaxanthin can be usually produced with high purity (>90%). Purity of astaxanthin may be compromised by accumulation of less oxygenated intermediates, such as adonixanthin, adonirubin, canthaxanthin, and echinenone.

References

1. Rohmer M, Seemann M, Horbach S, Bringer-Meyer S, Sahm H (1996) Glyceraldehyde 3-phosphate and pyruvate as precursors of isoprenic units in an alternative non-mevalonate pathway for terpenoid biosynthesis. J Am Chem Soc 118:2564–2566
2. Sandmann G (2001) Carotenoid biosynthesis and biotechnological application. Arch Biochem Biophys 385:4–12
3. Albrecht M, Misawa N, Sandmann G (1999) Metabolic engineering of the terpenoid biosynthetic pathway of *Escherichia coli* for production of the carotenoids β-carotene and zeaxanthin. Biotechnol Lett 21:791–795
4. Wang CW, Oh MK, Liao JC (1999) Engineered isoprenoid pathway enhances astaxanthin production in *Escherichia coli*. Biotechnol Bioeng 62:235–241
5. Kajiwara S, Fraser PD, Kondo K, Misawa N (1997) Expression of an exogenous isopentenyl diphosphate isomerase gene enhances isoprenoid biosynthesis in *Escherichia coli*. Biochem J 324:421–426
6. Fraser PD, Miura Y, Misawa N (1997) In vitro characterization of astaxanthin biosynthetic enzymes. J Biol Chem 272:6128–6135
7. Choi SK, Nishida Y, Matsuda S, Adachi K, Kasai H, Peng X, Komemushi S, Miki W, Misawa N (2005) Characterization of beta -carotene ketolases, CrtW, from marine bacteria by complementation analysis in *Escherichia coli*. Mar Biotechnol (NY) 7:515–522
8. Choi SK, Matsuda S, Hoshino T, Peng X, Misawa N (2006) Characterization of bacterial beta-carotene 3,3'-hydroxylases, CrtZ, and P450 in astaxanthin biosynthetic pathway and adonirubin production by gene combination in *Escherichia coli*. Appl Microbiol Biotechnol 72:1238–1246
9. Yuan LZ, Rouvière PE, LaRossa RA, Suh W (2006) Chromosomal promoter replacement of isoprenoid pathway for enhancing carotenoid production in *E. coli*. Metab Eng 1:79–90
10. Alper H, Stephanopoulos G (2008) Uncovering the gene knockout landscape for improved lycopene production in *E. coli*. Appl Microbiol Biotechnol 78:801–810
11. Sedkova N, Tao L, Rouvière EP, Cheng Q (2005) Diversity of carotenoid synthesis gene clusters from environmental *Enterobacteriaceae* strains. Appl Environ Microbiol 71:8141–8146
12. Tao L, Rouvière PE, Cheng Q (2006) A carotenoid synthesis gene cluster from a non-marine *Brevundimonas* that synthesizes hydroxylated astaxanthin. Gene 379:101–108
13. Tao L, Wilczek J, Odom JM, Cheng Q (2006) Engineering a β-carotene ketolase for astaxanthin production. Metab Eng 8:523–531

Chapter 8

Analysis of Canthaxanthin Production by *Gordonia jacobaea*

Patricia Veiga-Crespo, Teresa Vinuesa, Miquel Viñas, and Tomás G. Villa

Abstract

Commercial interest in the use of natural pigments isolated from microorganisms has increased in recent years; hence, molecules belonging to the polyisoprenoid group (i.e., β-carotene, astaxanthin, and canthaxanthin) have been the focus of much attention. The bacterium *Gordonia jacobaea* readily synthesizes and accumulates large amounts of canthaxanthin (β-β′-carotene-4,4′-dione), which is widely used in the food and cosmetics industries. In the present work, the integral process of canthaxanthin production by *G. jacobaea* is analyzed together with its application as natural sources for the industry. A great influence of culture media is observed on canthaxanthin levels. Also, the ability is found of extract the pigments with ethanol from bacteria. The concentration of the samples is a crucial point of the process, being mandatory to discard any process of heating the samples, because this provoked the pigment degradation. Despite this, the described method allows to consider *G. jacobaea* as a potential canthaxanthin producer for the industry.

Key words: *Gordonia jacobaea*, Canthaxanthin, Soy meal, Fermentor, Carotenoids

1. Introduction

The chemical synthesis of certain pigments such as those of polyprenoid origin has been an important advance in the past since it allowed an easy and cheap production. Their use in the food industry has been questioned, however, by consumer agencies on the grounds of their carcinogenic potential (1). This possible relationship between synthetic colorings and undesirable side effects has, therefore, increased the demand of the so-called natural pigments, carotenoids probably being the most important sources of natural pigments. These substances tend to be used as a substitutive of the synthetic red and yellow, which have been in many cases forbidden by the US Food and Drug Administration (FDA) (2).

José-Luis Barredo (ed.), *Microbial Carotenoids from Bacteria and Microalgae: Methods and Protocols*, Methods in Molecular Biology, vol. 892, DOI 10.1007/978-1-61779-879-5_8, © Springer Science+Business Media, LLC 2012

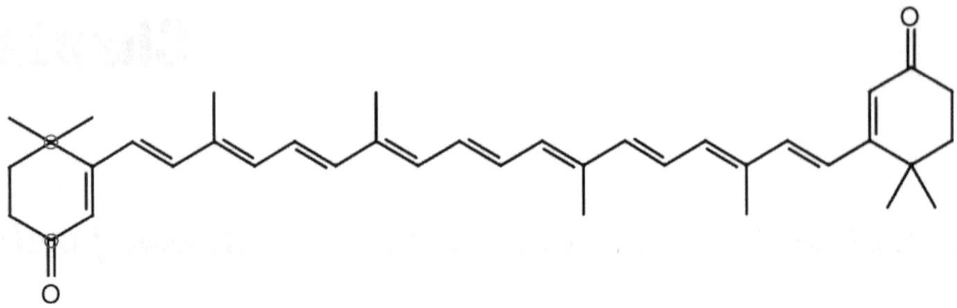

Fig. 1. Canthaxanthin structure.

Carotenoids are the most widely used pigments in the world. They are terpenoids with the isopentenyl-pyrophosphate (IPP) molecule as the basic unit and only plants and microorganisms can synthesize them, although they are present in animals by ingestion and are involved in multiple biologic functions in both plants and animals. They may be precursors of vitamins such as vitamin A, stimulate the immune response, and protect against free-radical-mediated damage, cardiovascular disease, and age-related macular degeneration (3–5). They can be used as feed supplements for poultry and in aquaculture (6).

Canthaxanthin (β-β′-carotene-4,4′-dione) is a keto-carotene pigment widely distributed in nature with industrial interest in several areas (7). It was first isolated from the mushroom *Cantharellus cinnabarinus.* It can be used as a food additive (E161g), and in cosmetics. Its chemical formula is $C_{40}H_{52}O_2$ (Fig. 1).

Different methods of chemical synthesis of astaxanthin and canthaxanthin have been described, i.e., the efficient synthesis based on sulfone compounds (8). The use of synthetic pigments has nevertheless several advantages over the natural ones. The chemical synthesis produces carotenoids of exceptional purity and consistency with lower production cost. On the other hand, the synthesis of some particular carotenoids may be a quite complex procedure, which increases the final cost of the product; and, in addition, the control of the final mixture of isomers is a hard task.

The preference for natural pigments over synthetic ones has stimulated research for the development of tests to determine whether the pigments are natural or synthetics to avoid fraudulent information from being provided to the final consumer. In the past, the natural sources of carotenoids were extracts from plants or animals, but those sources cannot cover the actual demands. Due to their ease of manipulation, microorganisms provide an excellent system for the large-scale production of carotenoids, as it has been shown by the yeast *Phaffia rhodozyma* (*Xanthophyllomyces dendrorhous*), the green algae *Haematococcus pluvialis* (9, 10), and the bacterium *Brevibacterium* KY-4313 (11). The use of microorganisms as a natural source of pigments can avoid difficulties, such as seasonal and geographic variability in their production. The two main problems associated to the natural synthesis of carotenoids are the ability of the cells to accumulate carotenoids and the ever-increasing costs of the downstream procedures.

Today, canthaxanthin is still mainly produced by chemical synthesis because of the low productivity and slow growth of *Brevibacterium* KY4313 and *Dictyococcus cinnabariunus* (12). The search for better candidates is still going on and many carotenoid-producing species have been described, such as *Thraustochytium* CHN-1 (an astaxanthin producer), *Gordonia jacobaea* MV1, or *Bradyrhizobium*, canthaxanthin producers (13–15) together with technical reports for improving the production rates.

Canthaxanthin is used in aquaculture for rainbow trout pigmentation because it has been demonstrated that it is cleared faster than astaxanthin from the rainbow trout plasma (16), possibly due to a higher rate of metabolic conversion of canthaxanthin (17).

The technical market research report from BCC Research about the Global Market for Carotenoids (FOD025C) expected an annual growth rate of 2.3% to reach a value of $919 million in 2015. The main carotenoids for the industry are thus far astaxanthin, canthaxanthin, and β-carotene.

As mentioned above, it has been suggested that carotenoids may play an important role as protective agents against several diseases. Thus, carotenoids seem to possess biological properties acting on the prevention of cancer and heart diseases. The protective action of carotenoids seems to be related to their antioxidant activity, acting as scavengers of free radicals (18). Canthaxanthin can be more potent as an antioxidant because it is able to participate in quenching reactions of reactive oxygen species (19). Based on its quenching activity, canthaxanthin has been proposed to reduce the risk of cancer.

Some studies have shown the potential antitumoral activity of canthaxanthin, and some of them strongly suggest its ability to protect against some types of cancer (20). For example, canthaxanthin was able to induce apoptosis in a prostate cancer cell line (LNCaP) because of its ability to interact with the retinoic acid receptor (21).

An undesirable side effect of canthaxanthin is its deposition as minute crystals in the macula of the eye leading to lower static luminance threshold, associated with retinopathy. However, it has been shown that this effect only takes place in overdosage cases and the crystals readily disappear by simple cessation of the canthaxanthin intake (22, 23).

In the European Union, "Regulation (EC) No 1831/2003" (24) regulates the use of additives in animal nutrition (2003). Canthaxanthin is furthermore allowed as an additive for the feed of dogs, cats, ornamental fish, birds, and pets (25). According to this, canthaxanthin can be used to a maximum content of 25 mg/kg of complete feeding stuff for poultry and salmonids, whereas the mixture of canthaxanthin with other carotenoids and xanthophylls is not permitted to exceed 80 mg/kg in the complete feeding stuff.

The genes involved in the synthesis of canthaxanthin in microorganisms have been characterized in several species (26, 27). The little knowledge about *crt* genes in microorganisms is due to their low rates of conserved evolution among species. For instance, when the active site of phytoene synthase is involved, only amino acids of precise given positions are conserved, whereas the rest of the residues are highly variable.

The basic subunit of the carotenoids is the isopentenyl diphosphate (IPP), a 5-carbon atoms molecule (5 C), and, through head/tail condensations, gives rise to geranylgeranyl pyrophosphate (GGPP) (20 C). The condensation of two molecules of GGPP gives rise to a phytoene molecule, which is the first 40 C compound of the pathway. After this, a number of cyclizations and desaturations take place. The pathway is complicated and may vary between different organisms (Fig. 2).

There are two possible strategies for increasing the natural carotenoid production in biological systems: (1) to increase the efficiency of biomass production or (2) to increase the efficiency of carotenoid biosynthesis. Indeed, a number of studies have demonstrated the influence of external factors, such as light, carbon source, growth temperature, and others (28).

The extraction yield differs significantly depending on the solvent used. According to their chemical structures, carotenoids are barely soluble in water but highly soluble in organic solvents, mainly hexane or acetone. The use of these solvents in the foodstuffs is not allowed, so it is mandatory to find out alternative methods of extraction.

As an example of the foregoing strategy, the recently isolated bacterium *G. jacobaea* MV-26 was used to optimize canthaxanthin production. A great influence of culture media was observed on canthaxanthin levels. Also, the ability of extract was found out the pigments with ethanol from bacteria. The concentration of the samples was a crucial point of the process, being mandatory to discard any process of heating the samples, because this provoked the pigment degradation. Despite this, the described method allows to consider *G. jacobaea* as a potential canthaxanthin producer for the industry.

2. Materials

2.1. Culture of G. jacobaea MV-26

1. *G. jacobaea* MV-26 (see Note 1) (29).
2. Tryptone Soy Meal Broth (TSB): 17 g/L of pancreatic digest of casein, 3 g/L of enzymatic digest of soybean meal, 5 g/L of sodium chloride, 2.5 g/L of dipotassium phosphate, and 2.5 g/L of dextrose. Adjust to pH 7.4. Sterilize at 121°C and 1.1 atm for 20 min.

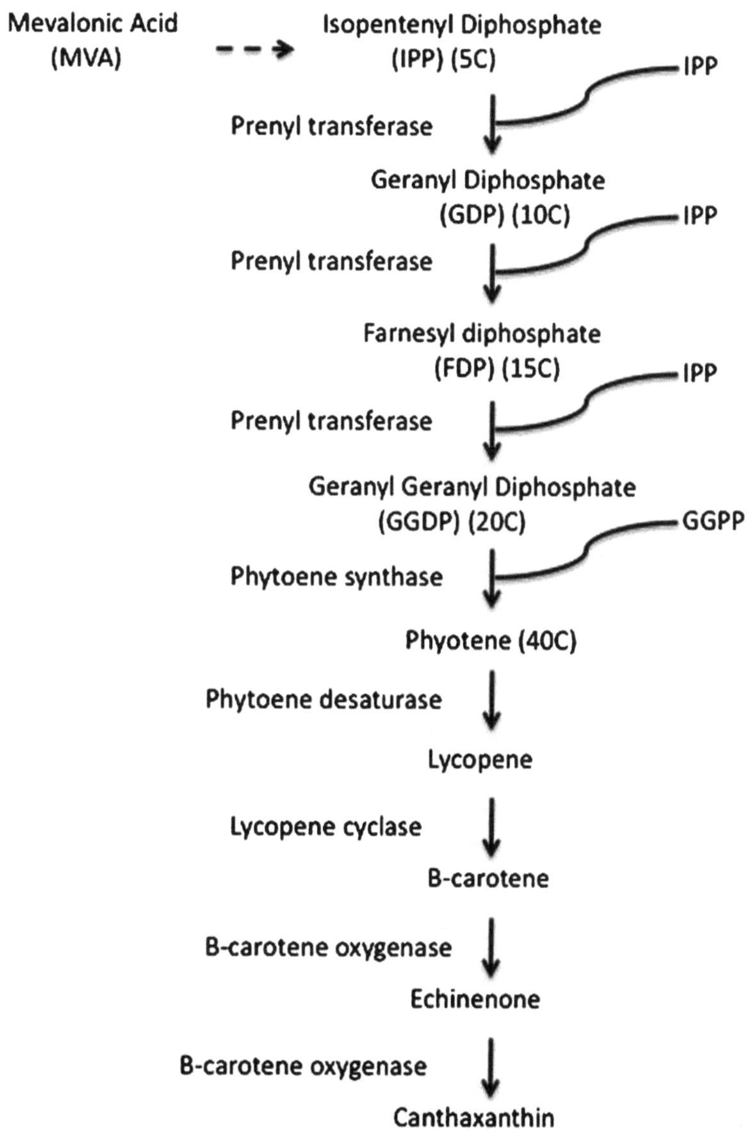

Fig. 2. Possible biosynthetic pathways for canthaxanthin formation. GGPS geranyl geranyl pyrophosphate synthase; CRTE bacterial GGPS; PSY phytoene synthase; CRTB bacterial PSY; PDS phytoene desaturase; CRTI bacterial PDS; LYCB lycopene cyclase; CRTY bacterial lycopene cyclase

3. TSB-agar: TSB and 2% agar.
4. Brain–heart infusion broth (BHIB): 12.5 g/L of brain infusion solids, 5 g/L of beef heart infusion solids, 10 g/L of proteose peptone, 2 g/L of glucose, 5 g/L of sodium chloride, and 2.5 g/L of disodium phosphate. Adjust to pH 7.4. Sterilize at 121°C and 1.1 atm for 20 min.
5. BHIB-agar: BHIB and 2% agar.

6. Yeast Peptone Broth (YPD): 10 g/L of bacto-yeast extract, 20 g/L of bacto-peptone, and 10 g/L of dextrose. Adjust to pH 7.4. Sterilize at 121°C and 1.1 atm for 20 min.
7. YPD-agar: YPD and 2% agar.
8. Spectrophotometer.
9. Soy meal-based medium 1: 100 g/L of glucose. Adjust to pH 7.4. Sterilize at 121°C and 1.1 atm for 20 min.
10. Soy meal-based medium 2: 5 g/L of soy-meal. Adjust to pH 7.4. Sterilize at 121°C and 1.1 atm for 20 min.
11. Soy meal-based medium 3: 50 g/L of glucose and 10 g/L of soy-meal. Adjust to pH 7.4. Sterilize at 121°C and 1.1 atm for 20 min.
12. Soy meal-based medium 4: 50 g/L of glucose and 10 g/L of soy-meal. Adjust to pH 7.4. Sterilize at 121°C and 1.1 atm for 20 min.
13. Soy meal-based medium 5: 15 g/L of glucose and 20 g/L of soy-meal glucose. Adjust to pH 7.4. Sterilize at 121°C and 1.1 atm for 20 min.
14. Soy meal-based medium 6: 10 g/L of glucose and 50 g/L of soy-meal. Adjust to pH 7.4. Sterilize at 121°C and 1.1 atm for 20 min.
15. Soy meal-based media-agar: Soy meal-based media and 2% agar.
16. Incubating shaker.
17. Braun BioStat C (B Braun Biotech International, Melgusen, Germany).
18. SPSS 12.0 software (IBM Corporation, Somers, NY, USA).

2.2. Extraction of the Pigments

1. Centrifuge.
2. Potassium phosphate buffer pH 7: 61.5 mL of 1 M K_2HPO_4 and 38.5 mL of 1 M KH_2PO_4. Distilled water is added to a final volume of 1 L and a final concentration of 0.1 M.
3. Teflon filters of 0.22-μm diameter of pore.

2.3. Analysis of the Pigments

1. Silica-gel column of 5-μm pore size, 25-cm length, and 45-mm diameter.
2. Mobile phase 1: Hexane–petroleum fraction:ethyl acetate 1:1 (v:v).
3. Mobile phase 2: Hexane–petroleum fraction:ethyl acetate 8:2 (v/v).
4. Standard of β-carotene (Sigma, St. Louis, MO, USA).
5. Standard of canthaxanthin (Sigma, St. Louis, MO, USA).
6. Chromatography paper 3 MM CHR.

2.4. Concentration of the Ethanol-Extracted Pigments

1. Rotavapor.
2. Lyophilizer.

3. Methods

3.1. Growth Curve of G. jacobaea MV-26

The way to establish the growth curve of *G. jacobaea* MV-26 when using TSB, BHIB, and YPD culture media is as follows.

1. Seed 250 mL of culture media with 500 μL from a culture of *G. jacobaea* MV-26 and incubate at 30 or 37°C for 8 days.
2. Take 1 mL of culture and measure OD_{600} using a spectrophotometer.
3. Take samples every 8 h during the exponential phase of growth, and every 24 h when the culture reaches the stationary phase.

In the case of the low-cost soy meal-based media, the way to establish the growth curve of *G. jacobaea* MV-26 is as follows.

1. Seed 250 mL of culture media with 500 μL from a culture of *G. jacobaea* MV-26 and incubate at 30 or 37°C for 8 days.
2. Take 1 mL of culture, and seed 100 μL in soy-meal-agar media.
3. Incubate for 24 h at 30°C.
4. Count the number of colony-forming units (CFUs).
5. Estimate the CFU with the formula: CFU/mL = number of CFU × dilution factor × 10 (see Note 2).
6. Take samples every 8 h during the exponential phase of growth, and every 24 h when the culture reaches the stationary phase.

3.2. Culture Conditions

1. Seed 250 mL of BHIB, TSB, YPD, and low-cost soy-meal-based media with 500 μL from a culture of *G. jacobaea* MV-26 and incubate at 30 or 37°C for 8 days in a rotary shaker (150 rpm), except for the YPD medium that the screening was implemented for 10 days.
2. Take 3 mL of culture for evaluating the levels of β-carotene and canthaxanthin by the hexane method (see Subheading 3.3) every 8 h during the exponential phase of growth, and every 24 h when the culture reaches the stationary phase (see Note 3).

After this first screening, the media that resulted in the highest levels of canthaxanthin production are used to scale up the cultures (see Note 4).

1. Seed 2 L of BHIB, TSB, YPD, Soy meal-based medium 1, and Soy meal-based medium 5 with 2 mL of a *G. jacobaea* MV-26

and incubate at 30°C for 8 days in a rotary shaker (150 rpm), except for the YPD medium that the screening is implemented for 10 days.

2. Take 3 mL of culture for evaluating the levels of β-carotene and canthaxanthin by the hexane method (see Subheading 3.3) every 8 h during the exponential phase of growth, and every 24 h when the culture reaches the stationary phase.

The results of the influence of the growth media on canthaxanthin production are subjected to statistical analysis with the SPSS 12.0 software (see Note 5).

3.3. Pigment Extraction

3.3.1. Hexane Method

1. Add 1 mL of 0.1 M potassium phosphate buffer (pH 7) and 3 mL hexane:petroleum fraction to 3 mL of culture.
2. Vortex vigorously for 10 min.
3. Centrifuge for 10 min at 5,000 × *g* to allow phase separation.
4. Recover the organic phase and filter across Teflon filters.
5. Evaluate the pigment content by HPLC (see Subheading 3.4).

3.3.2. Ethanol Method

1. Seed 2 L of BHIB, TSB, YPD, Soy meal-based medium 1, and Soy meal-based medium 5 with 2 mL of *G. jacobaea* MV-26 and incubate at 30°C until the culture reaches the maximum productivity of canthaxanthin.
2. Harvest the biomass at 4°C by means of continuous-flow centrifugation at 30,000 × *g*.
3. Resuspend the cells in 1 vol of pure ethanol.
4. Vortex vigorously during 10 min.
5. Centrifuge for 10 min at 5,000 × *g* and recover the ethanol.
6. Discard the first extraction (see Note 6).
7. Repeat the process until the pellet lose its characteristic orange color (see Note 7).

3.4. Analysis of Pigments

3.4.1. Analysis of Pigments by HPLC

1. Use the following parameters: (a) flow of the mobile phase across the column 1 mL/min at 0.4 kpsi, (b) temperature 25°C, and (c) wavelength of 480 nm.
2. Inject 30 μL of standard solutions of canthaxanthin and β-carotene in order to determine the retention times of both compounds. Both standards are suspended in hexane:petroleum fraction and filtered across teflon filters of 0.22 μm prior to their use.
3. Make a calibration curve to determine the concentration of the pigments in the samples by injecting 30 μL of standard solutions of known concentration. The calibration curves are done with 50, 100, 250, and 500 μg/mL of β-carotene and 100, 250, 500, and 1,000 μg/mL of canthaxanthin.

4. Inject 30 μL of the pigments extracted by hexane method.
5. Repeat all the measurements by triplicate in an independent manner.

3.4.2. Analysis of Pigments by TLC

1. Deposit the mobile phase in the chromatography jar and let the atmosphere be sutured.
2. Spot the samples on the plate and dry them under cool airflow.
3. Spot standards of β-carotene and canthaxanthin in order to identify the bands corresponding to these carotenoids.
4. Put the plate in the chromatography chamber.
5. Develop the TLC until the front reaches the top of the plate.

3.5. Fermentor Growth

1. Seed 20 L of YPD medium with 3 L of a culture in the exponential phase of *G. jacobaea* MV-26.
2. Incubate the culture for 8 days at 30°C, 200 rpm, and airflow of 15 L/min.
3. Take 1 mL every 8 h during the exponential phase of growth and every 24 h during the stationary one for elaborating the growth curve in the fermentor by spectrophotometry.
4. Take 3 mL every 8 h during the exponential phase of growth and every 24 h during the stationary one for elaborating the carotenoid production curve in the fermentor by the hexane and HPLC methods (see Note 8).
5. Stop the fermentor when the culture reaches the maximum productivity of canthaxanthin.
6. Centrifuge the culture in the continuous-flow centrifuge.
7. Recover the pellet and extract the pigment by the ethanol method.

3.6. Concentration of the Ethanol-Extracted Pigments

3.6.1. Vacuum Concentration

1. Protect the samples against the light during the whole concentration process.
2. Put the ethanol-extracted pigments coming from the fermentor in the rotavapor.
3. Heat the ethanol to facilitate the dryness of the samples at different temperatures (35, 50, 72, and 80°C) to test the influence of the temperature in the process.
4. Concentrate the samples' volume by ten times.
5. Analyze the integrity of the samples by TLC as previously described (see Subheading 3.4).

3.6.2. Lyophilization

1. Freeze the ethanol-extracted samples coming from the fermentor (see Subheading 3.5) at −70°C for 24 h.
2. Place them in the lyophilizer and protect against the light.

3. Keep the samples under vacuum at –50°C.
4. Concentrate the samples' volume by ten times and analyze their integrity by TLC (see Note 9).

About 3.2 g of dry pigment can be obtained from the fermentor (see Subheading 3.5) after 7 days of culture of *G. jacobaea* MV-26 in YPD medium followed by ethanol extraction and concentration in the lyophilizer (see Note 10).

4. Notes

1. *G. jacobaea* MV-26 was obtained in our lab by random mutagenesis of wild-type *G. jacobaea* MV-1 (29).
2. The elaboration of the growth curve by spectrophotometry is not possible because of the solid particles present in the soy meal-based media. Molasses-based media (from 10 to 30 g/L molasses) were also checked with *G. jacobaea* MV-26, but it was not able to grow up.
3. From data of the growth curve of *G. jacobaea* MV-26 and the peaks of canthaxanthin and β-carotene production, it can be observed that the production of β-carotene is inversely proportional to canthaxanthin accumulation (Fig. 3). It is necessary

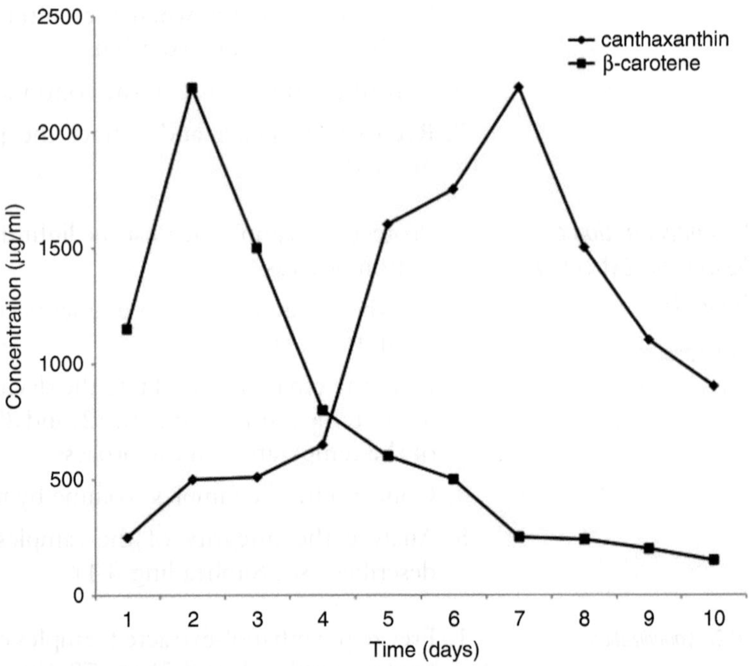

Fig. 3. Evolution of the two main pigments (canthaxanthin and β-carotene) produced by *G. jacobaea* MV-26 on YPD medium.

Table 1
Production of canthaxanthin using different culture media

Media	Concentration of canthaxanthin (μg/mL)
YPD	2,140
BHIB	2,489
TSB	1,790
Soy meal-based medium 1	1,000
Soy meal-based medium 2	2,340
Soy meal-based medium 3	1,000
Soy meal-based medium 4	2,650
Soy meal-based medium 5	3,440
Soy meal-based medium 6	13,373

that *G. jacobaea* reaches the stationary phase to accumulate canthaxanthin. During the exponential phase, the amount of canthaxanthin increases as the bacterial mass do. Another relevant aspect is the fact that canthaxanthin and β-carotene show an inverse pattern of accumulation. So far, this has two important implications: on the one hand, it demonstrates that β-carotene is an intermediate in the keto-carotenoids' synthetic pathway in *G. jacobaea* and, on the other hand, it would be possible to obtain both pigments from *G. jacobaea* according to industrial requirements by simply varying the harvesting time. The production of canthaxanthin and β-carotene is different depending on the media. In all the cases, the maximum productivity of canthaxanthin is obtained at the stationary phase of growth, but the timing and the length of production are different (Table 1). The negative controls show that all the differences come from *G. jacobaea* cells and not from media-containing carotenoids.

4. The soy-meal-based media show high rates of production, being the Soy meal-based medium 6 (10 g/L of glucose and 50 g/L of soy-meal) the best one. However, it had to be discarded because of its hard manipulation. It is mandatory to explore new soluble compounds extracted from soy-meal in order to be applicable in fermentation technology. The ability of the soy-meal to increase the *G. jacobaea* canthaxanthin could be related with the presence of carotenoidic precursors, such as the mevalonic acid. This, plus the extra amounts of glucose, would increase the canthaxanthin production at the beginning

of the stationary phase. This effect could be also observed on BHI broth, a terpenoid-rich medium. However, the BHI high cost makes difficult its application in an industrial process. There are no differences in the rates of canthaxanthin production of *G. jacobaea* MV-26 at different temperatures. This latter aspect is relevant in an industrial process, where increase of only 1°C has large economic impact in the whole process.

5. Statistical analysis confirms the influence of culture media composition on canthaxanthin production of *G. jacobaea* MV-26.
6. The analysis of the different steps of extraction by TLC shows that the first extraction must be discarded because of the altered pattern of carotenoidic bands, which could be related to the rest of bacterial cell walls and cellular membranes. On the contrary, in the successive steps, bands corresponding to canthaxanthin and β-carotene are clear.
7. The direct extraction of the pigments with ethanol is an interesting result since ethanol by itself is unable to extract pigments from other carotenoid producers, such as *X. dendrorhous* (*P. rhodozyma*) or *H. pluvialis*. Although the extraction with hexane shows higher yields than with ethanol, its use is not permitted for foodstuffs because of its potential toxicity.
8. Soy-meal-based media are also eliminated due to difficulties in manipulation. The culture of *G. jacobaea* in a semi-industrial scale is carried out reproducing the previous results obtained for lower volumes. The growth curve and the production curve match up perfectly. An associated problem in these experiments is the amount of foam in some parts of the growth cycle of *G. jacobaea* MV-26; however, it is easily controlled by the addition of industrial antifoam.
9. The pigments are sensitive to temperature when using the rotavapor-based procedure to concentrate the ethanol-extracted pigments. However, the samples concentrated by lyophilization give the correct pattern of bands by TLC. The temperature shows a relevant influence on the integrity of carotenoids. When the rotavapor is used, a degradation of the extracted carotenoids occurs. On the other hand, when the ethanol is removed at low temperatures, the integrity of canthaxanthin and β-carotene is maintained, but more time is spent for pigment preparation.
10. Two aspects must be considered to calculate the yield of pigments extraction: (a) canthaxanthin represents 80% of the total pigments from *G. jacobaea* MV-26 if the culture is harvested at the moment of the maximum productivity and (b) the highest authorized concentration of canthaxanthin by law in the trout feeds is 100 mg/kg combined with astaxanthin or 25 mg/kg of pure canthaxanthin. Taking those data into account, 3.2 g

of dry pigment from *G. jacobaea* MV-26 contains 2.8 g of canthaxanthin, which means that ca 100 kg of trout feed could be elaborated using exclusively canthaxanthin. Only 50 kg of feed could be prepared if canthaxanthin is combined with astaxanthin in a proportion 1:1.

Acknowledgment

Dr. Veiga-Crespo wants to acknowledge to la Xunta de Galicia for the fellow "Anxeles Alvariño" and to the Spanish Government for the "Jose Castillejo" program.

References

1. Newsome RL (1986) Food colors. Food Technology 40(7):49–56
2. Tafoya A, García F (1993) Colorantes. In: Biotecnología alimentaria. García Garibay, Quintero Ramírez y López Murguía. Ed. Limusa Noriega Editores, pp 479–517
3. Seddon JM, Ajani UA, Sperduto RD, Hiller R, Blair N, Burton TC, Farber MD, Gragoudas ES, Haller J, Miller DT (1994) Dietary carotenoids, vitamins A, C and E and advance age related macular degeneration. JAMA 272:1413–1420
4. Willet WC, Trichopoulos D (1996) Nutrition and cancer: a summary of evidence. Cancer Causes Control 7:178–180
5. Shahaidi F, Metusalach S, Brown JA (1998) Carotenoid pigments in seafoods and aquaculture. Crit Rev Food Sci Nutr 38:1–67
6. Johnson EA, An GH (1999) Astaxanthin from microbial sources. Critical Rev Biotechnol 11:297–326
7. Hannibal L, Lorquin J, D'Ortoli NA, García N, Chaintreuil C, Masson-Boivin C, Dreyfus B, Giraud E (2000) Isolation and characterization of canthaxanthin biosynthesis genes from the photosynthetic bacterium *Bradyrhizobium* sp. Strain ORS278. J Bacteriol 182: 3850–3853
8. Choi S, Koo S (2005) Efficient syntheses of the keto-carotenoids canthaxanthin, astaxanthin and astacene. J Org Chem 70: 3328–3331
9. Johnson EA, Villa TG, Lewis MJ (1980) *Phaffia rhodozyma* as an astaxanthin source in salmonid diets. Aquaculture 20:123–134
10. Lorentz RT, Cysewski GR (2000) Commercial potential for *Haematococcus* microalgae as a natural source of astaxanthin. Trends Biotechnol 18:160–167
11. Nelis HT, de Leenher AP (1991) Microbial sources of carotenoid pigments used in food and feed. J Appl Bacteriol 70:181–191
12. Lorquin J, Molouba F, Dreyfus BL (1997) Identification of the carotenoid pigment canthaxanthin from photosynthetic *Bradyrhizobium* strains. Appl Environ Microbiol 63:1151–1154
13. de Miguel T, Sieiro C, Poza M, Villa TG (2000) Isolation and taxonomic study of a new canthaxanthin-containing bacterium, *Gordonia jacobaea* MV-1 sp. nov. Int Microbiol 3:107–111
14. Carmona ML, Naganuma T, Yamaoka Y (2003) Identification by HPLC-MS of carotenoids of the Thraustochytrium CHN-1 strain isolated from the Seto Inland Sea. Biosci Biotechnol Biochem 67:884–888
15. Gobantes I, Choubert G, Laurentie M, Milicua JCG, Gomez R (1997) Astaxanthin and canthaxanthin kinetics after ingestion of individual doses by immature rainbow trout *Oncorhynchus mykiss*. J Agric Food Chem 45:454–458
16. Salvador AM, Alonso-Damian A, Choubert G, Milicua JC (2007) Effect of soybean phospholipids on canthaxanthin lipoproteins transport, digestibility, and deposition in Rainbow trout (*Oncorhynchus mykiss*) muscle. J Agric Food Chem 55:9202–9207
17. Kleinova M, Hewitt M, Brezova V, Madden JC, Cronin MTD, Valko M (2007) Antioxidant properties of carotenoids: QSAR prediction of their redox potentials. Gen Physiol Biophys 26:97–103
18. Edge R, McGarvey DJ, Truscott TG (1997) The carotenoids as anti-oxidants – a review. J Photochem Photobiol B 41:189–200

19. Peto R, Doll R, Buckley JD, Sporn MD (1981) Can dietary beta-carotene materially reduce human cancer rates? Nature 290:201–208
20. Kumaresan N, Sanjay KR, Venkatesh KS, Kadeppagari R-K, Vijayalakshmi G, Umesh-Kumar S (2008) Partially saturated canthaxanthin purified from *Aspergillus carbonarius* induces apoptosis in prostate cancer cell line. Appl Microbiol Cell Physiol 80: 467–473
21. Harnois C, Samson J, Malenfant M, Rousseau A (1989) Canthaxanthin retinophaty. Anatomic and functional reversibility. Arch Ophthalmol 107:538–540
22. Baker RTM (2002) Canthaxanthin in aquafeed applications: is there any risk? Trends Food Sci Technol 12:240–243
23. Regulation (EC) 1831/2003 of the European Parliament and the Council of 22 September 2003 on additives for use in animal nutrition. Official Journal of the European Union, L268/29, 18.10.2003
24. Commission Regulation (EC) No 880/2004 of 29 April 2004 authorising without time limit the use of beta-carotene and canthaxanthin as additive in feedingstuffs belonging to the group of colouring matters including pigments. Official Journal of the European Union, L162/68, 30.4.2004
25. Krubasik P, Sandmann G (2000) A carotenogenic gene cluster from *Brevibacterium linens* with novel lycopene cyclase genes involved in the synthesis of aromatic carotenoids. Mol Gen Genet 263:423–432
26. Cantera JJL, Kawasaki H, Seki T (2002) Farnesyl diphosphate synthase gene of three phototrophic bacteria and its use as a phylogenetic marker. Int J Syst Evol Microbiol 52:1953–1960
27. Sieiro C, Poza M, de Miguel T, Villa TG (2003) Genetic basis of microbial carotenogenesis. Int Microbiol 6:11–16
28. Veiga-Crespo P, Blasco L, Poza M, Villa TG (2005) Influence of culture conditions of *Gordonia jacobaea* MV-26 on canthaxanthin production. Intern Microbiol 8:55–58
29. de Miguel T, Sieiro C, Poza M, Villa TG (2001) Analysis of canthaxanthin and related pigments from *Gordonia jacobaea* mutants. J Agric Food Chem 48:1200–1202

Chapter 9

Isolation and Light-Stimulated Expression of Canthaxanthin and Spirilloxanthin Biosynthesis Genes from the Photosynthetic Bacterium *Bradyrhizobium* sp. Strain ORS278

Eric Giraud and André Verméglio

Abstract

Some aerobic photosynthetic bacteria produce a cocktail of carotenoids, some of them being of a high economic value. A good example is the photosynthetic *Bradyrhizobium* sp. strain ORS278, which synthesizes, in addition to the photosynthetic carotenoid spirilloxanthin, large amounts of canthaxanthin. Here, we describe the procedures that have been successfully used to isolate the different *crt* genes involved in the synthesis of both carotenoids in this bacteria. The synthesis of these carotenoids is stimulated under far-red light by a bacteriophytochrome. The procedure we developed to study the effect of the light on carotenoids synthesis is also described. Finally, we describe a procedure to genetically transform photosynthetic *Bradyrhizobium* strain ORS278 for improvement of canthaxanthin production.

Key words: Canthaxanthin, Spirilloxanthin, Carotenoid, *crt* genes, Photosynthetic *Bradyrhizobium*, Bacteriophytochrome, LED, Light regulation, Genetic engineering

1. Introduction

Aerobic phototrophic bacteria display the remarkable feature to synthesize an unusually diverse variety of carotenoids. Besides the synthesis of photosynthetic carotenoids like spirilloxanthin or spheroidenone, they often produce large amount of bicyclic carotenoids (β-carotene and hydroxyl derivatives), which act probably as oxygen singlet scavengers (1, 2). Among them, the *Bradyrhizobium* sp. ORS278 strain is certainly one of the most studied at the genetic and physiological levels. In addition to spirilloxanthin, this bacterium produces important quantities of canthaxanthin (3). This orange-red-colored pigment is used as a food

José-Luis Barredo (ed.), *Microbial Carotenoids from Bacteria and Microalgae: Methods and Protocols*, Methods in Molecular Biology, vol. 892, DOI 10.1007/978-1-61779-879-5_9, © Springer Science+Business Media, LLC 2012

additive (4, 5). In cosmetology and pharmacology, it is also combined with β-carotene for use as a dermal photo-protector (6).

The first three steps of canthaxanthin and spirilloxanthin biosynthesis pathways are common up to the formation of lycopene through the successive action of CrtE, CrtB, and CrtI enzymes. The synthesis of canthaxanthin from lycopene necessitates two additional enzymes (7): CrtY which catalyzes the cyclization of lycopene leading to β-carotene, and CrtW which oxygenates β-carotene to form canthaxanthin (Fig. 1). The synthesis of spirilloxanthin involves three additional enzymes CrtC, CrtD, and CrtF which catalyze the successive reactions of hydratation, desaturation, and methylation of lycopene (Fig. 1) (8, 9).

Our research over the past decade has permitted to characterize the different *crt* genes involved in canthaxanthin and spirilloxanthin biosynthesis and to understand the molecular mechanism of their regulation by light. Unexpectedly, ORS278 strain contains two distinct *crt* gene clusters, one dedicated for spirilloxanthin synthesis and the other for canthaxanthin synthesis (10, 11). Interestingly, each cluster contains the first genes, *crtE*, *crtI*, and *crtB*, necessary for carotenoids' synthesis (Fig. 1). Despite this redundancy, the biosyntheses of both carotenoids are strongly interconnected at the level of the common precursor lycopene (11). Indeed, the up-regulation of *crt* spirilloxanthin genes by light stimulates the synthesis of both carotenoids. This light control is due to the action of a bacteriophytochrome, which, depending on the light conditions, antagonizes the repressive action of the photosynthesis-specific regulator PpsR (11–13).

Here, we describe the strategies we used to isolate and characterize the two *crt* gene clusters present in ORS278 strain. A method to precisely study the effect of the light quality on carotenoids' biosynthesis is also shown. Finally, we propose a procedure to genetically transform ORS278 strain to stimulate the production of canthaxanthin, a valuable carotenoid.

2. Materials

2.1. Isolation of Spirilloxanthin and Canthaxanthin Biosynthesis Genes

1. *Bradyrhizobium* sp. strain ORS278 (LMG P-12187).
2. crtIf primer: 5′-GTNGGNGCRGGCACNCAYCC-3′.
3. crtBr primer: 5′-TCGCGRGCRATRTTSGTSARRTG-3′.
4. pufLf primer: 5′-TTYTAYGTNGGNTTYTTYGG-3′.
5. pufMr primer: 5′-CCCATNGTCCANCGCCARAA-3′.
6. SuperCos1 cosmid vector kit (Stratagene, La Jolla, CA).

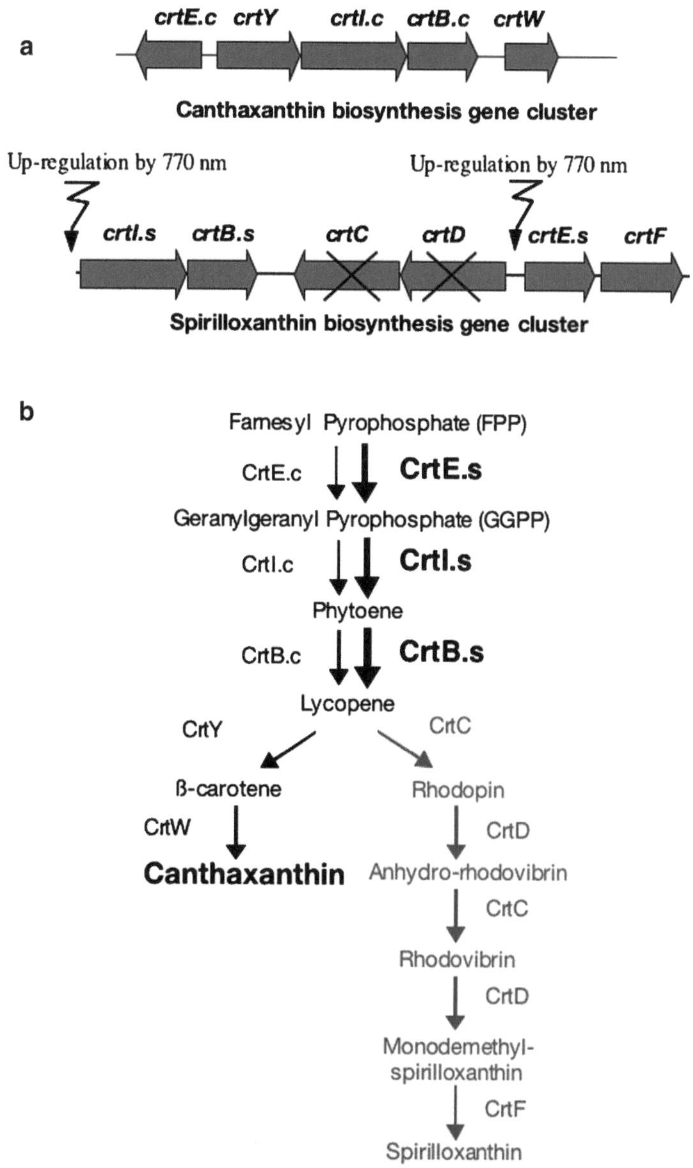

Fig. 1. General strategy to improve canthaxanthin production in *Bradyrhizobium* sp. strain ORS278. Note that the *crtE*, *crtI*, and *crtB* genes are present in each cluster. To distinguish them, we designate the genes from the spirilloxanthin *crt* cluster *crtE.s*, *crtI.s*, and *crtB.s* and those from the canthxanthin gene cluster *crtE.c*, *crtI.c*, and *crtB.c*. To stimulate the production of canthaxanthin in ORS278 strain, the *crtC* and *crtD* genes are deleted and the expression of the *crtE.s*, *crtI.s*, and *crtB.s* is up-regulated by far-red light illumination at 770 nm. (**a**) Organization of the canthaxanthin and spirilloxanthin gene clusters of ORS278 strain. (**b**) Canthaxanthin and spirilloxanthin biosynthesis pathways. In grey is presented the part of the spirilloxanthin pathway that is shut down. The lycopene produced principally through the successive action of CrtE.s, CrtI.s, and CrtB.s enzymes is rerouted towards the CrtY and CrtW enzymes for a maximal canthaxanthin biosynthesis.

7. Gigapack® III Gold Packaging Extract kit (Stratagene, La Jolla, CA).
8. QIAquick PCR extraction kit (QIAGEN GmbH, Hilden, Germany).
9. Minipreps DNA Purification kit (Promega, Madison, WI).
10. Luria–Bertani (LB) broth: 10 g/L tryptone, 5 g/L yeast extract, and 10 g/L NaCl.
11. LB-agar: LB and 16 g/L agar.
12. Kanamycin stock solution: 50 mg/mL.
13. Omnitrays Numc (Thermo Fisher Scientific, Rochester, NY).
14. TSP Numc (Thermo Fisher Scientific, Rochester, NY).

2.2. Light Control of Carotenoids' Biosynthesis

1. YM-modified medium: 2.5 g sodium glutamate, 2 g yeast extract, 0.66 g K_2HPO_4, 0.05 g NaCl, 0.1 g $MgSO_4{\cdot}7H_2O$, 0.04 g $CaCl_2$, and 0.004 g $FeCl_3$. Add water to a volume of 900 mL. Mix and adjust to pH 6.8. Make up to 1 L with water.
2. YM-modified agar medium: YM-modified medium and 16 g/L agar.
3. Illumination apparatus. Light-Emitting Diodes (LEDs) are made by a large number of manufacturers. Emission wavelengths can be found from 370 to 950 nm, the LED's bandwidth usually lies between 20 and 30 nm, and intensities up to 100 $\mu E/m^2/s$ can be obtained.
4. Acetone/methanol: 7 mL acetone and 2 mL methanol.
5. Hypersil C18 column: 5 μm, 250 by 4.6 mm (Phenomenex, Torrance, CA).
6. Acetonitrile/methanol/isopropanol: 40 mL acetonitrile, 50 mL methanol, and 10 mL isopropanol.
7. Photodiode array detector 996 (Waters, Milford, MA).
8. Canthaxanthin standard (Aventis, Paris, France).

2.3. Engineering of Bradyrhizobium ORS278 Strain to Improve Canthaxanthin Production

1. crtDf primer: 5′-TAGTCGACGCAATGGCGCGCCACGATCTATC-3′.
2. crtCr primer: 5′-ACAGTCGACCGGTCTTGGAGCGGTGATAATG-3′.
3. pGEM-T vector (Promega, Madison, WI, USA).
4. pKOK5 (14).
5. pJQ200mp18 (15).
6. *E. coli* S17-1 λpir (Biomedal, Sevilla, Spain).
7. Nalidixic acid stock solution: 35 mg/mL.

3. Methods

3.1. Isolation of Spirilloxanthin and Canthaxanthin Biosynthesis Genes

The level of identity among *crt* genes is relatively low (for example, the CrtE, CrtI, and CrtB paralogs identified in ORS278 display between 35 and 50% of amino acid identity). This could constitute an obstacle for the isolation of the *crt* genes, in particular for a bacterial species phylogenetically distant from those already characterized. In the case of the photosynthetic *Bradyrhizobium* strain ORS278, two different strategies were used for the isolation of spirilloxanthin and canthaxanthin *crt* gene clusters. The first one is based on the use of degenerated primers defined from consensus sequences identified from the alignment of CrtI and CrtB proteins (see Note 1). The second is based on the observation that, in purple bacteria, the *crt* genes involved in the synthesis of photosynthetic carotenoids have been always found associated to other photosynthesis genes (*bch*, *puf*, etc.) in the so-called photosynthesis gene cluster (PGC) of around 45 kb (16). Among them, the *pufLM* genes that encode the major polypeptides of the photosynthetic reaction center are well conserved. They could, therefore, be used as a probe to identify cosmids or clones containing the PGC and consequently the *crt* genes. The following protocol, designed for the isolation of the two *crt* gene clusters from *Bradyrhizobium* ORS278 strain, could be a useful starting point for designing isolation strategies for *crt* genes from other species. Depending on the photosynthetic character or not of the carotenoid studied, one or the other strategy should be privileged to obtain a specific probe of the corresponding *crt* genes.

1. Amplify by polymerase chain reaction (PCR) a specific region of *crtI/crtB* genes or *pufL/pufM* genes using the degenerated primers crtIf/crtBr or pufLf/pufMr, respectively. Use a touchdown PCR program as follows: initial denaturation at 94°C for 5 min followed by 20 cycles consisting of a 30-s denaturation at 94°C, 30-s annealing temperature from 60 to 50°C, and 1-min primer extension at 72°C, followed by 15 cycles consisting of a 30-s denaturation at 94°C, 30-s annealing temperature at 50°C, and 1-min primer extension at 72°C.
2. Purify the corresponding PCR product using a QIAquick PCR extraction kit and sequence it.
3. Design specific primers from the obtained sequence. These primers will be used to screen by PCR a DNA genomic library.
4. Construct a DNA genomic library of the considered strain using the SuperCos1 cosmid vector (see Note 2). Briefly, the genomic DNA is partially digested with *Sau*3AI, dephosphorylated with Calf Intestinal Alkaline Phosphatase (CIAP), and ligated into the SuperCos1 vector previously double digested

with *Xba*I/*BamH*I. The resulting high-molecular-weight concatemers are encapsidated using the Gigapack III Gold Packaging Extract kit.

5. Pick the clones and array them into 96-well plates, each plate containing 120 μL of LB medium supplemented with kanamycin (50 μg/mL) (see Note 3). After overnight culture at 37°C, add 50 μL of 70% glycerol per well and store the library at −80°C.
6. Subculture each 96-well plate onto an LB-agar plate (Omnitrays Numc) with the help of a 96-pin lid (96 plots TSP Numc).
7. Pool the colonies from an agar plate with 10 mL of water and extract the plasmids using the Minipreps DNA Purification kit.
8. Screen by PCR the different plasmid pools using the specific primers defined in step 3 in order to identify the positive plates (see Note 4).
9. Subculture the positive plates and screen again by PCR directly on colonies in order to identify a positive clone containing the wanted *crt* genes (see Note 5).
10. Sequence completely the cosmid or a subclone containing the *crt* genes identified by PCR or Southern-blot hybridization in order to characterize the *crt* gene cluster.

3.2. Light Control of Carotenoids' Biosynthesis

Light stimulation of carotenoids' biosynthesis has been reported in numerous organisms, including plants, fungi, and bacteria (17). In higher plants, this regulation is mediated by a phytochrome that controls the level of the phytoene synthase expression (18). This class of light sensor has been identified in photosynthetic and non-photosynthetic bacteria (19). Phytochromes and bacteriophytochromes are light sensor chromoproteins that display two photo-interconvertible forms, a red-absorbing form (Pr) and a far-red absorbing form (Pfr) (20). Depending on the (bacterio)phytochromes, the Pr or the Pfr form could be the active form (20). It has been also reported that (bacterio)phytochromes could control carotenoids' biosynthesis, such as deinoxanthin in the non-photosynthetic bacteria *Deinococcus radiodurans* (19), and spirilloxanthin and canthaxanthin in ORS278 strain (11). The simplest way to determine whether a carotenoid biosynthesis is under the control of a bacteriophytochrome is to check the effect of light quality in the red/far-red region on its synthesis. Below, we describe the procedure that we used for ORS278 strain.

1. Prepare a fresh culture of the ORS278 strain. Grow the bacteria in Falcon tubes of 50 mL containing 20 mL of YM-modified medium. Incubate the culture in the dark at 37°C under agitation (170 rpm) for 5 days.
2. Inoculate homogeneously the bacteria onto round Petri dishes (diameter 9 cm) containing YM-modified agar medium. Prepare one Petri dish per each light condition tested.

3. Place each Petri dish under an illumination apparatus equipped of LED of different peak wavelengths. The LED is fixed on the cover of a Petri dish placed on the top of a plastic pipe whose diameter is slightly larger than the diameter of the Petri dishes. This pipe of 20-cm high is coated with a reflecting paint to funnel the light and its extremity covered by a piece of tracing paper to homogenize the illumination. The wavelength peaks of the LEDs are chosen in order to span the two spectral forms of bacteriophytochromes, typically between 550 and 800 nm (see Note 6).
4. After 1 week of incubation at 37°C, resuspend the cells grown at the surface of the Petri dishes in 6 mL of water containing 9 g/L NaCl.
5. Harvest the cells by centrifugation (3,200 × *g* for 30 min at 4°C) and resuspend the pellets with 1 mL of cold acetone/methanol to extract carotenoids.
6. Place the samples in a rotary agitator and incubate for 30 min at 4°C in the dark.
7. Centrifuge (3,200 × *g* for 30 min at 4°C), keep the supernatant, and repeat twice the extraction procedure on the pellets.
8. Pool the supernatants and analyze the carotenoids' content by HPLC. For sipirilloxanthin and canthaxanthin analysis, use the following conditions: Hypersil C18 column, acetonitrile/methanol/isopropanol as eluent, and a flow rate of 0.8 mL/min. Monitor the eluted fractions with a photodiode array detector, scanning from 270 to 600 nm every 2 s. Identify carotenoids by their retention times and by comparison of the spectral features with those of pure compounds or with reported data. Determine the amount of canthaxanthin from the area of the peak detected at 480 nm using a calibration curve obtained with a canthaxanthin standard. If you lack of spirilloxanthin standard, estimate the amount of spirilloxanthin from the area of the peak detected at 494 nm using the canthaxanthin correlation coefficient. Determine the normalized quantity of carotenoid with respect to the dry weight of the pellet.
9. Plot the amount of carotenoids produced according to the illumination wavelengths (see Notes 7 and 8).

3.3. Engineering of *Bradyrhizobium* ORS278 Strain to Improve Canthaxanthin Production

As previously stated, a cross talk between the canthaxanthin and spirilloxanthin biosynthesis pathways exists at the level of the common precursor lycopene. The stimulation of canthaxanthin production by far-red light observed in the ORS278 strain results, in fact, to the overexpression of the spirilloxanthin *crt* genes cluster. This enhances the production of lycopene and, as a consequence, the canthaxanthin synthesis. A strategy to further increase canthaxanthin production is to construct a deletion mutant in the *crtC* and *crtD* genes that act downstream in the spirilloxanthin biosynthesis pathway. This completely reroutes the lycopene towards the CrtY and CrtW enzymes for maximal canthaxanthin synthesis (Fig. 1). Such engineering strategy to improve the synthesis of the desired carotenoid could be applied to other aerobic photosynthetic bacteria that are known to produce a cocktail of photosynthetic and non-photosynthetic carotenoids. However, the characterization of the different *crt* genes as described in Subheading 3.1 is needed. Below, we describe the methodology for the construction of a *crtCD* deletion mutant of the ORS278 strain and its use to overproduce canthaxanthin.

1. Amplify by PCR the DNA region (3.5 kb) encompassing the *crtC* and *crtD* genes using the primers crtDf and crtCr.
2. Clone the PCR product into the pGEM-T vector.
3. Delete a 2-kb region inside the *crtC* and *crtD* genes by *Xho*I digestion and insert in place the 4.7-kb *Sal*I pKOK5 cassette containing a kanamycin resistance gene and a *lacZ* reporter gene (14) (see Note 9).
4. Remove the DNA region containing the mutated *crt* genes by *Sal*I digestion and clone the fragment into the *Xho*I site of the pJQ200mp18 suicide vector (15) (see Note 10).
5. Transfer by electroporation the pJQ200mp18 plasmid containing the mutated *crt* genes into *E. coli* S17-1 λpir to permit its mobilization by bipartite conjugation in the ORS278 strain.
6. Grow *E. coli* S17-1 λpir containing the pJQ200mp18 plasmid overnight at 37°C in LB medium supplemented with kanamycin (50 μg/mL). Grow ORS28 strain for 5 days at 37°C in YM-modified medium.
7. Harvest 2.5 mL of *E. coli* S17-1 λpir and 10 mL of ORS278 cells by centrifugation (3,200 × *g* for 30 min at 4°C).
8. Wash the pellets twice with 9 g/L NaCl.
9. Resuspend each pellet in 200 μL of 9 g/L NaCl.
10. Gently mix the two pellets and transfer to a YM-modified agar plate. Incubate for 3 days at 37°C.
11. Resuspend the mating in 1 mL of YM-modified medium, make serial dilutions until 10^{-4} into 9 g/L NaCl, and spread 100 μL of each dilution onto YM-modified agar plates containing kanamycin (50 μg/mL) and nalidixic acid (35 μg/mL) (see Note 11).
12. Incubate the plates at 37°C for at least 10 days (see Note 12).
13. Screen the colonies by PCR to check that the first event of recombination occurred.
14. Grow one recombinant strain for 1 week at 37°C in YM-modified medium supplemented with 50 μg/mL kanamycin and 35 μg/mL nalidixic acid.

15. Make serial dilutions until 10^{-4} into 9 g/L NaCl and spread 100 μL of each dilution onto YM-modified agar plates containing sucrose (7%), kanamycin (50 μg/mL), and nalidixic acid (35 μg/mL) (see Note 13).
16. Incubate the plates at 37°C for at least 1 week.
17. Screen the colonies by PCR to check that the second event of recombination occurred.
18. Store the mutant strain in glycerol (25% v/v) at −80°C.
19. Cultivate the mutant in YM-modified medium under 770-nm illumination as previously described in Subheading 3.2.
20. Extract the carotenoids as previously described in Subheading 3.2 (see Note 14).

4. Notes

1. The genes *crtI* and *crtB* encoding, respectively, phytoene synthase and phytoene desaturase, two enzymes involved in the initial steps of carotenoid biosynthesis (Fig. 1), have been isolated and characterized from various microorganisms. In all of the cases, these genes were found adjacent and with the same direction of transcription. Comparison of the deduced amino acid sequences of CrtI and CrtB proteins from *Erwinia uredovora*, *Erwinia herbicola*, *Flavobacterium* sp. ATCC21588, *Rhodobacter sphaeroides*, and *Agrobacterium aurantiacum* revealed well-conserved domains at the C-terminal end of CrtI (LVGAGTHPG) and in the central region of CrtB (QLTNIARD). These motifs were chosen for designing the degenerated primers crtIf and crtBr.
2. This is a critical step. It is recommended to follow precisely the protocols furnished by Stratagene.
3. Considering an insert size from 30 to 42 kb, about 2,000 clones are estimated to be enough to cover the entire genome of the ORS278 strain (about 7.2 Mb).
4. In our case, it was not possible to identify a positive clone by its color. This was not due to a problem of functionality of *Bradyrhizobium* CRT enzymes in *E. coli*, but to a problem of promoter recognition preventing *crt* genes' expression (10).
5. In order to limit the number of PCR reactions, the colonies present in the same line and the same column could be pooled. This diminishes the number of PCR reactions from 96 to 20.
6. To examine the light effect on carotenoids' biosynthesis in ORS278 strain, we use six different wavelengths (590, 660,

735, 770, 810, and 850 nm). They cover the absorption range of both the Pr and Pfr forms of a bacteriophytochrome.

7. Light control of carotenoid biosynthesis by a bacteriophytochrome typically should result in a bell curve with an optimum around 650 nm if Pfr is the active form, or an optimum around 750 nm if Pr is the active form. In the case of the ORS278 strain, we observed that the production of canthaxanthin increased twice under 770-nm light, whereas the production of spirilloxanthin increased three times.
8. Alternatively, the same procedure could be used to examine the effect of the light on *crt* genes' expression. In this last case, we constructed a transformant strain, where a reporter gene (such as *lacZ*, *gusA*, etc.) is under the control of the promoter region(s) of the *crt* genes. The expression of the reporter gene is then analyzed according to the different light conditions tested. According to this method, we determined that the *crt* spirilloxanthin gene cluster is only up-regulated by far-red light, whereas the synthesis of both carotenoid is stimulated (11).
9. The mutant construction strategy is designed so that the flanking regions of the cassette, which correspond to the 5′-end of *crtD* and 3′-end of *crtC*, have a size of around 700 bp in order to promote the double-crossing-over event.
10. The use of the pJQ200mp18 suicide plasmid, which does not replicate in ORS278 strain, permits to integrate the DNA region containing the deleted *crtC* and *crtD* genes into the chromosomal *crtC* and *crtD* genes due to the similarity between one of the flanking regions of the insert and the genomic DNA.
11. The ORS278 strain is resistant to nalidixic acid. The addition of nalidixic acid to the medium inhibits the growth of *E. coli* cells. The addition of kanamycin permits, on the other hand, to select the ORS278 recombinant cells in which the pJQ200mp18 vector is integrated into the chromosome.
12. *Bradyrhizobium* strains grow very slowly, the first colonies appearing between 7 and 10 days.
13. Clones in which double events of recombination occurred are selected by plating on sucrose-containing medium. Cells retaining the vector backbone are lost since the *sacB* gene on the pJQ200mp18 backbone converts sucrose to toxic levansucrose.
14. In such condition, the total amount of canthaxanthin produced by the *crtCD* mutant reached 1.5 mg of canthaxanthin per g of dry cells, against 0.3 mg/g for the wild-type strain grown in the dark (11).

References

1. Takaichi S, Shimada K, Ishidsu JI (1990) Carotenoids from the aerobic photosyntehtic bacterium, *Erythrobacter longus*: β-carotene and its hydroxyl derivates. Arch Microbiol 153: 118–122
2. Yurkov VV, Beatty JT (1998) Aerobic anoxygenic bacteria. Microbiol Mol Biol Rev 62: 695–724
3. Lorquin J, Molouba F, Dreyfus BL (1997) Identification of the carotenoid canthaxanthin from photosynthetic *Bradyrhizobium* strains. Appl Environ Microbiol 63:1151–1154
4. Kläui H (1982) Industrial and commercial uses of carotenoids. In: Britton G, Goodwin TW (eds) Carotenoid chemistry and biochemistry. Pergamon, Oxford, pp 309–328
5. Simpson KL, Katayama T, Chichester CO (1981) Carotenoids in fish feeds. In: Bauernfeind JC (ed) Carotenoids as colorants and vitamin A precursors. Academic, New York, pp 463–538
6. Camera E, Mastrofrancesco A, Fabbri C, Daubrawa F, Picardo M, Sies H, Stahl W (2009) Astaxanthin, canthaxanthin and beta-carotene differently affect UVA-induced oxidative damage and expression of oxidative stress-responsive enzymes. Exp Dermatol 18: 222–231
7. Misawa N, Satomi Y, Kondo K, Yokoyama A, Kajiwara S, Saito T, Ohtani T, Miki W (1995) Structure and functional analysis of a marine bacterial carotenoid biosynthesis gene cluster and astaxanthin biosynthetic pathway proposed at the gene level. J Bacteriol 177:6575–6584
8. Takaichi S (1999) Carotenoids and carotenogenesis in anoxygenic photosynthetic bacteria, p 40-69. In: Frank HA, Young AJ, Britton G, Cogdell RJ (eds) The photochemistry of carotenoids. Kluwer Academic, The Netherlands
9. Pinta V, Ouchane S, Picaud M, Takaichi S, Astier C, Reiss-Husson F (2003) Characterization of unusual hydroxy- and keto-carotenoids in *Rubrivivax gelatinosus*: involvement of enzyme CrtF or CrtA. Arch Microbiol 179:354–362
10. Hannibal L, Lorquin J, Angles d'Ortoli N, Garcia N, Chaintreuil C, Masson-Boivin C, Dreyfus B, Giraud E (2000) Isolation and characterization of canthaxanthin biosynthesis genes from the photosynthetic Bradyrhizobium sp. strain ORS278. J Bacteriol 182:3850–3853
11. Giraud E, Hannibal L, Fardoux J, Jaubert M, Jourand P, Dreyfus B, Sturgis J, Verméglio A (2004) Two distinct *crt* gene clusters for two different functional classes of carotenoid in *Bradyrhizobium*. J Biol Chem 279: 15076–15083
12. Giraud E, Fardoux J, Fourrier N, Hannibal L, Genty B, Bouyer P, Dreyfus B, Verméglio A (2002) Phytochrome controls the photosystem synthesis in anoxygenic bacteria. Nature 417: 202–205
13. Jaubert M, Zappa S, Fardoux J, Adriano JM, Hannibal L, Elsen S, Lavergne J, Verméglio A, Giraud E, Pignol D (2004) Light and redox control of photosynthesis gene expression in *Bradyrhizobium*: dual roles of two PpsR. J Biol Chem 279:44407–44416
14. Kokotek W, Lotz W (1989) Construction of a *lacZ*-kanamycine–resistance cassette, useful for site-directed mutagenesis and as a promoter probe. Gene 84:467–471
15. Quandt J, Hynes MF (1993) Versatile suicide vectors which allow direct selection for gene replacement in gram-negative bacteria. Gene 127:15–21
16. Alberti M, Burke DH, Hearst JE (1995) Structure and sequence of the photosynthesis gene cluster. In: Blankenship RE, Madigan MT, Bauer CE (eds) Anoxygenic photosynthetic bacteria. Kluwer Academic, Dordrecht, pp 775–805
17. Bramley PM, Mackenzie A (1988) Regulation of carotenoid biosynthesis. Curr Top Cell Regul 29:291–343
18. Von Lintig J, Welsch R, Bonk M, Giuliano A, Kleinig H (1997) Light-dependent regulation of carotenoid biosynthesis occurs at the level of phytoene synthase expression and is mediated by phytochrome in *Sinapsis alba* and *Arabidopsis thaliana* seedlings. Plant J 12:625–634
19. Davis SJ, Vener AV, Vierstra RD (1999) Bacteriophytochromes: phytochrome-like photoreceptors from nonphotosynthetic eubacteria. Science 286:2517–2520
20. Giraud E, Verméglio A (2008) Bacteriophytochromes in anoxygenic photosynthetic bacteria. Photosynth Res 97:141–153

Chapter 10

Construction of Carotenoid Biosynthetic Pathways Through Chromosomal Integration in Methane-Utilizing Bacterium *Methylomonas* sp. Strain 16a

Rick W. Ye and Kristen Kelly

Abstract

Methylomonas sp. strain 16a is an obligate methanotrophic bacterium that uses methane or methanol as the sole energy and carbon source. In order to engineer a stable strain to produce carotenoids, integration of genes or gene clusters in various nonessential locations in the chromosome is used. Construction of a canthaxanthin-producing strain involves the integration of canthaxanthin biosynthetic genes including the *crtW* gene for the β-carotenoid ketolase. Addition of the *crtZ* gene that encodes the β-carotenoid hydroxylase in this strain leads to the production of astaxanthin. Further increase in titer and yield for astaxanthin is obtained by integration of another set of astaxanthin biosynthetic gene cluster in a separate location of the chromosome.

Key words: Methane, Methanotroph, Pathway engineering, Chromosomal integration, Conjugation, Homologous recombination, Carotenoids, Canthaxanthin, Astaxanthin

1. Introduction

Obligate methanotrophs are those bacteria that can only grow on methane or methanol. However, examples of engineering this group of organisms to produce commercial products remain scarce. It can be difficult to introduce a complex pathway into methanotrophs since these organisms are not well studied and the genetic systems are less developed as compared to those for *E. coli* and other industrial organisms. In addition, delivering methane gas to the organism as carbon source is not as convenient as using sugars as substrates. Despite these drawbacks, methanotrophs have been considered as production hosts due to the natural abundance of methane. Furthermore, methane can be generated from biomass or biowaste using microbial processes.

José-Luis Barredo (ed.), *Microbial Carotenoids from Bacteria and Microalgae: Methods and Protocols*, Methods in Molecular Biology, vol. 892, DOI 10.1007/978-1-61779-879-5_10, © Springer Science+Business Media, LLC 2012

Canthaxanthin and astaxanthin are carotenoids that have been widely used commercially (1). Biosynthesis of carotenoids is derived from the isoprenoid pathway. β-carotene can be made from farnesyl pyrophosphate by expression of *crtE*, *crtY*, *crtI*, and *crtB* genes (2–4). The two β-ionone rings in β-carotene are modified by ketolases and hydroxylases to produce canthaxanthin and astaxanthin. In some bacterial systems, the β-carotene hydroxylase CrtZ introduces the hydroxyl groups on the β-ionone rings, while the ß-carotene ketolase CrtW catalyzes the reaction of adding two keto-groups (5, 6). Only the CrtW ketolase is necessary for the production of canthaxanthin, while a combination of both CrtZ hydroxylase and CrtW ketolase is required for the biosynthesis of astaxanthin. Based on color intensity of bacterial colonies, improvement of canthaxanthin and astaxanthin production can be achieved by screening mutation libraries of ketolase CrtW (7, 8).

To demonstrate the feasibility and the commercial potential of engineering complex pathways in methanotrophs, here we describe a procedure for the construction of canthaxanthin- and astaxanthin-producing strains in *Methylmonas* sp. strain 16a. To maintain strain stability, which is required for large-scale fermentation process, construction of these strains is achieved through integration of the carotenoid pathway genes in the chromosome and manipulation of the expression level with appropriate promoters and gene copy numbers.

2. Materials

In addition to the routine techniques to handle microorganisms and molecular biological materials, care must be taken to manipulate methane gas due to its explosive nature.

2.1. Strains and Plasmids

Additional information about strains, plasmids, and gene clusters can be found in Table 1.

1. *Methylomonas* sp. 16a ATCC PTA-2402 (ATCC, Manassas, VA, USA).
2. *Methylomonas* MWM1200 ATCC PTA-6887 (ATCC, Manassas, VA, USA) (see Note 1).
3. *E. coli* pRK2013 ATCC 37159 (ATCC, Manassas, VA, USA).
4. pSUKSM (9) (see Note 2).
5. pSUKSMori (9).
6. pSUKSMorihps (9).
7. pSUKSMald (9).
8. pSUKSMaldhps (9).

Table 1
Bacterial strains and plasmids

Strains or plasmids	Relevant characteristics	References
Strains		
Methylomonas sp. 16a	WT	ATCC PTA-2402
MWM1200	Non-pigmented *Methylomonas*	ATCC PTA-6887
E. coli with pRK2013	Helper strain for conjugation	ATCC 37159
Plasmids		
pSUKSM	Km^r, an integration vector derived from pACYC containing *sacB* and *ori*T	(9, 13)
pSUKSMori	Derived from pSUKSM; contains a noncoding region for homologous recombination	(9)
pSUKSMorihps	Derived from pSUKSMori; P_{hps}	(9)
pSUKSMald	Derived from pSUKSM; contains the *ald* region for homologous recombination	(9)
pSUKSMaldhps	Derived from pSUKSMald; P_{hps}	(9)
pDCQ333	Derived from pBHR1, Km^r; contains *crtWE idi crtYIB*. *crtW* from *Paracoccus* sp. N81106 (GenBank: D58420); *crtE idi crtYIB* cluster from strain DC404 (GenBank: DQ090834)	(9)
pDCQ334	Derived from pDCQ333, Km^r; contains *crtWZE idi crtYIB* with *crtZ* from *Paracoccus* sp. N81106 (GenBank: D58420)	(10)
pDCQ392	Derived from pBHR1, Km^r; contains *crtWZE idi crtYIB* with both *crtW* and *crtZ* from DC263 (GenBank: DQ309446) and *crtE idi crtYIB* from DC413 (GenBank: DQ090836)	(10)

9. pDCQ333 (10).
10. pDCQ334 (9).
11. pDCQ392 (9).

2.2. Growth Medium for Methylomonas and E. coli

1. Luria–Bertani (LB): Add 10 g of bacto-tryptone, 5 g of yeast extract, and 10 g of NaCl into 1 L of deionized water. Autoclave at 121°C for 15 min.
2. LB-agar: LB and 15 g/L agar. Autoclave at 121°C for 15 min.
3. Kanamycin (Kan) stock solution (50 mg/mL).
4. BTZ medium: 0.537 g NH_4Cl, 0.5 g KH_2PO_4, 0.5 g Na_2SO_4, 0.2 g $MgCl_2·6H_2O$, 0.1 g $CaCl_2·2H_2O$, 50 mL HEPES (1.0 M pH 7.0), and 10 mL trace mineral stock solution. Bring with distilled water to 1 L. Autoclave at 121°C for 15 min.
5. BTZ-agar: BTZ and 15 g/L agar. Autoclave at 121°C for 15 min.

6. Trace mineral stock solution: 12.8 g nitriloacetic acid, 0.0254 g $CuCl_2 \cdot 2H_2O$, 0.3 g $FeCl_2 \cdot 4H_2O$, 0.1 g $MnCl_2 \cdot 4H_2O$, 0.312 g $CoCl_2 \cdot 6H_2O$, 0.10 g $ZnCl_2$, 0.01 g H_3BO_3, 0.01 g $Na_2MoO_4 \cdot 2H_2O$, and 0.184 g $NiCl_2 \cdot 6H_2O$. Bring with distilled water to 1 L. Adjust the pH of the solution to pH 7.0 with NaOH solution. Autoclave at 121°C for 15 min.

2.3. Growth Apparatus for Methylomonas

1. Serum bottles (150 or 500 mL) sealed with PTFE backed butyl rubber stoppers (Wheaton Scientific; Wheaton, IL, USA) (see Note 3).
2. Syringe (see Note 3).
3. Gas jar or gas box (see Note 4).

2.4. Primer List

1. A: 5′-GCTCTAGAATTGGTAATCTTCTGTTATTTATTG-3′.
2. B: 5′-CGGAATTCTCACGCCGTTTCGGCTGGTTGAATG-3′.
3. C: 5′-CATTCAACCAGCCGAAACGGCGTGAGAATTCCT GATGTAGTTCAAACG-3′.
4. D: 5′-GAAGATCTTCGCCGTCCGCC ATGCGCTAGCGGC-3′.
5. E: 5′-ATGACAATTGGTCGACGCGCTAAGGATTGGGGT GCGTCG-3′.
6. F: 5′-ATGACAATTGCCTAGGGAATTCTGTACAGTGATG TGCTCCGAAAGTTT-3′.
7. G: 5′-TTCTAGAAAAGCCAAAGCCTGAGTATGACGA-3′.
8. H: 5′-CGCAATTGAATTCGTTTAAACAGTACTTCATTAG TCATCCCGTGTCCAAGAA-3′.
9. I: 5′-TTCTTGGACACGGGATGACTAATGAAGTACTGTT TAAACGAATTCAATTGA CTCAAATGACAACCAACGCGTGATC-3′.
10. J: 5′-GAAGATCTCCCGGA CAGCGTCACCATCGGCATG-3′.
11. K:5′-GGCCATGCTGTACATCTAGAAAGGAGGAATAAAC CATGACCA-3′.
12. L: 5′-CGCGTACGCCTAGGTTAGGTGCGTTCTTGGGCT TCGGCA-3′.
13. M:5′-GCATGCTAGCCGACGGCTGATCGCATGCTGGCT TATCA-3′.
14. N: 5′-GCATACTAGTTACGCGCGCCAACACCGTTGACA TGAAA-3′.
15. O: 5′-GCATGAATTCGGGTAGAGCCGCGTAATGTGCCG ACC-3′.
16. P:5′-GCTAAGATCTGATGATCGCCTGTCTATCCTCGGG AGCA-3′.

3. Methods

For industrial applications where a large volume of product is made, it is desirable to have a stable strain. One approach to construct such strain is through the integration of various carotenoid genes or gene clusters in different locations of the chromosome. These locations have to be nonessential, but allow an adequate level of expression of the integrated genes.

Integration of recombinant genes into *Methylomonas* is obtained using an integration vector which cannot replicate in this organism. Classical triparental mating is used to transfer DNA from *E. coli* donor to *Methylomonas*. *Methylomonas* sp. Strain 16a is sensitive to Kan and thus the Kan-resistant gene can be used as the selection marker. After conjugation, the first event of homologous recombination is selected based on resistance to Kan due to the integration of the vector by single crossover. Addition of sucrose in the growth medium selects for the second recombination event which results in the excision of the plasmid from the chromosome (11, 12).

3.1. Cloning of the Canthaxanthin Gene Cluster in the Integration Vector

1. The first step to construct the integration vector is to select a DNA region for homologous recombination. In this case, a noncoding region located 13.6 kb downstream from the putative origin of replication of the genome is chosen to integrate the canthaxanthin cluster through homologous recombination (see Note 5).
2. Amplify the 1.1-kb homologous upstream region with primers A and B. An XbaI site is incorporated into the primer A, while an EcoRI site is introduced in the primer B.
3. Amplify the 1.3-kb downstream region with primers C and D. The primer C contains an EcoRI site while primer D has a BglII site.
4. Clone the two fragments into the XbaI and BglII sites of the integration vector pSUSMK. The resulting plasmid is pSUSMKori (Fig. 1), which has a single EcoRI site in the middle of the noncoding region available for cloning of a promoter and the carotenoid gene cluster for integration.
5. Amplify the *hps* promoter (see Note 6) as an MfeI and EcoRI fragment with primers E and F.
6. Clone the PCR fragment into the EcoRI site of pSUSMKori vector. The resulting plasmid is pSUMKorihps, which has a single EcoRI site downstream from the *hps* promoter.
7. Digest pDCQ333 with EcoRI to obtain a fragment containing the canthaxanthin biosynthetic cluster *crtEidiYIB* (Table 1).

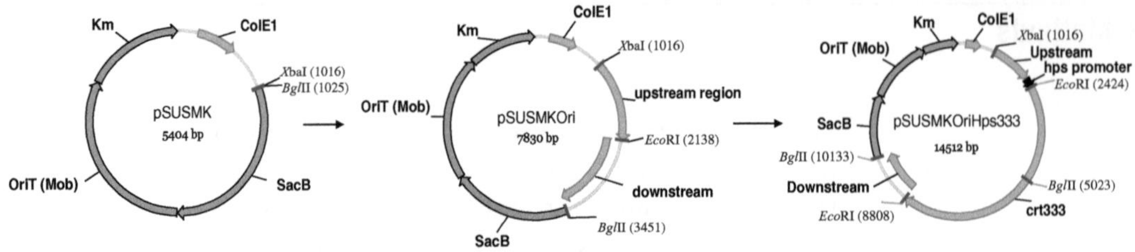

Fig. 1. Construction of integration vector for expression of carotenoid genes in the chromosome. The first step is the cloning of the genomic regions targeted for integration. The second step is the introduction of the *crt* genes between these two regions.

8. Clone this fragment into the EcoRI site of the integration vector pSUMKorihps. The resulting plasmid with proper orientation to the *hps* promoter is pSUSMKorihps333 (Fig. 1).

3.2. Integration of the Canthaxanthin Pathway Through Conjugation and Homologous Recombination

The procedure is outlined in Fig. 2.

1. Start overnight cultures of the helper (*E. coli* strain with pRK2013) and the donor (*E. coli* strain with the integration plasmid pSUSMKorihps333) in LB medium supplemented with Kan (50 μg/mL), and incubate the culture at 37°C.
2. Start a culture of the non-pigmented MWM1200 recipient strain in BTZ medium and incubate the culture at 30°C. It may be necessary to start the culture for MWM1200 a day earlier since it grows slower as compared to *E. coli* strains.
3. For conjugation experiment, the ratio of these three strains is 1:2:6 based on volume (see Note 7).
4. Harvest the overnight cultures by centrifugation and wash them with BTZ medium before combining them together.
5. Spot the pellet containing all three strains onto a nonselective BTZ agar plate supplemented with 0.5% yeast extract.
6. Incubate the plate in a jar or box filled with 25% methane for 3 days at 30°C.
7. Spread the cultures onto BTZ agar plates supplemented with Kan (50 μg/mL) for selection.
8. Place the plates in a jar or box with methane gas mixture and incubate them at 30°C until single colonies with orange color appear.
9. Streak these colonies onto new BTZ plates.
10. Repeat this process to purify the *Methylomonas* exconjugants away from the *E. coli* cells.
11. Culture the selected exconjugants at 30°C in liquid BTZ medium without Kan in serum bottles. Carry out passage of the grown cultures three times.

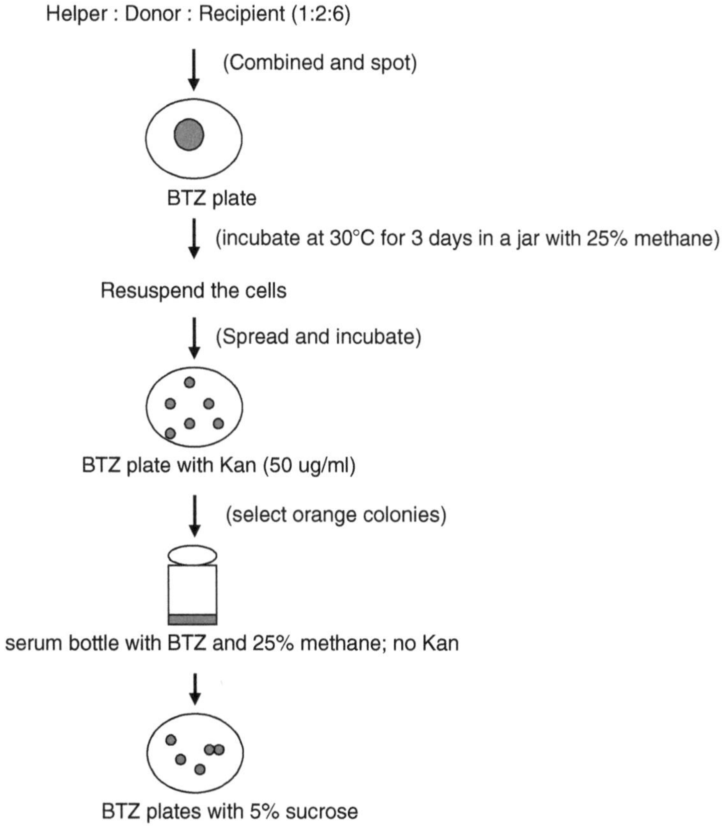

Fig. 2. Outline of procedures for integration of *crt* genes in the chromosome of *Methylomonas* sp. strain 16a (see Subheading 3.2).

12. Plate these cultures on BTZ plates supplemented with sucrose (5% final concentration) (see Note 8).
13. Test the growth of the individual colonies with orange color on BTZ agar plates supplemented with Kan (50 μg/mL).
14. Select strains that are sensitive to Kan.
15. Confirm the integration of the cluster via double crossover with two outside primer sets (see Note 9).
16. Name the selected strain as Orihps333. Extract the canthaxanthin from Orihps333 and analyze the canthaxanthin content by HPLC (8).

3.3. Integration of the crtZ Gene for Astaxanthin Production

The basic strategy and procedure are similar to the steps outlined in Subheadings 3.1 and 3.2.

1. The DNA region containing *crtN, ald,* and *crtNb* genes (see Note 10) is chosen for the integration of *crtZ* gene for astaxanthin biosynthesis.

2. Amplify the 5′ DNA region (upstream) for homologous recombination with primer set G and H, and the 3′ DNA region (downstream) with primers I and J.
3. To amplify the entire region, carry out overlapping PCR with both fragments as the template using the forward primer of the upstream region (primer G) and the reverse primer of the downstream region (primer J).
4. Digest PCR product with XbaI and BglII.
5. Clone the fragment into the integration vector pSUSMK, resulting in the construction of pSUSMKald.
6. Insert the *hps* promoter to the plasmid, which results in the construct pSUSMKaldhps.
7. Amplify the *crtZ* from pDCQ334 plasmid with primers K and L.
8. Clone the *crtZ* gene into pSUSMKaldhps to obtain the plasmid pSUSMKaldhps-crtZ.
9. Transfer the plasmid pSUSMKaldhps-crtZ into the canthaxanthin-producing strain Orihps333 through conjugation as described in Subheading 3.2. The integrated strain obtained is Ax6-5.
10. Extract and analyze the carotenoid content of Ax6-5 by HPLC (8). The strain obtained should produce astaxanthin (see Note 11).

3.4. Integration of an Additional Astaxanthin Gene Cluster

To avoid the unstability of integrated strains, a different astaxanthin biosynthetic gene cluster *crtWZEidiYIB* from pDCQ392 (Table 1) is used for integration into strain Ax6-5. The DNA region *ccp* encoding the cytochrome *C* peroxidase (see Note 12) is chosen as the integration site.

1. Amplify the upstream region including the *ccp* promoter with primers M and N.
2. Amplify the downstream region with primers O and P.
3. Clone the upstream region as an NheI and SpeI fragment into the XbaI site of the integration vector pSUSMK.
4. Clone the downstream region as an EcoRI and BglII fragment, generating the construct pSUSMKccp.
5. Digest the gene cluster *crtWZEidiYIB* as an EcoRI fragment from pDCQ392 and clone it in the EcoRI site of pSUSMKccp.
6. Transfer the resulting plasmid into strain Ax6-5 via conjugation as described in Subheading 3.2. After completing the double crossover selection, name the resulting strain as Ax392 (see Note 13).
7. Analyze the astaxanthin content of Ax392 by HPLC (8).

Although the experiment starts with integrating the canthaxanthin first, direct integration of the astaxanthin clusters in multiple locations is desirable if that is the product of interest. In this experiment, two sets of astaxanthin biosynthetic genes from different organisms are chosen to reduce the likelihood of instability due to homologous recombination.

4. Notes

1. The colorless MWM1200 strain derived from the wild-type *Methylomonas* is used for integration experiments.
2. The integration plasmid pSUKSM has a Kan-resistant marker for selection in MWM1200, an *oriT* region for conjugal transfer, and an *sacB* region for counter selection for double crossovers. The *oriT* region is required for transfer of the plasmid from *E. coli* to recipient bacteria. However, this plasmid does not have regions for replication from broad host range plasmids and thus cannot replicate in *Methylmonas.*
3. For growing the strain in liquid medium, serum bottles with 150- or 500-mL volume are used. The bottles are sealed with PTFE backed butyl rubber stoppers. The gas/liquid ratio in the bottle is at least 8:1 in all experiments to allow the presence of sufficient oxygen and methane gas. Deliver methane with a syringe into the serum bottle so that the gas phase in the serum bottle contains 25% methane by volume.
4. For incubation of agar plates, place plates in a gas jar or gas box. Flush the jar or the gas box with methane/air mixture for 40 min before sealed. The gas mixture consisted of 25% methane and 75% air by volume.
5. The putative origin of replication is next to the *dnaA* gene which encodes DnaA, a replication initiation factor. A region close to the origin of replication is chosen for integration with the assumption that the integrated gene will be expressed at higher level due to the increase in copy number during replication. However, this assumption has not been proven. For homologous recombinations, the two homologous regions flanking the DNA to be integrated have to be around 0.8 kb or longer to obtain good efficiency of recombination.
6. The *hps* gene encodes the 3-hexulose-6-phosphate synthase, which catalyzes the first step in the RuMP pathway for formaldehyde assimilation. The *hps* promoter is used to express the canthaxanthin gene cluster. This is one of the genes that is highly expressed based on DNA microarray experiment (9).

7. We normally use 5 mL of helper strain, 10 mL of the donor strain (with the integration vector), and 30 mL of Kan-sensitive recipient *Methylomonas* host.
8. Strains with single crossover still have the vector backbone which contains the *sacB* gene and thus are sensitive to sucrose. The purpose of passage is to enrich those that have undergone double crossover. Those strains growing on sucrose will be either the parent strain or the strains with integration via double crossover. The sucrose has to be freshly prepared.
9. Four primers are needed for confirmation. Design two outside primers based on genomic regions outside the integration region and two internal primers in opposite direction based on the *crt* region.
10. This DNA region is involved in the native pigment production and has been shown to be nonessential for cell growth. Mutations in this region render *Methylomonas* sp. strain 16a colorless and thus useful for the introduction of carotenoid gene cluster for visual selection based on color formation.
11. We observed that the strain with a single copy of the *crtZ* gene produced both astaxanthin and the mono-ketolated intermediate adonixanthin. The astaxanthin consisted of 64% of the total carotenoid.
12. The *ccp* region has been shown to support a high level of expression of *crt* genes (13).
13. In strain Ax392, the expression of *crtWZEidiYIB* gene cluster is under the control of the native promoter for the *ccp* gene. This final strain should produce a higher amount of astaxanthin. We observed that the strain with a second complete astaxanthin cluster produces up to 95% of the total carotenoid in the culture as astaxanthin. Among the different forms of astaxanthin isomers, the all *E*-isomer is 89%. In addition, the strain proved to be very stable under fermentation conditions. To achieve a high level of astaxanthin production, it is important to maintain a high oxygen level. Lower oxygen level often results in accumulation of intermediates and thus lower selectivity for astaxanthin production for engineered strains (14).

References

1. Bhosale P, Bernstein PS (2005) Microbial xanthophylls. Appl Microbiol Biotechnol 68:445–455
2. Misawa N, Satomi Y, Kondo K, Yokoyama A, Kajiwara S, Saito T, Ohtani T, Miki W (1995) Structure and functional analysis of a marine bacterial carotenoid biosynthesis gene cluster and astaxanthin biosynthetic pathway proposed at the gene level. J Bacteriol 177:6575–6584
3. Miura Y, Kondo K, Saito T, Shimada H, Fraser PD, Misawa N (1998) Production of the carotenoids lycopene, β-carotene, and astaxanthin in the food yeast *Candida utilis*. Appl Environ Microbiol 64:1226–1229

4. Mochimaru M, Masukawa H, Takaichi S (2005) The cyanobacterium *Anabaena* sp. PCC 7120 has two distinct beta-carotene ketolases: CrtO for echinenone and CrtW for ketomyxol synthesis. FEBS Lett 579: 6111–6114
5. Fraser PD, Miura Y, Misawa N (1997) In vitro characterization of astaxanthin biosynthetic enzymes. J Biol Chem 272:6128–6135
6. Sieiro C, Poza M, de Miguel T, Villa TG (2003) Genetic basis of microbial carotenogenesis. Int Microbiol 6:11–16
7. Tao L, Wilczek J, Odom JM, Cheng Q (2006) Engineering a beta-carotene ketolase for astaxanthin production. Metab Eng 8: 523–531
8. Ye RW, Stead KJ, Yao H, He H (2006) Mutational and functional analysis of the beta-carotene ketolase involved in the production of canthaxanthin and astaxanthin. Appl Environ Microbiol 72:5829–5837
9. Ye RW, Yao H, Stead K, Wang T, Tao L, Cheng Q, Sharpe PL, Suh W, Nagel E, Arcilla D, Dragotta D, Miller ES (2007) Construction of the astaxanthin biosynthetic pathway in a methanotrophic bacterium *Methylomonas sp.* strain 16a. J Ind Microbiol Biotechnol 34:289–299
10. Tao L, Rouviere PE, Cheng Q (2006) A carotenoid synthesis gene cluster from a non-marine *Brevundimonas* that synthesizes hydroxylated astaxanthin. Gene 379:101–108
11. Fu R, Voordouw G (1997) Targeted gene-replacement mutagenesis of dcrA, encoding an oxygen sensor of the sulfate-reducing bacterium *Desulfovibrio vulgaris Hildenborough.* Microbiology 143:1815–26
12. Schweizer HP (1992) Allelic exchange in *Pseudomonas aeruginosa* using novel ColEl-type vectors and a family of cassettes containing a portable oriT and the counter-selectable *Bacillus subtilis sacB* marker. Mol Microbiol 6:1195–1204
13. Sharpe PL, Dicosimo D, Bosak MD, Knoke K, Tao L, Cheng Q, Ye RW (2007) Use of transposon promoter-probe vectors in the metabolic engineering of the obligate methanotroph *Methylomonas sp.* strain 16a for enhanced C40 carotenoid synthesis. Appl Environ Microbiol 73:1721–1728
14. Tao L, Sedkova N, Yao H, Ye RW, Sharpe PL, Cheng Q (2007) Expression of bacterial hemoglobin genes to improve astaxanthin production in a methanotrophic bacterium *Methylomonas* sp. Appl Microbiol Biotechnol 74:625–633

Chapter 11

Genetic Modification in *Bacillus subtilis* for Production of C30 Carotenoids

Isamu Maeda

Abstract

C30 carotenoids, which have shorter backbones than C40 carotenoids, are known to be produced in the pathogenic bacterium *Staphylococcus aureus* that causes opportunistic infection. The first committed enzyme in the C30 carotenoid synthetic pathway is dehydrosqualene synthase CrtM. CrtM converts farnesyl pyrophosphate to dehydrosqualene. Dehydrosqualene desaturase CrtN then converts dehydrosqualene to the yellow C30 carotenoid, 4,4′-diaponeurosporene. This chapter describes a method to synthesize C30 carotenoids in *Bacillus subtilis*, which is generally recognized as a safe (GRAS) organism. Introduction of *S. aureus crtM* and *crtN* genes into *B. subtilis* results in yellow pigmentation. The *B. subtilis* transformant accumulates two C30 carotenoids, 4,4′-diapolycopene and 4,4′-diaponeurosporene. Furthermore, together with *crtMN*, introduction of *S. aureus crtP* and *crtQ* genes, which encode mixed function oxidase and glycosyltransferase, respectively, donates the ability to produce glycosylated C30 carotenoic acid. Thus, carotenoid biosynthesis genes of *S. aureus* is applicable to genetically modify *B. subtilis* in order to construct a safe organism producing C30 carotenoids.

Key words: *Bacillus subtilis*, *crtM*, Diapolycopene, Diaponeurosporene, Glycosyl diaponeurosporenoate

1. Introduction

Carotenoids that possess the C30 backbone such as staphyloxanthin are found in a small group of bacteria, including *Staphylococcus aureus* (1) and *Heliobacterium* sp. (2). Five or six genes have been reported to be involved in the staphyloxanthin biosynthetic pathway (3, 4). The first gene *crtM*, which encodes dehydrosqualene synthase CrtM, is involved in the conversion of farnesyl pyrophosphate to dehydrosqualene. The second gene *crtN*, which encodes dehydrosqualene desaturase CrtN, is responsible for the conversion of dehydrosqualene to the yellow carotenoid 4,4′-diaponeurosporene or the red carotenoid 4,4′-diapolycopene. The terminal methyl group of 4,4′-diaponeurosporene is then oxidized to form

José-Luis Barredo (ed.), *Microbial Carotenoids from Bacteria and Microalgae: Methods and Protocols*, Methods in Molecular Biology, vol. 892, DOI 10.1007/978-1-61779-879-5_11, © Springer Science+Business Media, LLC 2012

4,4′-diaponeurosporenoic acid by CrtP, encoded by the *crtP* gene. The carboxyl group of 4,4′-diaponeurosporenoic acid is esterified with glucose to yield glycosyl 4,4′-diaponeurosporenoate by the glycosyltransferase CrtQ, encoded by the *crtQ* gene. Finally, glycosyl 4,4′-diaponeurosporenoate is acylated with 12-methyltetradecanoyl to form staphyloxanthin by the acyl transferase CrtO, encoded by the *crtO* gene. The carotenoids in the staphyloxanthin biosynthetic pathway are shown to possess antioxidative effects (5, 6), and could be potentially applicable as food/feed additives and preservatives.

S. aureus is known to be a pathogenic bacterium that poses a serious public health threat (5), and therefore it is not suitable for industrial production of fine chemicals such as carotenoids. *Heliobacterium* sp. is an obligate anaerobe (7), and industrial application of this bacterium may accompany with difficulty to maintain anaerobic conditions in a large-scale culture. The Gram-positive bacterium, *Bacillus subtilis*, has long been used for the industrial production (8), and is generally recognized as a safe (GRAS) organism (9). Yoshida et al. have succeeded in forcing *B. subtilis* to produce 4,4′-diaponeurosporene and 4,4′-diapolycopene by introducing *S. aureus crtMN* genes (6). Furthermore, this chapter shows that glycosylated carotenoic acid is produced in the bacterium by introducing *S. aureus crtMNPQ* genes. Thus, *S. aureus* carotenoid biosynthetic genes and *B. subtilis* are potentially applicable as transforming DNA and as host organism to produce the C30 carotenoids of staphyloxanthin biosynthesis pathway.

2. Materials

2.1. Bacterial Strains

1. *B. subtilis* JH642 (Dr. Tsutomu Sato, Hosei University, Tokyo, Japan) (see Note 1).
2. *S. aureus* NCTC 50581 (National Collection of Type Cultures, London, UK).
3. *Escherichia coli* DH5α (Takara Bio, Shiga, Japan).

2.2. Culture Media

1. Tryptone soya broth (TSB): 17.0 g/L pancreatic digest of casein, 3.0 g/L papaic digest of soybean meal, 5.0 g/L sodium chloride, 2.5 g/L di-basic potassium phosphate, and 2.5 g/L d-glucose; adjust to pH 7.3 before autoclaving at 121°C for 15 min (see Note 2).
2. Luria–Bertani (LB) medium: 10.0 g/L tryptone, 5.0 g/L yeast extract, and 5.0 g/L sodium chloride; adjust to pH 7.5 before autoclaving at 121°C for 15 min (see Note 3).

3. Tetracycline (2.0 mg/mL): Filter through 0.2-μm filter, and add to the culture medium after autoclaving at a final concentration of 20 μg/mL when selective pressure is required.
4. Solution A: 10% (v/v) glycerol, 0.5 M sorbitol, and 0.5 M mannitol.

2.3. DNA Manipulation

1. pHY300PLK (Takara Bio, Shiga, Japan) (see Note 4).
2. pGEM-T (Promega KK, Tokyo, Japan) (see Note 5).
3. PCR primer crtM-F: 5′-GC*GGATCC*GT<u>AAGAGAGG</u>ACTAGTATGAC (the introduced *Bam*HI site is shown in italics and the ribosome-binding site (RBS) is underlined).
4. PCR primer crtN-R: 5′-CGCC*GTCGAC*CGTTATGTTCAACAG (the introduced *Sal*I site is shown in italics).
5. PCR primer crtP-F: 5′-CA*GTCGAC*GT<u>AAGAGAGG</u>TGAACGTCATG (the introduced *Sal*I site is shown in italics and RBS is underlined).
6. PCR primer crtQ-R: 5′-CGCCAGGC*TCTAGA*TTTATTGTTCT (the introduced *Xba*I site is shown in italics).
7. DNeasy tissue kit (Qiagen, Tokyo, Japan).
8. Pfx50 DNA polymerase (Invitrogen, Tokyo, Japan).
9. MinElute PCR purification kit (Qiagen, Tokyo, Japan).
10. MinElute reaction purification kit (Qiagen, Tokyo, Japan).
11. Rapid DNA dephos and ligation kit (Roche Diagnostics Japan, Tokyo, Japan).
12. Gene pulser cuvette (Bio-rad, Hercules, California, USA).
13. Gene pulsar II (Bio-rad, Hercules, California, USA).

2.4. Analyses of Carotenoids

1. HPLC equipment Waters 1525 (Waters, Milford, MA, USA).
2. C18 column (μBondapak C18, 3.9×300 mm, 125-Å pore size) (Waters, Milford, MA, USA).
3. Photodiodearray detector Waters 2996 (Waters, Milford, MA, USA).
4. Acetonitrile–methanol–distilled water (47.5:47.5:5.0; v/v/v).
5. HLC-Disk 3 (0.45-μm pore size, polytetrafluoroethylene) (Kanto Chemical, Tokyo, Japan).

3. Methods

3.1. DNA Construction

1. Incubate *S. aureus* in LB medium at 37°C and 150 rpm overnight.
2. Centrifuge the culture containing 2×10^9 cells at $5{,}000 \times g$ and discard supernatant.

3. Extract DNA from the pellet with proteinase K and buffer AL in DNeasy tissue kit.
4. Purify DNA using DNeasy Mini spin column.
5. Extract DNA from DNeasy Mini spin column with 200 μL of buffer AE in DNeasy tissue kit.
6. Amplify open reading frames (ORFs) of *crtM* with the designed RBS and *crtN* with its RBS from *S. aureus* total DNA using crtM-F and crtN-R primers and Pfx50 DNA polymerase.
7. Prepare a PCR mixture as follows: 5 μL of 10× Pfx50 PCR mix, 1.5 μL of 10 mM dNTP mix, 1.5 μL of 10 μM crtM-F, 1.5 μL of 10 μM crtN-R, 1 μL of *S. aureus* total DNA, 1 μL of Pfx50 DNA polymerase, and 38.5 μL of distilled sterilized water.
8. Set a thermal program as follows: 1 step of 94°C 2 min; 35 cycles of 94°C 15 s, 60°C 15 s, and 68°C 2 min; and 1 step of 68°C 5 min.
9. Place the PCR mixture on a thermal cycler and run the program.
10. When the 35 cycles complete, add 0.5 μL of ExTaq DNA polymerase to the mixture for an adenine nucleotide tailing during the next 5-min step.
11. Purify the PCR fragment with the MinElute PCR purification kit.
12. Ligate the purified PCR fragment into the pGEM-T cloning vector to construct pGEMcrtMN.
13. Digest pGEMcrtMN with *Bam*HI and *Sal*I to generate the fragment crtMN.
14. Digest pHY300PLK with *Bam*HI and *Sal*I.
15. Purify the fragment crtMN and digested pHY300PLK with MinElute reaction purification kit.
16. Ligate the fragment crtMN and digested pHY300PLK with the *Bam*HI and *Sal*I sites to generate pHYcrtMN, a plasmid for carotenoid production.
17. Amplify ORFs of *crtP* with the designed RBS and *crtQ* with its RBS from *S. aureus* total DNA as described above. Use crtP-F and crtQ-R primers instead of crtM-F and crtN-R primers.
18. Ligate the purified fragment into pGEM-T cloning vector to construct pGEMcrtPQ.
19. Digest pGEMcrtPQ with *Sal*I and *Xba*I to generate the fragment crtPQ. Digest pHYcrtMN with *Sal*I and *Xba*I.
20. Purify the fragment crtPQ and digested pHYcrtMN with MinElute reaction purification kit.
21. Ligate the fragment crtPQ and digested pHYcrtMN with the *Sal*I and *Xba*I sites of pHYcrtMN to construct pHYcrtMNPQ, another plasmid for carotenoid production (Fig. 1).

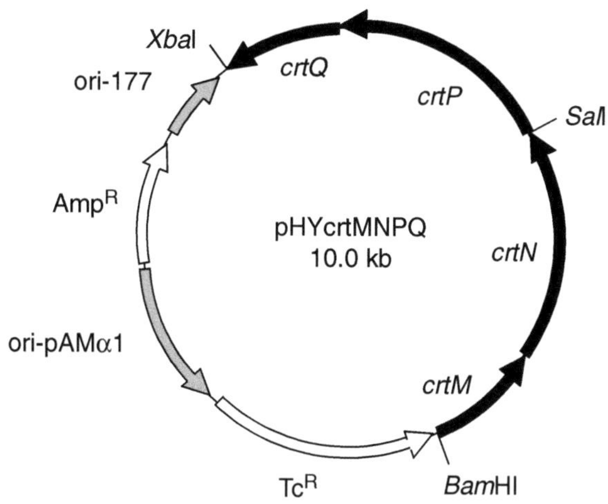

Fig. 1. Structure of pHYcrtMNPQ, a plasmid for carotenoid production. *Staphylococcus aureus crtMNPQ* genes are inserted into the *Bam*HI/*Xba*I sites of pHY300PLK. Arrows indicate directions of transcription or replication. ori-177, replication origin of plasmid pACYC177; ori-pAMα1, replication origin of plasmid pAMα1; Amp^R, ampicillin-resistance gene; Tc^R, tetracycline-resistance gene.

3.2. Introduction of DNA into B. subtilis by Electroporation

1. Grow an overnight culture of *B. subtilis* at 150 rpm and 37°C.
2. Inoculate 2 mL of *B. subtilis* overnight culture into 32 mL LB medium containing 0.5 M sorbitol.
3. Incubate at 150 rpm and 37°C until OD_{600} becomes 0.85–0.95 (see Note 6).
4. Transfer the culture on ice for 10 min to cool down.
5. Collect cells by centrifugation and resuspend the pellet in 10 mL ice-cold solution A (see Note 7).
6. Collect cells again by centrifugation and resuspend the pellet in 0.8 mL ice-cold solution A (see Note 8).
7. Add 200 ng of plasmid DNA to 70 μL of the cell suspension in an ice-cold electroporation cuvette (see Note 9) and mix well by the top of tip.
8. Electroporate cells with parameter setting at 25 μF, 2.0 kV/mm, and 300 Ω.
9. Transfer the cells to 1 mL LB medium containing 0.5 M sorbitol and 0.38 M mannitol immediately and cultivate for 3 h.
10. Dilute the culture and spread it onto LB plates containing 20 μg/mL tetracycline.
11. Pick up colonies and confirm that transformants contain the expected plasmid by PCR using several sets of primers (see Note 10).

3.3. Extraction of Carotenoids

1. Inoculate a single colony of transformants into 50 mL TSB and cultivate at 150 rpm and 37°C for 24 h.
2. Collect cells by centrifugation, wash once with distilled water, resuspend with 1 mL distilled water, and transfer the cell suspension to a 7-mL glass tube with screw cap (see Note 11).
3. Remove water from the cell suspension by lyophilization.
4. Pour 1 mL of methanol to the glass tube containing lyophilized cells.
5. Extract carotenoids with methanol at 55°C for 20 min. Use a vortex mixer for shaking. Shake the glass tubes for 2 min, heat for 10 min, shake for 2 min, heat for 10 min, and finally shake for 2 min.
6. Set the tubes in a swing bucket rotor, and centrifuge at 3,000 × *g* for 5 min (see Note 12).
7. Filter the supernatants through HLC-Disk 3 and collect the filtrates as carotenoid extracts.

3.4. Analyses of Carotenoids

1. Set carotenoid extracts in a reverse-phase HPLC system equipped with a C18 column.
2. Elute carotenoids with acetonitrile–methanol–distilled water at a flow rate of 1.0 mL/min.
3. Monitor a line of HPLC with A_{460} with a photodiode array detector (Fig. 2a). Set the absorption spectrum range from 230 to 800 nm.
4. Identify carotenoids from their absorption spectra (Fig. 2b, c). For example, *B. subtilis* harboring pHYcrtMN produces two major carotenoids, 4,4′-diaponeurosporene (see Note 13) and 4,4′-diapolycopene (see Note 14), which are the third and last products of reactions catalyzed by CrtN, respectively (Fig. 3). *B. subtilis* harboring pHYcrtMNPQ produces two major carotenoids, one of which is considered to be glycosylated C30 carotenoic acid (retention time = 10.0 min) (Fig. 2b) (see Note 15). Another carotenoid is 4,4′-diaponeurosporene (retention time = 14.4 min) (Fig. 2c).

4. Notes

1. It is used as a host strain for C30 carotenoid production.
2. Use TSB for cultivation of *B. subtilis* to produce carotenoids.
3. Use LB medium for cultivation of *S. aureus*, *B. subtilis*, and *E. coli* to isolate recombinant strain and to prepare DNA source.

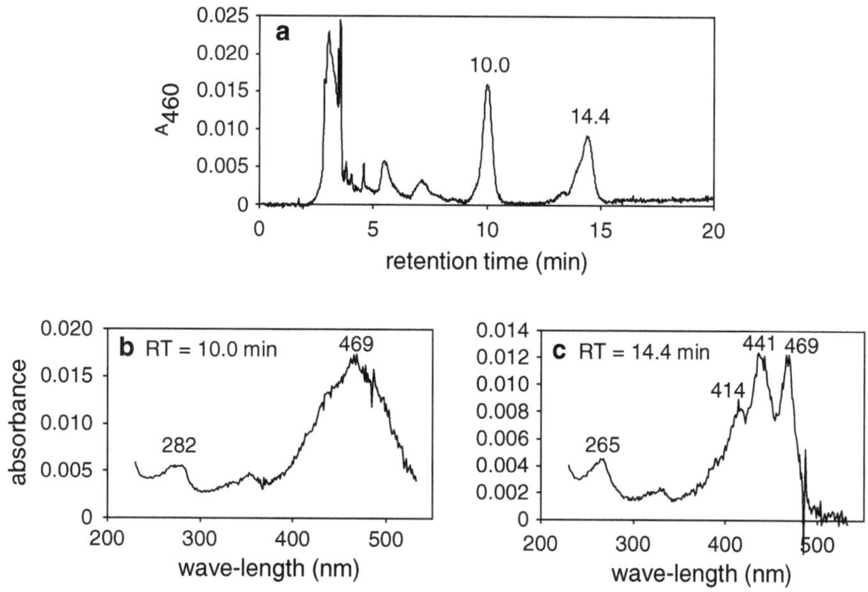

Fig. 2. Analyses of C30 carotenoids produced by *B. subtilis* harboring pHYcrtMNPQ. (**a**) Chromatogram of a cell extract. Retention time (min) is indicated above each peak. (**b**, **c**) Absorption spectra of the peaks eluted at 10.0 and 14.4 min. Wavelength of absorption maximum (nm) is indicated above each peak.

4. *B. subtilis–E. coli* shuttle vector: Construct plasmids for C30 carotenoid production based on this plasmid.
5. TA cloning vector: Use this plasmid or other TA cloning vectors for the cloning of PCR fragments amplified with *Taq* polymerase. Check nucleotide sequences of PCR-amplified fragments in TA cloning vector.
6. Use conventional spectrophotometer in combination with cuvette with a light path length of 1.0 cm.
7. This step should be repeated more than three times to wash out electrolytes from medium.
8. Cell suspension can be kept on ice within a day of preparation.
9. Stand gene pulser cuvette with 1-mm electrode gap on crashed ice.
10. Pick up a colony and wash the cells by sterilized distilled water. Finally, resuspend the cells in 50 μL sterilized distilled water and divide the suspension into two portions. Add one portion directly to PCR reaction solution and keep another portion at 4°C for inoculation.
11. Tubes must be sealed with inner septum made of silicon.
12. Adjust the speed so that the glass tubes do not break.

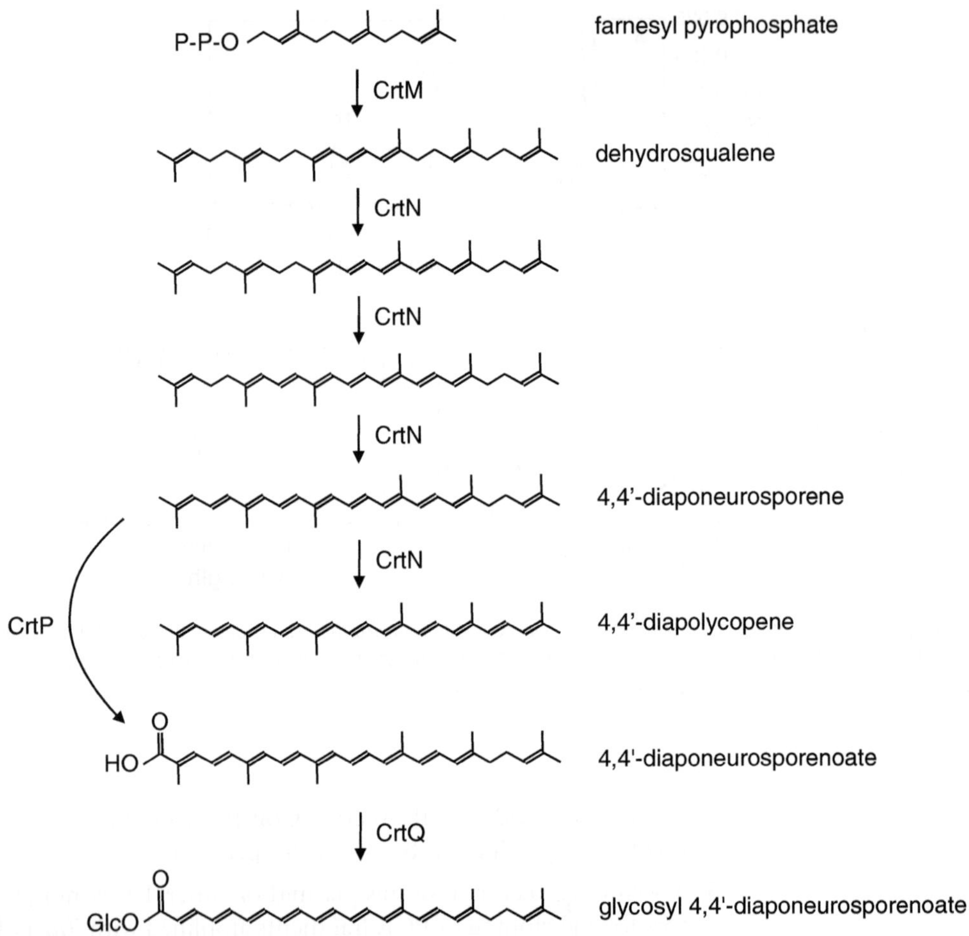

Fig. 3. Schematic representation of enzymes and molecular structures of intermediate in the staphyloxanthin biosynthetic pathway of *S. aureus*.

13. This spectrum contains absorption maxima of 265, 414, 441, and 469 nm, which are identical to those of 4,4′-diaponeurosporene (2).
14. This spectrum contains absorption maxima of 293, 443, 472, and 501 nm, which are identical to those of 4,4′-diapolycopene (10).
15. Absorption maximum in this spectrum is 469 nm, whereas absorption maximum for glycosyl 4,4′-diaponeurosporenoate has been reported to be 460 nm (4).

Acknowledgment

This work was supported by a grant-in-aid 05A22703a from the Industrial Technology Research Program in 2005 of NEDO Japan to I.M.

References

1. Marshall JH, Wilmoth GJ (1981) Pigments of *Staphylococcus aureus*, a series of triterpenoid carotenoids. J Bacteriol 147:900–913
2. Takaichi S, Inoue K, Akaike M, Kobayashi M, Oh-oka H, Madigan MT (1997) The major carotenoid in all known species of heliobacteria is the C30 carotenoid 4,4′-diaponeurosporene, not neurosporene. Arch Microbiol 168:277–281
3. Köcher S, Breitenbach J, Müller V, Sandmann G (2009) Structure, function and biosynthesis of carotenoids in the moderately halophilic bacterium *Halobacillus halophilus*. Arch Microbiol 191:95–104
4. Pelz A, Wieland KP, Putzbach K, Hentschel P, Albert K, Götz F (2005) Structure and biosynthesis of staphyloxanthin from *Staphylococcus aureus*. J Biol Chem 280:32493–32498
5. Liu CI, Liu GY, Song Y, Yin F, Hensler ME, Jeng WY, Nizet V, Wang AH, Oldfield E (2008) A cholesterol biosynthesis inhibitor blocks *Staphylococcus aureus* virulence. Science 319:1391–1394
6. Yoshida K, Ueda S, Maeda I (2009) Carotenoid production in *Bacillus subtilis* achieved by metabolic engineering. Biotechnol Lett 31:1789–1793
7. Bryantseva IA, Gorlenko VM, Kompantseva EI, Tourova TP, Kuznetsov BB, Osipov GA (2000) Alkaliphilic heliobacterium *Heliorestis baculata* sp. nov. and emended description of the genus *Heliorestis*. Arch Microbiol 174:283–291
8. Sauer U, Cameron DC, Bailey JE (1998) Metabolic capacity of *Bacillus subtilis* for the production of purine nucleosides, riboflavin, and folic acid. Biotechnol Bioeng 59: 227–238
9. Widner B, Behr R, Von Dollen S, Tang M, Heu T, Sloma A, Sternberg D, Deangelis PL, Weigel PH, Brown S (2005) Hyaluronic acid production in *Bacillus subtilis*. Appl Environ Microbiol 71:3747–3752
10. Takaichi S (2000) Characterization of carotenes in a combination of a C18 HPLC column with isocratic elution and absorption spectra with a photodiode-array detector. Photosynth Res 65:93–99

Chapter 12

Carotenoids' Production from Halophilic Bacteria

María de Lourdes Moreno, Cristina Sánchez-Porro, María Teresa García, and Encarnación Mellado

Abstract

Carotenoids have received considerable attention due to their interesting industrial applications and, more importantly, their potential beneficial effects on human health. Halophiles comprise a heterogeneous group of microorganisms that need salts for optimal growth. The pigments produced by these halophilic organisms comprise phytoene, β-carotene, lycopene, derivatives of bacterioruberin, and salinixanthin. Here, we describe the procedure to obtain salinixanthin from the extremely halophilic bacterium *Salinibacter ruber*. Additionally, we describe the expression of the β-carotene biosynthetic genes *crtE*, *crtY*, *crtI*, and *crtB* from *Pantoea agglomerans* in the moderately halophilic bacterium *Halomonas elongata* obtaining a strain able to produce practically pure β-carotene. Thus, the use of these halophilic microorganisms as a source of carotenoids constitutes an important commercial alternative in the production of carotenoids from biological sources.

Key words: Carotenoid, Salinixanthin, β-Carotene, *crt* genes, Halophiles, *Salinibacter ruber*, *Halomonas elongata*, Genetic engineering

1. Introduction

Hypersaline environments constitute typical examples of environments with extreme conditions, not only due to their high salinity but also because they may be exposed to high or low temperatures, low oxygen conditions, and in some cases high pH values. Hypersaline environments can be classified as athalassohaline and thalassohaline. Thalassohaline waters are originated by seawater condensation while the athalassohaline by evaporation of inland surface water (1). The predominant groups of microorganisms inhabiting these hypersaline environments are moderately halophilic and extremely halophilic microorganisms. The extreme halophiles are well represented among the *Halobacteriaceae*, although some extremely halophilic bacteria have been described (1). In the last years, an increased

José-Luis Barredo (ed.), *Microbial Carotenoids from Bacteria and Microalgae: Methods and Protocols*, Methods in Molecular Biology, vol. 892, DOI 10.1007/978-1-61779-879-5_12, © Springer Science+Business Media, LLC 2012

interest in the extreme halophiles has been detected, probably due to their potential biotechnological uses (2, 3).

Saltern ponds are frequently colored due to the presence of pigmented microorganisms, including *Dunaliella*, rich in β-carotene, Haloarchaea producing mainly bacterioruberin, and halophilic Bacteria, such as *Salinibacter ruber* that produces the carotenoid called salinixanthin.

The halophilic green algae accumulate high concentrations of carotenoids, such as β-carotene, astaxanthin, and canthaxanthin. Among them, the best studied is *Dunaliella salina*, a flagellate species that grows in saline media and tolerates high temperature (up to about 40°C). These extreme growth conditions facilitate the cultivation system, being an attractive candidate for commercial production of β-carotene (4, 5). The main carotenoid accumulated by *D. salina* is the β-carotene, which accounted for about 20% of the total carotenoids and it is accumulated in globules, located in the interthylacoid space within the chloroplast of the cell (6).

The family Halobacteriaceae contains exclusively halophiles that live in natural environments containing high salt concentrations (as high as 5 M NaCl) and grow optimally at 3.4–5.1 M NaCl. Most haloarchaea are brightly red-orange due to a large amount of a red membrane consisting of carotenoid pigments (7). Among the accumulated carotenoids, the most represented are C-50 carotenoids, mainly bacterioruberin and its derivatives (monoanhydrobacterioruberin and bisanhydrobacterioruberin) (8). Other carotenoids found in minor amounts in Haloarchaea are lycopersene, *cis*- and *trans*-phytofluene, and *cis*- and *trans*-phytoene (9). The function of these pigments seems to be related to a protective effect against light damage in the visible and ultraviolet range of the spectrum and a reinforcement of the cell membrane. It has been demonstrated that the red bacterioruberin-containing wild type of *Halobacterium salinarum* survives better than the colorless wild type in strong light (10). In the genus *Halobacterium*, it is described a variation of pigmentation according to the salt concentration of the medium (11).

S. ruber is a halophilic bacterium present in significant number in the communities identified in NaCl-saturated saltern crystallizer ponds, coexisting with members of the family *Halobacteriaceae*. This bacterium contains a high proportion of carotenoids in its membrane; however, the carotenoid identified (salinixanthin) is a C-40 acyl glycoside carotenoid, not related to the C-50 bacterioruberins identified in haloarchaea (12, 13).

Both haloarchaea and extremely halophilic bacteria grow under non-aseptic conditions due to the high concentrations of salts necessary for cultivation, which prevent contamination by other microorganisms. On the other hand, the process to obtain the carotenoids is simple, being extracted directly from the cells without any mechanical operation, because NaCl concentration below 15% induces cell lysis (14). Thus, the use of these halophilic microorganisms as a source of carotenoids constitutes an important commercial alternative in the production of carotenoids from biological sources.

2. Materials

1. *S. ruber* M31^T (CECT 5946) (Spanish Type Culture Collection, Valencia, Spain).
2. *Halomonas elongata* ATCC 33173^T (American Type Culture Collection, Manassas, VA, USA).
3. *Escherichia coli* DHα (15).
4. pAC-BETA-04 (16, 17) (see Note 1).
5. pBBR1MCS-2 (18) (see Note 2).
6. pRK600 (19) (see Note 3).
7. Saline medium: 195 g/L NaCl, 25 g/L $MgSO_4 \cdot 7H_2O$, 16.3 g/L $MgCl_2 \cdot 6H_2O$, 1.25 $CaCl_2 \cdot 2H_2O$, 5.0 g/L KCl, 0.25 g/L $NaHCO_3$, 0.625 g/L NaBr, and 1.0 g/L yeast extract. Adjust to pH 7.0. Autoclave at 121°C for 15 min.
8. SW30 medium: 234 g/L NaCl, 39.0 g/L $MgCl_2 \cdot 6H_2O$, 61.0 g/L $MgSO_4 \cdot 7H_2O$, 6.0 g/L KCl, 0.7 g/L NaBr, 0.2 g/L $NaHCO_3$, 1.0 g/L $CaCl_2 \cdot 2H_2O$, and 0.5% (w/v) yeast extract. Autoclave at 121°C for 15 min.
9. Luria–Bertani (LB): Add 10 g of bacto-tryptone, 5 g of yeast extract, and 10 g of NaCl into 1 L of deionized water. Autoclave at 121°C for 15 min.
10. LB-agar: LB and 20 g/L agar. Autoclave at 121°C for 15 min.
11. Orbital shaker.
12. Centrifuge Sorvall Evolution RC Superspeed (Thermo Scientific, Waltham, MA, USA).
13. SpeedVac™ concentrator (Thermo Scientific, Waltham, MA, USA).
14. Lyophilizer.
15. Water bath.
16. Heater block.
17. Buffer A: Methanol/acetone (3:7.3).
18. Buffer B: Acetone/hexane (1:1).
19. Buffer C: Acetone/hexane (2:3).
20. Buffer D: Methanol/ethylacetate/petrol (25:25:50).
21. Buffer E: Acetonitrile/methanol/chloroform (47:47:6).

22. Buffer F: Phenol:chloroform (24:24).
23. Buffer G: Methanol/dichloromethane (50:50).
24. Antibiotics: Rifampicin, kanamycin, and cloramphenicol.
25. *Eco*RI and appropriate buffer.
26. T4 DNA Ligase and appropriate buffer.
27. Spectrophotometer.
28. TLC LuxPlate® Silicagel 60 F (Merck, Darmstadt, Germany).
29. High-Performance Liquid Chromatography (HPLC) equipment.
30. Column Hypersil ODS (5 μm, 100× 4.6 mm) (Thermo Scientific, Waltham, MA, USA).
31. Carotenoids standards (Sigma, St. Louis, MO, USA).
32. Membrane filters (0.45 μm) (Millipore, Bedford, MA, USA).

3. Methods

Here, we describe two methods: (1) a method to obtain salinixanthin from an extremely halophilic bacterium (see Subheadings 3.1–3.3) and (2) a method to produce β-carotene in a recombinant moderately halophilic bacterium (see Subheadings 3.4–3.9).

3.1. Production of Salinixanthin by the Extremely Halophilic Bacterium *S. ruber* M31^T

S. ruber M31^T contains a high proportion of carotenoids in its membrane, producing colonies of similar appearance to haloarchaea (20). It has been demonstrated that the carotenoids in haloarchaea contribute to survive better in strong light in comparison to colorless wild type. The principal (>96% of total) carotenoid found in *S. ruber* M31^T is salinixanthin, a C-40 acyl glycoside carotenoid chemically related to the carotenoids found in *Rhodothermus marinus* and different from C-50 carotenoids produced by haloarchaea (21).

1. To grow *S. ruber* M31^T, inoculate homogeneously the bacterium onto round Petri dishes containing saline medium (see Note 4) (21) and incubate at 37°C during 7 days to obtain isolated colonies.
2. Inoculate liquid saline medium (2-L Erlenmeyer flasks containing 1 L of saline medium) with a single colony and incubate at 37°C on a rotary shaker (200 rpm) for 1 day.
3. Harvest the cells in the late-exponential growth phase (see Note 5) by centrifugation (30 min, 5,500 ×*g*, 4°C).
4. Dry the pellet in a SpeedVac™ concentrator.
5. When necessary, preserve samples by lyophilization and/or store in the dark at −20°C.

3.2. Extraction of Salinixanthin

General precautions for work with carotenoids need to be taken (22).

1. Mix 9.4 g of lyophilized cells from *S. ruber* with 15 mL of water to facilitate cell lysis.
2. Extract the carotenoids with 150 mL of solution A.
3. Dry the solution at reduced pressure.
4. Precipitate white contaminants with cold acetone (see Note 6).
5. Repeat this step until all water-soluble pigments are into the acetone.
6. Elute the carotenoids with solution B (23).
7. Take the solution almost to dryness (reduced pressure, room temperature).
8. Achieve preparative TLC (see Note 7) in solution C eluted with methanol. Fat-soluble pigments remain in the hexane.

3.3. Analysis of Carotenoids

A conclusive identification is a prerequisite in obtaining reliable data (see Note 8).

1. Absorption spectrum UV/visible: The carotenoid extracts are analyzed by scanning the absorbance in the wavelength region of 400–600 nm using a spectrophotometer. The total carotenoid content in the hexane extract is estimated by measuring the absorbance at 490 nm. The results are given as OD per 100 mL culture (24).
2. TLC analysis: The extracts are purified by two TLC steps on silica gel involving first 35% acetone in petrol to remove the carotenes and then solution D (25).
3. HPLC analysis: The composition of carotenoids in the resulting extracts is analyzed with HPLC equipped with diode array detector with detection wavelengths set to 450 and 480 nm using a Hypersil ODS column (5 μm, 100× 4.6 mm) and solution E with a flow rate of 1 mL/min as the mobile phase (26, 27). The standards are commercial carotenoids in ethylacetate for chromatography.

3.4. Construction of pBBR1BETA, Expressing the Gene crtEYIB and the cDNA Encoding IPP Isomerase

Engineering halophilic bacteria to produce carotenoids is a subject of great scientific and commercial interest. The aim of this method is the expression of a biosynthetic pathway in the moderately halophilic bacterium *H. elongata* ATCC 33173^T to produce the high-value product β-carotene. A gene cluster, consisting of the β-carotene biosynthetic gene crt*EYIB* from *Pantotea agglomerans*, and the cDNA encoding IPP isomerase from *Haematococcus pluvialis* (16, 17) are expressed in *H. elongata* ATCC 33173^T to produce β-carotene.

1. Digest the vector pBBR1MCS-2 (5,144 bp; Km^R) (10 μg) with *Eco*RI to prepare for directional cloning.

2. Digest the plasmid pAC-BETA-04 (~12.6 kb) with *Eco*RI to obtain a 10.1-kb fragment containing all five carotenoids' biosynthetic genes.
3. Purify the digested DNA by extraction with solution F and standard ethanol precipitation.
4. Prepare the ligation mixture (10 μL total volume): pBBR1MCS-2 digested with *Eco*RI, 10.1-kb *Eco*RI-fragment from pAC-BETA-04, 10× ligation buffer, DNA-free water, and T4 DNA ligase (see Note 9).
5. Incubate the reaction mixture overnight at 16°C or for 4 h at 20°C (see Note 10).

3.5. Expression of the Plasmid pBBR1BETA in E. coli

1. Add the transforming DNA (10 μL ligation) in 50 μL of the *E. coli* competent cells. Swirl the tubes gently several times to mix their contents. Store the tubes on ice for 30 min (see Note 11).
2. Transfer the tubes to a rack placed in preheated 42°C circulating water. Store the tubes in the rack for exactly 45 s. Do not shake the tubes (see Note 12).
3. Rapidly transfer the tubes to an ice bath. Allow the cells to cool for 1–2 min.
4. Add 800 μL of LB medium to each tube.
5. Warm the cultures to 37°C in a water bath, and then transfer the tubes to a shaking incubator set at 37°C.
6. Incubate the cultures for 45 min to allow the bacteria to recover and to express the antibiotic resistance marker encoded by the plasmid (see Note 13).
7. Transfer the appropriate volume (50, 100, 200, and 300 μL) of transformed *E. coli* competent cells onto agar LB medium containing the appropriate antibiotic.
8. Store the plates at room temperature until the liquid has been absorbed.
9. Invert the plates and incubate them at 37°C. Transformed colonies should appear in 12–16 h.
10. Check the transformants (see Note 14).
11. The resultant plasmid was designated pBBR1BETA (Fig. 1).

3.6. Expression of the β-Carotene Genes in H. elongata ATCC 33173^T

To express the β-carotene genes in *H. elongata* ATCC 33173^T, the plasmid pBBR1BETA was introduced into a spontaneous rifampicin-resistant (Rf^R) mutant of *H. elongata* ATCC 33173^T (see Note 15) by conjugal transfer using pRK600 (Cm^R) as a helper plasmid (28). Conjugal transfer of plasmid is performed by filter mating on solid media.

1. Incubate overnight the donor strain (*H. elongata*/pBBR1BETA), the recipient strain (*H. elongata* Rf^R), and the helper strain

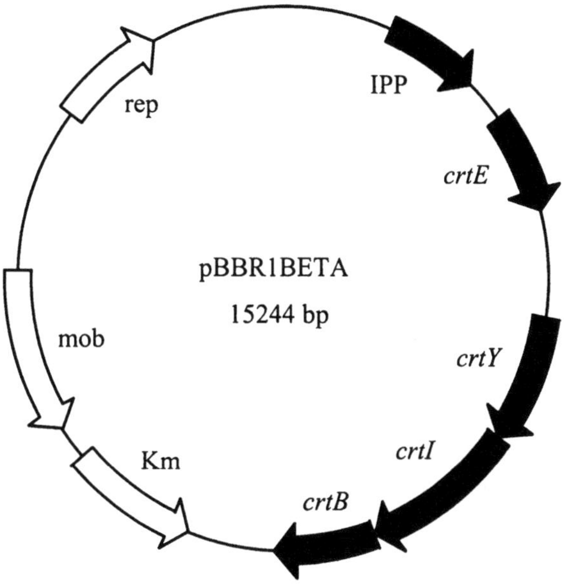

Fig. 1. Restriction map of the plasmid pBBR1BETA containing the β-carotene biosynthesis genes *crtE*, *crtY*, *crtI*, and *crtB* from *Pantoea agglomerans* and the gene encoding *Haematococcus pluvialis* IPP isomerase downstream of the strong bacterial *trc* promoter. The ORFs and the direction of transcription are shown by the large arrows. *Km* kanamycin resistance gene, *mob* mobilization gene, *rep* replication origin.

(*E. coli*/pRK600) in their respective medium and antibiotics: LB Km (50 μg/mL), SW2 Rf (25 μg/mL), and LB Cm (30 μg/mL), respectively.

2. Mix the culture of bacteria in proportion 1:1:1.
3. Centrifuge for 1 min at 13,000 × *g* to precipitate the cells.
4. Wash twice with 500 μL of SW2 medium to eliminate the rest of the antibiotics and resuspend in 150 μL of SW2.
5. The mating mixture is taking place on the surface of a sterile 0.45-μm pore-membrane filter onto a plate of SW2 medium.
6. Incubate at 37°C for 10–12 h.
7. Resuspend the cells of the filter in 1 mL of 20% (w/v) sterile glycerol and after appropriate dilutions, plate onto SW2 medium containing rifampicin (25 μg/mL), to counterselect the donor strain, and kanamycin (50 μg/mL), to select for *H. elongata* transconjugants carrying the plasmid pBBR1BETA (see Note 16).
8. Select one of the transconjugants of *H. elongata*. These colonies can be distinguished from the parental unconjugated strain by a subtle difference in the color. Colonies presumably containing β-carotene are deep orange-yellow in color.

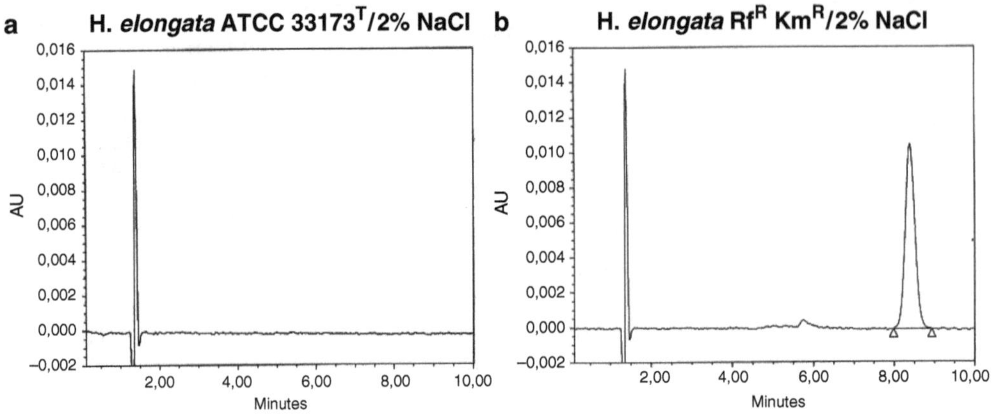

Fig. 2. HPLC profile of the carotenoids biosynthesized after flask fermentation by (**a**) *Halomonas elongata* TCC33173^T in SW with 2% (w/v) NaCl and (**b**) *H. elongata* Rf^R Km^R in SW with 2% (w/v) NaCl.

3.7. Culture of H. elongata for Optimal β-Carotene Production

Successful expression of the β-carotene biosynthetic pathway in *H. elongata* ATCC 33173^T opens the possibility of engineering halophilic bacteria for carotenoids' production. β-carotene is strongly dependent on NaCl concentration in the culture medium, with the highest production in the presence of 2% NaCl (Fig. 2).

1. Seed a Petri dish containing SW2 RfKm medium.
2. Inoculate 5 mL of SW2 RfKm medium with a single colony of the selected transconjugant *H. elongata* Rf^R Km^R.
3. Incubate the culture at 37°C.
4. Inoculate 50 mL of SW2 RfKm medium with 100 μL of *H. elongata* Rf^R Km^R.
5. Incubate overnight at 37°C under aerobic conditions with continuous shaking for 40 h.
6. Harvest the cells by centrifugation (10 min, 5,000 × *g*, 4°C).

3.8. Extraction of β-Carotene

General precautions for work with carotenoids need to be taken (22).

1. After harvesting the cells, wash the cell pellet with sterilized water, and extract it with 1 volume of solution G.
2. Dissolve the carotenoids in acetone and analyze by HPLC with a Hypersil ODS column (5 μm, 100× 4.6 mm) and buffer E as the mobile phase using a flow rate of 1 mL/min.

3.9. Analysis of β-Carotene

Perform the analysis of β-carotene following the method described in Subheading 3.3.

4. Notes

1. Plasmid pAC-BETA-04 (~12.6 kb) is constructed by the combination of plasmid pAC-BETA (10.6 kb) that includes the β-carotene biosynthetic genes *crtE, crtY, crtI*, and *crtB* (crt*EYIB)* from *P. agglomerans* (16) and a fragment ~2.0 kb containing the gene encoding *H. pluvialis* IPP isomerase downstream of the strong bacterial *trc* promoter (17).
2. Plasmid pBBR1MCS-2 (5,144 bp; Km^R) has several advantages over many of the currently available broad host range vectors: (1) it is relatively small (<5.3 kb); (2) it possesses an extended multiple cloning site (MCS); (3) it allows direct selection of recombinant plasmid molecules in *E. coli* via disruption of the LacZ alpha peptide; (4) it is mobilizable when the RK2 transfer functions are provided in *trans*; and (5) it is compatible with IncP, IncQ, and IncW group plasmids, as well as with ColE1- and P15a-based replicons (18).
3. pRK600 (Cm^R) is used as a helper plasmid (19).
4. To prepare the saline medium, use hot water and add the salt gradually. The last solution to be added must be $CaCl_2$ in order to avoid the precipitation of the solution.
5. To optimize the extraction of carotenoids, cells are grown until optical density 1.5 (late exponential phase).
6. As an alternative to precipitation in cold acetone, the white contaminants can be removed by this procedure: Add water to the raw extract dissolved in benzene to give two phases. Add methanol to precipitate the white contaminants in the hypophase. Remove the hypophase, repeat the procedure with increasing amounts of hexane in the epiphase, and stop before the carotenoid turned hypophasic.
7. Thin-Layer Chromatography (TLC) is a chromatographic technique used to separate mixtures. The process is similar to paper chromatography with the advantages of faster runs, better separations, and the choice between different stationary phases. Because of its simplicity and speed, TLC is often used for monitoring chemical reactions and for the qualitative analysis of reaction products.
8. In some studies using HPLC, the peaks of the chromatograms are designated as the signals for different carotenoids without concrete evidence of identification. This has led the world authorities establish minimum criteria for carotenoids' identification: (1) absorption spectrum UV/visible according to the suggested chromophore and (2) chromatographic properties consistent in two different systems: TLC and HPLC, including co-chromatography with authentic standard.

9. The molar ratio of plasmid vector to insert DNA fragment should be ~1:1 in the ligation reaction. The final DNA concentration should be ~10 ng/μL.
10. Depending on the commercial company could be differences, so the recommendation is to follow the indications of the manufacturer.
11. In the different experiments, it is essential to include positive controls to measure the efficiency of transformation, and negative controls to eliminate the possibility of contamination and to identify the potential causes of failure. Negative control: An aliquot of competent cells to which no DNA is added should be carried through the transformation experiment. The entire aliquot should be plated on a single agar plate containing the appropriate antibiotic used to select transformants. No bacterial colonies should grow on these plates or on a selective plate that received no bacteria. Positive control: An aliquot of competent cells should be transformed with a known amount of a standard preparation of circular superhelical plasmid DNA.
12. Heat shock is a crucial step. It is very important the cells to be raised to exactly the right temperature at the correct rate.
13. To maximize the efficiency of transformation, gently agitate (<225 cycles/min) the cells during the recovery period.
14. To check the transformant colonies, a commercial kit for plasmid DNA extraction can be used.
15. To obtain *H. elongata* spontaneous Rf^R mutants, *H. elongata* ATCC 33173^T has to be cultivated in SW2 rifampicin (25–100 μg/mL) plates and incubated at 37°C overnight. Colonies that appear in these plates correspond to *H. elongata* Rf^R mutants.
16. As a control for spontaneous mutations, plate both parental strains on the SW2 Rf Km medium.

References

1. Ventosa A (2006) Unusual micro-organisms from unusual habitats: hypersaline environments. In: Logan NA, Lappin-Scott HM, Oyston PCF (eds) Prokaryotic diversity: mechanisms and significance. Cambridge University Press, Cambridge, pp 223–253
2. Ventosa A, Nieto JJ (1995) Biotecnological applications and potentialities of halophilic microorganisms. World J Microbiol Biotecnol 11:85–94
3. Oren A (2010) Industrial and environmental applications of halophilic microorganisms. Environ Technol 31:825–834
4. Borowitzka LJ, Borowitzka MA, Moulton T (1984) Mass culture of *Dunaliella*: from laboratory to pilot plant. Hydrobiology 117:115–121
5. Hosseini TA, Shariati M (2010) *Dunaliella* biotechnology: methods and applications. J Appl Microbiol 107:14–35
6. Ben-Amotz A (1999) *Dunaliella* β-carotene. From science to commerce. In: Seckbach J (ed) Enigmatig microorganisms and life in extreme environments. Kluwer Academic, Dordrecht, The Netherlands, pp 399–410

7. Oren A, Rodríguez-Valera F (2001) The contribution of *Salinibacter* species to the red coloration of saltern crystallizar ponds. FEMS Microbiol Ecol 36:123–130
8. Oren A (2009) Saline lakes around the world: unique systems with unique values. Nat Res Environ Issues 15:246–255
9. Tindall BJ (1992) The family Halobacteriaceae. In: Balows A, Trüper HG, Dworkin M, Harder W, Schleifer KH (eds) The prokaryotes: a handbook on the biology of bacteria. Ecophysiology, isolation, identification and applications. Springer, New York, NY, pp 768–810
10. Dundas ID, Larsen H (1963) A study on the killing by light of photosensitized cells of *Halobacterium salinarium*. Arch Mikrobiol 46:19–28
11. Rodriguez-Valera F, Ventosa A, Juez G, Imhoff JF (1985) Variation of environmental features and microbial populations with salt concentrations in a multi-pond saltern. Microbiol Ecol 11:107–115
12. Anton J, Oren A, Benlloch S, Rodriguez-Valera F, Amann R, Rossello-Mora R (2002) *Salinibacter ruber* gen. nov., sp. nov., a novel, extremely halophilic member of the Bacteria from saltern crystallizer ponds. Int J Syst Evol Microbiol 52:485–491
13. Lutnaes BF, Strand A, Petursdottir SK, Liaaen-Jensen S (2004) Carotenoids of thermophilic bacteria *Rhodothermus marinus* from submarine Icelandic hot springs. Biochem Syst Ecol 32:455–468
14. Asker D, Ohta Y (2002) Production of canthaxanthin by *Haloferax alexandrinus* under non-aseptic conditions and a simple, rapid method for its extraction. Appl Microbiol Biotechnol 6:743–750
15. Hanahan D (1983) Studies on transformation of *Escherichia coli* with plasmids. J Mol Biol 166:557–580
16. Cunningham FX, Pogson B, Sun Z, McDonald KA, DellaPenna D, Gantt E (1996) Functional analysis of the beta and epsilon lycopene cyclase enzymes of *Arabidopsis* reveals a mechanism for control of cyclic carotenoid formation. Plant Cell 8:1613–1626
17. Sun Z, Gantt E, Cunningham FX (1996) Cloning and functional analysis of the beta-carotene hydroxylase of *Arabidopsis thaliana*. J Biol Chem 271:24349–24352
18. Kovach ME, Elzer PH, Hill DS, Robertson GT, Farris MA, Roop RM, Peterson KM (1995) Four new derivatives of the broad-host-range cloning vector pBBR1MCS, carrying different antibiotic-resistance cassettes. Gene 166:175–176
19. Ditta G, Stanfield S, Corbin D, Helinski DR (1980) Broad host range DNA cloning system for Gram-negative bacteria: construction of a gene bank of *Rhizobium meliloti*. Proc Natl Acad Sci U S A 77:7347–7351
20. Mongodin EF, Nelson KE, Daugherty S, DeBoy RT, Wister J, Khouri H et al (2005) The genome of *Salinibacter ruber*: Convergence and gene exchange among hyperhalophilic bacteria and archaea. Proc Natl Acad Sci 102:18147–18152
21. Lutnaes BJ, Oren A, Liaaen-Jensen S (2002) New C40-carotenoid Acyl glycoside as principal carotenoid in *Salinibacter ruber*, an extremely halophilic eubacterium. J Nat Prod 65:1340–1343
22. Schiedt K, Liaaen-Jensen S (1995) Carotenoids. In: Britton G, Liaaen-Jensen S, Pfander H (ed) Birkhaüser, Basel, p 81–107
23. Calo P, de Miguel T, Sieiro C, Velazquez JB, Villa TG (1995) Ketocarotenoids in halobacteria: 3-hydroxy-echinenone and *trans-astaxanthin*. J Appl Microbiol 79:282–285
24. Asker D, Ohta Y (1999) Production of Canthaxanthin by extremely halophilic bacteria. J Bios Bioeng 88:617–621
25. Taylor RF, Davies BH (1983) The triterpenoid carotenoids and related terpenoids in *Staphylococcus aureus* 209P. Can J Biochem Cell Biol 91:892–905
26. Rodríguez-Sáiz M, Paz B, de la Fuente JL, López-Nieto MJ, Cabri W, Barredo JL (2004) Genes for carotene biosynthesis from *Blakeslea trispora*. Appl Environ Microbiol 70:5589–5594
27. Rudakov OB, Perikova LI, Bolotov VM, Stashina GA (2004) Chromatographic determination of natural and artificial carotenoids in foodstuff. Vestn Beloruss Gos Univ Ser Khim Biol
28. Vargas C, Coronado MJ, Ventosa A, Nieto JJ (1997) Host range, stability, and compatibility of broad host-range plasmids and a shuttle vector in moderately halophilic bacteria. Evidence of intragenic and intergenic conjugation in moderate halophiles. Syst Appl Microbiol 20:173–181

Chapter 13

Construction and Utilization of Carotenoid Reporter Systems: Identification of Chromosomal Integration Sites That Support Suitable Expression of Biosynthetic Genes and Pathways

Pamela L. Sharpe and Deana J. DiCosimo

Abstract

In order to metabolically engineer microorganisms to produce compounds of interest, it is often desirable to integrate foreign genes into the chromosome of the host. However, the consequences of these genetic alterations are not always predictable. The use of a reporter system can often assist in determining chromosomal locations for optimal expression of foreign biosynthetic genes. The method described here involves the construction and utilization of promoterless carotenoid transposons, which provides a colorimetric screen for identifying the best chromosomal integration sites for the expression of the genes of interest. The transposons (pUTmTn5::392W and pUTmTn5::392) contain the carotenoid genes required for the production of canthaxanthin and astaxanthin, respectively. Thus, when promoterless transposons insert into the host's genome, the color of the colonies will vary based on their chromosomal location. There is a correlation between the color intensity of the colonies and the expression of the carotenoid transposon. The transposon insertion site can be determined via direct chromosomal sequencing. This sequence information is used to guide the site-specific integration of biosynthetic genes and pathways of interest.

Key words: Reporter systems, Promoterless transposons, Carotenoids, Canthaxanthin, Astaxanthin, Metabolic engineering

1. Introduction

Several of the key C_{40} carotenoids, such as lycopene, β-carotene, canthaxanthin, and astaxanthin, are being produced at commercial scale. These C_{40} carotenoids are being marketed to various industries, including food additives, colorants for beverages, and feed for animal and aquaculture (1). Currently, the majority of carotenoids being sold is made through chemical synthesis or is extracted from various plant sources. However, in recent years, the interest in

José-Luis Barredo (ed.), *Microbial Carotenoids from Bacteria and Microalgae: Methods and Protocols*, Methods in Molecular Biology, vol. 892, DOI 10.1007/978-1-61779-879-5_13, © Springer Science+Business Media, LLC 2012

producing these commercially important C_{40} carotenoids via metabolic engineering of microbial hosts has increased significantly (2). Examples of both bacteria, primarily *Escherichia coli* and *Methylomonas* sp., and yeast (*Yarrowia lipolytica*) have been genetically engineered to produce several different carotenoids (3, 4).

The genetic engineering of microbial systems for carotenoid production requires the stable expression of key biosynthetic genes and usually involves the chromosomal integration of pathway genes. However, not every chromosomal location supports gene expression to the same degree. Here, in this work, we describe a method for constructing a carotenoid reporter system (promoterless carotenoid transposon) that enhances the identification of locations within the host's chromosome that supports optimal expression of key biosynthetic genes and pathway (2).

2. Materials

2.1. Sources of C_{40} Carotenoid Gene Clusters

1. *crtEidiXYIBZ* gene cluster from Enterobacteriaceae bacterium DC413 (GenBank accession number DQ090836).
2. *crtW* gene from *Brevundimonas vesicularis* DC263 (GenBank accession number DQ309446).
3. *crtZ* gene from *B. vesicularis* DC263 (GenBank accession number DQ309446).

2.2. DNA Primers

1. Primer MCS.F: 5′AATTCCCGGGACTAGTACGCGTGCGG CCGCCCATGGCATATGTTCGAACCCGGGTACC3′ (see Note 1).
2. Primer MCS.R: 5′GGTACCCGGGTTCGAACATATGCC ATGGGCGGCCGCACGCGTACTAGTCCCGGGAATT3′ (see Note 1).
3. Primer <u>Kn</u>AvrI<u>I</u>Kpn<u>I</u>Bst<u>BI.R:</u> 5′-ATGCTTCGAACGGGTA CCTAGGATGGATGCGTGATCTGATCC-3′ (see Note 2).
4. Primer <u>Kn</u>Spe<u>I.F:</u> 5′-TGGCTTCGAACGATGAATTGTG TCTC-3′ (see Note 2).
5. Primer <u>392KnSeq.R</u>: 5′-GGGACGGCGGCTTTGTTGAA TAAATCG-3′ (see Note 3).
6. Primer <u>392KnSeq.F</u>: 5′-GTAAGCATCCTGTTCGCTTGCC GCATGG-3′ (see Note 3).

2.3. Strains and Plasmids for Triparental Mating

1. *E. coli* SY327λ *pir* [Δ(*lac-pro*) *arg*E(Ap) *rec*A56 *nal*A Rfr (λ *pir*) π protein for R6Kγ *ori*] (5).
2. *E. coli* DH5α [F⁻ φ80*lac*ZΔM15 Δ(*lac*ZYA-*arg*F)U169 *rec*A1 *hsd*R17(r_k^-, m_k^+) *pho*A *sup*E44 *thi*-1 *gyr*A96 *rel*A1 λ⁻] (Invitrogen, Carlsbad, CA, USA).

3. *E. coli* DH10B [F^- *mcr*A Δ(*mrr-hsd*RMS-*mcr*BC) φ80*lac*ZΔM15 Δ*lac*X74 *rec*A1 *end*A1 *ara*D139 Δ(*ara-leu*)7697 *gal*U *gal*K λ^- *rps*L *nup*G] (Invitrogen, Carlsbad, CA, USA).
4. pUTmTn5*gfptet* (2).
5. pCR2.1 (Invitrogen, Carlsbad, CA, USA).

2.4. Medium and Antibiotics for Growing Strains

1. LB (Luria–Bertani): Add 10 g of bacto-tryptone, 5 g of yeast extract, and 10 g of NaCl into 1 L of deionized water. Autoclave at 121°C for 15 min.
2. LB-agar: LB and 20 g/L agar. Autoclave at 121°C for 15 min.
3. LB/Kan^{50} agar plates (Teknova, Hollister, CA, USA).
4. LB/Cm^{25} agar plates (Teknova, Hollister, CA, USA).
5. LB/Str^{50} agar plates (Teknova, Hollister, CA, USA).
6. LB/Kan^{50}/Str^{50} (Teknova, Hollister, CA, USA).
7. LB/Amp^{100} (Teknova, Hollister, CA, USA).

2.5. Reagents, Kits, and Equipment

1. Restriction enzymes: *Xma*I, *Spe*I, *Nhe*I, *Bsp*EI, *Xho*I, *Bst*BI, *Mlu*I.
2. X-gal (5-bromo-4-chloro-3-indolyl-b-galactopyranoside) (Invitrogen, Carlsbad, CA, USA).
3. Shrimp Alkaline Phosphatase (SAP) (Affymetric USB Products, Santa Clara, CA, USA).
4. Zymoclean™ Gel DNA Recovery Kit (The Epigenetic Company, Irvine, CA, USA).
5. QIAquick Nucleotide Removal Kit (Qiagen Inc., Valencia, CA, USA).
6. QIAquick Purification Kit (Qiagen Inc., Valencia, CA, USA).
7. QIAprep Spin Miniprep Kit (Qiagen Inc., Valencia, CA, USA).
8. End It™ DNA End Repair Kit (Epicentre Biotechnologies, Madison, Wisconsin, USA).
9. Blue Juice™ Gel Loading Buffer (Invitrogen, Carlsbad, CA, USA).
10. High DNA Mass Ladder (Invitrogen, Carlsbad, CA, USA).
11. Epicentre EZ::TN<Kan-2>(Epicentre Biotechnologies, Madison, WI, USA).
12. Fast-Link™ DNA Ligation Kits (Epicentre Biotechnologies, Madison, WI, USA).
13. 0.7% TBE agarose gels.
14. 0.8% TBE agarose gels.
15. 10× TBE buffer (pH 8.3): Weight 86.4 g of Tris, 44 g boric acid, and 7.36 g of EDTA. Add water and dissolve. Bring to

800 mL with water. For 1× working solution, dilute stock solution 1:10 with sterilized water.

16. Single edge razor.
17. Gel pulser cuvettes (1 mm).
18. Falcon 2059 test tube 14-mL polypropylene round bottom (Becton Dickinson Labware, Franklin Lakes, NJ, USA).
19. Screw-capped freezer vials.
20. Ice/Ice bucket.
21. Sorvall Super T21 with ST-H750 rotor (Sorvall, Newport Pagnell, Buckinghamshire, England).
22. NanoDrop 1000 Spectrophotometer (Thermo Fisher Scientific, Waltham, MA, USA).
23. Gradient PCR machine.
24. PCR machine.
25. Gene Pulser.
26. Low UV transilluminator.
27. Incubator.
28. Shaking incubator.
29. Water bath.
30. Microcentrifuge.
31. Gel apparatus.
32. Power supply.

3. Methods

Promoter probe transposons containing various reporter genes, such as *lacZ*, *lux*AB, and *xylE* have been used to study the expression of foreign genes insert randomly into its host genome (6). To metabolically engineer *E. coli* for the production of C_{40} carotenoids, a promoterless transposable element based on the Tn5 minitransposon (7, 8) that contain the β-carotene synthetic genes *crtEYIB* (including the isoprenoid gene *idi*) and the β-modification genes *crtWZ* was constructed and used to determine the which chromosomal locations support the best expression of the carotenoid gene cluster (2, 3, 9, 10). The following method can be used to construct the promoter-probe vectors, pUTmTn5-392WKn and pUTmTn5-392Kn, which contain the *crtWEidiYIB* and *crtWZEidiYIB* gene clusters, respectively. The expression of these clusters is necessary for the production of canthaxanthin and astaxanthin.

3.1. Cloning a Multiple Cloning Site into pUTmTn5 Transposon Vector

To increase the number of restriction sites available for cloning into the pUTmTn5*gfptet* transposon vector, add a multiple cloning site (MCS) between the inside end (IE) and outside end (OE) (Fig. 1).

1. Digest pUTmTn5 with *Xma*I using the following digestion mixture: 10 μL of plasmid DNA (~50 ng/μL), 7.5 μL of 10× buffer, 0.75 μL of BSA, 3.0 μL of *Xma*I, and 53.75 μL of sterile dH_2O in a total reaction volume 75 μL.
2. Incubate the digestion mixture at 37°C in a water bath for ~2 h.
3. To dephosphorylate, add 1 U of SAP for every 1 pmol of DNA ends directly to the *Xma*I digestion reaction.
4. Incubate the SAP reaction at 37°C for 60 min.
5. Inactivate/stop the dephosphorylation reaction by heating the reaction at 65°C for 15 min.

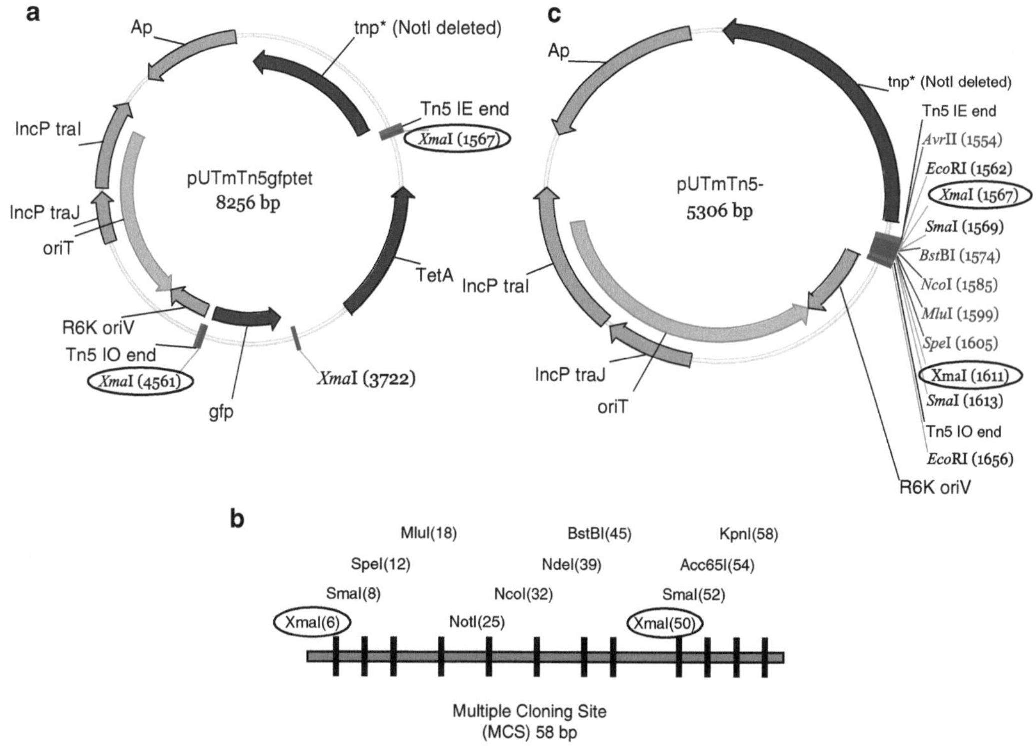

Fig. 1. Preparation of the pUTmTn5 vector backbone for the construction of the promoterless carotenoid transposon. (**a**) Digestion of pUTmTn5gfptet with *Xma*I releases the *gfp* and *tet*A genes from between the Tn5 transposon ends (Tn5 OE and Tn5 IE). (**b**) Formation of a Multiple Cloning Site (MCS) by annealing two 58-nucleotide primers. Twelve restriction sites will make-up the MCS. The restriction sites include *Xma*I, *Sma*I, *Spe*I, *Mlu*I, *Not*I, *Nco*I, *Nde*I, *Bst*BI, *Xma*I, *Sma*I, *Acc*65I, and *Kpn*I. The MCS is digested with *Xma*I. The *Xma*I sites are circled. (**c**) Ligation of the MCS with the *Xma*I-digested pUTmTn5gfptet vector DNA. The resulting vector is pUTmTn5–.

6. Separate the dephosphorylated DNA fragments on a 0.7% TBE agarose gel. As a digestion control, include a lane on the gel of the undigested pUTmTn5 vector.
7. Stain the agarose gel for about 10 min in an ethidium bromide bath.
8. Excise the digested vector fragment from the gel using a single edge razor and a low UV transilluminator to visualize the DNA fragment.
9. To purify the *Xma*I-digested pUTmTn5 using the Zymoclean™ Gel DNA Recovery Kit, add 3 volumes of Agarose Dissolving Buffer to the *Xma*I-digested pUTmTn5 DNA gel sample. Process the DNA sample using a 1.5-mL microcentrifuge tube.
10. Incubate the sample at 55°C for 10 min or until the agarose gel slice dissolves.
11. Transfer the DNA sample to a Zymo-Spin™ Column in a Collection Tube.
12. Centrifuge the spin column at 10,000 × *g* in a microcentrifuge for 30 s.
13. Discard the flowthrough.
14. Add 200 μL of Wash Buffer to the spin column.
15. Centrifuge the spin column at 10,000 × *g* in a microcentrifuge for 30 s.
16. Repeat the wash step.
17. Add 6 μL of water to the column matrix.
18. Transfer the column to a new 1.5-mL microcentrifuge tube.
19. Elute the DNA by centrifugation at 10,000 × *g* for 30 s.
20. Determine the DNA concentration using NanoDrop 1000 spectrophotometer.

To expand the restriction sites available for cloning, add a multiple cloning site (MCS) with eight additional restriction sites to pUTmTn5.

1. Adjust the primers MCS.F and MCS.R concentrations to 100 pmol/μL.
2. To anneal the primers, add 20 μL of each primer to a PCR microfuge tube.
3. Add 60 μL sterile dH_2O to the mixture.
4. Heat the primer–water mixture to 100°C for 5 min.
5. Gradually cool the mixture to 40°C over a 20-min period (after heating to 100°C, turn off the heat block).
6. Transfer the tube to ice.
7. To digest the annealed DNA primers with *Xma*I, prepare the following mixture: 20 μL of annealed DNA primers, 45 μL of

sterile dH_2O, 7.5 μL 10× buffer, 0.75 μL BSA, and 1.0 μL *Xma*I, in a total reaction volume of 75 μL.

8. Incubate the digestion reaction at 37°C in a water bath for about 3 h.
9. To purify the *Xma*I-digested MCS DNA fragment using the QIAquick Nucleotide Removal Kit, add 10 volumes of Buffer PN to 1 volume of the reaction sample.
10. Mix the solution by moving the micro-centrifuge tube back and forth.
11. Place a QIAquick spin column in the 2-mL collection tube provide in the kit.
12. Apply the sample to the QIAquick column.
13. Centrifuge the column for 60 s at 6,000×*g* in a microcentrifuge.
14. Discard the flowthrough.
15. Put the QIAquick column back into the same collection tube.
16. Wash the column using 750 μL of Buffer PE.
17. Discard the flowthrough.
18. Place the QIAquick column back into the empty collection tube.
19. Centrifuge the column for an additional 60 s at 13,000×*g* in a microcentrifuge.
20. Place the QIAquick column into a clean 1.5-mL microcentrifuge tube.
21. Add 50 μL of Buffer EB to the center of the QIAquick membrane to elute the DNA.
22. Let the column stand for 60 s.
23. Centrifuge the column for 60 s at 13,000×*g* in a microcentrifuge.
24. Determine the DNA concentration using NanoDrop 1000 spectrophotometer.
25. To ligate the MCS into the *Xma*I site of pUTmTn5, set up several ligation reactions with various insert to vector ratios. Reaction #1: 9.75 μL of H_2O, 1.5 μL of ATP, 1.5 μL of ligase buffer, 0.25 μL of vector DNA (40 ng/μL), 0.25 μL of insert DNA (10 ng/μL), and 1.0 μL of ligase in a total volume of 15.0 μL. Reaction #2: 8.75 μL of H_2O, 1.5 μL of ATP, 1.5 μL of ligase buffer, 0.25 μL of vector DNA (40 ng/μL), 2.0 μL of insert DNA (10 ng/μL), and 1.0 μL of ligase in a total volume of 15.0 μL. Reaction #3: 10.75 μL of H_2O, 1.5 μL of ATP, 1.5 μL of ligase buffer, 0.25 μL of vector DNA (40 ng/μL), 0.0 μL of insert DNA (10 ng/μL), and 1.0 μL of ligase in a total volume of 15.0 μL.

26. Incubate the ligation reactions for 5 min at room temperature.
27. Inactivate the ligation reaction by heating the mixture for 15 min at 70°C.

3.2. Preparation of E. coli SY327 Electrocompetent Cells

The pUTmTn5 vector is a conditional replicating plasmid. It has a R6K origin of replication, thus an *E. coli* host that expresses the lambda *pir* gene is required for propagation of the vector. The *E. coli* strain used in this work is strain SY327.

1. Inoculate a single colony into 6 mL of LB broth to generate the seed culture.
2. Use 3 mL of the seed culture to inoculate 200 mL of LB broth in a 500-mL shake flask.
3. Incubate the SY327 cells at 37°C with aeration.
4. Harvest cells at an OD_{600} between 0.5 and 0.7 by centrifugation at 4°C at 4,000 × *g*.
5. Discard the supernatant.
6. Resuspend the cells into 100 mL of ice-cold H_2O, centrifuge at 4,000 × *g* for 15 min, and discard the supernatant.
7. Resuspend the cells into 50 mL of ice-cold H_2O containing glycerol H_2O, centrifuge at 4,000 × *g* for 15 min, and discard the supernatant.
8. Resuspend the cells into 10 mL of ice-cold H_2O containing glycerol H_2O, centrifuge at 4,000 × *g* for 15 min, and discard the supernatant.
9. Resuspend the cells into 1 mL of ice-cold H_2O containing glycerol H_2O, centrifuge at 4,000 × *g* for 15 min, and discard the supernatant.
10. Transfer 150 μL aliquots of the cells into screw-capped freezer vials.
11. Quick-freeze the competent cells using liquid nitrogen and store at −80°C.

3.3. Transformation of the pUTmTn5 + MCS Ligation Reaction into E. coli SY327 Electrocompetent Cells

1. Thaw tubes of competent cells onto ice.
2. Place the 1-mm Gene Pulser cuvettes and the 1.5-mL microcentrifuge tubes onto ice for at least 10 min.
3. Transfer 0.5 μL and 1.0 μL of the DNA ligation reaction from Subheading 3.2 to two mircrocentrifuge tubes.
4. Add 30 μL of the SY327 cells into the tubes containing the ligation reaction. Mix gently by flicking the side of the microcentrifuge tubes. Don't generate bubbles.
5. Transfer the mixture of cells and DNA into the ice-cold cuvette.
6. Set the metrics [voltage (1.8 kV), resistance (200 ohms (Ω)) and capacitance (25 micro Faradays (μF))] of the electroporator.

7. Electroporate the cells. Press both button of the pulser at the same time for ~10 s.
8. Immediately following electroporation, add 400 μL of the SOC recovery medium into the cuvette. Mix by pipetting.
9. Transfer the cells into a 14-mL polypropylene round bottom test tube (Falcon 2059) and incubate the cells at 37°C with aeration (250 rpm) for 90 min.
10. Plate 50 μL and 100 μL the recovered cells onto LB + Amp^{100} agar plates.
11. Incubate the plates at 37°C for several days.

3.4. Evaluation of Colonies (Transformants) Having the pUTmTn5 Vector with the Correct Insert DNA Fragment

1. To isolation of plasmid DNA using the QIAprep Spin Miniprep Kit, streak or patch isolated colonies from Subheading 3.3 onto fresh LB + Amp^{100} agars plates and incubate plates at 37°C for several days.
2. Isolate the plasmid DNA for 5–10 patches (isolates) using the QIAprep Spin Miniprep Kit.
3. Use an inoculating loop to transfer and resuspend a large loop of cells into a 1.5-mL microcentrifuge tube containing 250 μL of Buffer P1. Vortex the cells well to ensure that no clumps of cells remain.
4. Add 250 μL of Buffer P2 and mix well by inverting the tubes 4–6 times. Mix by inverting the tube. Do not vortex at this step, which will result in shearing of the DNA. (Note at this step of the reaction should not exceed 5 min).
5. Add 350 μL of Buffer N3. Mix well and immediately by inverting the tubes 4–6 times.
6. Centrifuge the solution at 13,000 × *g* in the microcentrifuge for 10 min.
7. Transfer the supernatants to a QIAprep spin column.
8. Centrifuge at 13,000 × *g* for 60 s. Discard the flowthrough.
9. Add 500 μL of Buffer PE to the QIAprep spin column to wash the DNA on the column. Centrifuge the column for 60 s at 13,000 × *g*. Discard the flowthrough.
10. Centrifuge the column again for 60 s to remove any remaining wash buffer. This is an important step because any residual ethanol may interfere with subsequent enzymatic reactions.
11. Transfer the QIAprep columns to clean 1.5-mL microcentrifuge tubes.
12. Add 50 μL of Buffer EB to the center of each QIAprep spin column.
13. Let the column stand for 60 s and then centrifuge the QIAprep column for 60 s to elute the plasmid DNA.

14. To digest the plasmid DNA with *Spe*I and *Nhe*I, prepare the following digestion mixture: 5 μL of plasmid DNA, 2.0 μL of 10× buffer, 0.2 μL of BSA, 0.5 *Spe*I (20/μL), 0.5 μL of *Nhe*I (20U/ μL), and 11.8 μL sterile dH_2O to a total reaction volume of 20 μL.
15. Incubate the reaction for 1.5 h at 37°C in a water bath.
16. Separate the DNA fragments on a 0.7% TBE agarose gel. As a digestion control, include a lane on the gel of the undigested pUTmTn5 vector.
17. Stain the agarose gel for ~10 min in an ethidium bromide bath.
18. Excise the digested vector fragment from the gel using a single edge razor (use a low UV transilluminator to visualize the DNA fragment).
19. There should be two DNA fragments with a predicted size of ~4.2 kb and 1.1 kb.
20. To confirm the orientation of the MCS via DNA sequencing, use DNAsequencingprimerspUTmTn5Seq.FandpUTmTn5Seq.R to determine the DNA sequence of two or three candidate plasmids that have the correct restriction patterns.
21. Name the vector pUTmTn5+ or pUTmTn5− depending on the orientation of the MCS.

3.5. Cloning the Promoterless Carotenoid Gene Cluster from pDCQ392W and pDCQ392 into pUTmTn5 Transposon Vector

Please refer to Subheadings 3.3 and 3.4 of Chapter 7 for the methods used to assemble gene clusters pDCQ392W and pDCQ392.

1. To prepare the carotenoid plasmids (pDCQ393W and pDC392) ligation into the pUTmTn5 transposon vector, digest the plasmids with *Bsp*EI and *Spe*I (Fig. 2a).
2. To digest pDCQ392W and pDCQ392 with *Bsp*EI, set up the following digestion reaction (see Note 4). 1× Reaction: 5.0 μL of plasmid DNA (50 ng/ μL), 2.0 μL of 10× buffer, 0.2 μL of BSA, 1.0 μL *Bsp*EI (10 U/μL), and 11.8 μL of sterile dH_2O in a total reaction volume of 20.0 μL. 10× Reaction: 50.0 μL of plasmid DNA (5.0 ng/ μL), 20.0 μL of 10× buffer, 2.0 μL of BSA, 10.0 μL *Bsp*EI (10 U/μL), and 118.0 μL of sterile dH_2O in a total reaction volume of 200.0 μL.
3. Incubate the reaction at 37°C in a water bath for 2 h.
4. To purify the *Bsp*EI digested plasmids (pDCQ392W and pDCQ392), add 400 μL of Buffer PBI from the QIAquick purification kit to the 200 μL *Bsp*EI digestion reaction. Pipette to mix (see Note 5).
5. Put the QIAquick spin column into a 2-mL collection tube that is provided in the kit.
6. Apply the sample to the column and centrifuge for 60 s. The DNA will bind to the QIAquick column.

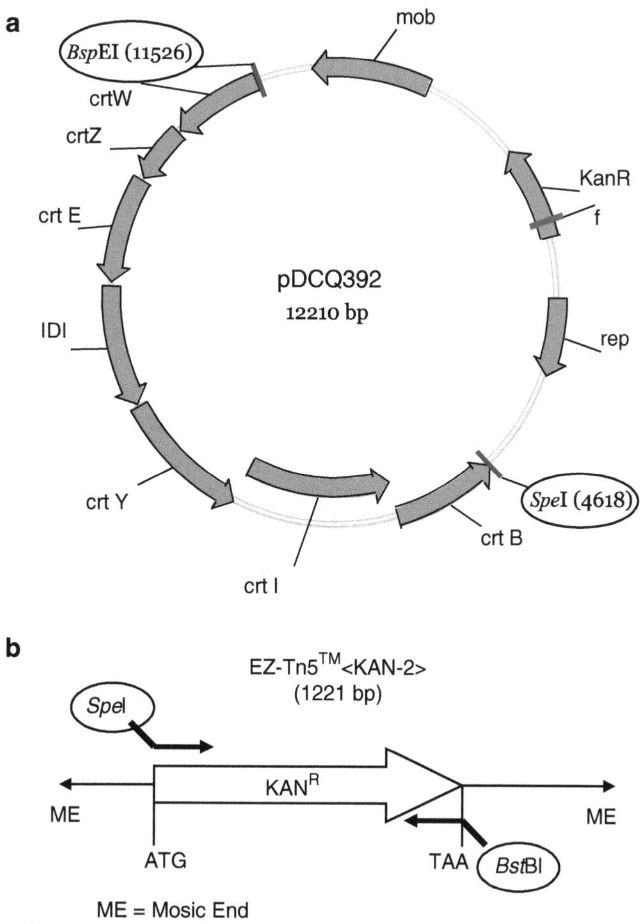

Fig. 2. Preparation of the crt gene cluster for the construction of the promoterless carotenoid transposon. (**a**) The promoterless carotenoid gene cluster is provided by the digestion of pDCQ392 with *Bsp*EI and *Spe*I. The genes present within the carotenoid gene cluster are *crt*W, *crt*Z, *crt*Y, *crt*I, and *crt*B. The isoprenoid gene, *idi*, is also present within the cluster. (**b**) The kanamycin resistance gene is provided through the PCR amplification of the kanamycin gene from the EZ-Tn5™ <KAN-2> transposon. The PCR primers used are KnBstB.R and KnSpeI.F.

7. Discard the flowthrough.
8. Place the column back into the 2-mL collection tube.
9. Add 750 μL of Buffer PE to the QIAquick column to wash the DNA.
10. Centrifuge the sample for 60 s.
11. Discard the flowthrough.
12. Place the column back into the 2-mL collection tube and centrifuge for an additional 60 s (see Note 6).
13. Place the QIAquick column into a clean 1.5-mL microcentrifuge tube.

14. Elute the DNA from column. Add 30 μL of TE buffer to the center of the QIAquick member and let the column stand for 60 s.
15. Centrifuge the column for 60 s.
16. To blunt-end the *Bsp*EI digested DNA, set-up the following DNA End Repair reaction in a 1.5-mL microcentrifuge tube: 28.0 μL of the *Bsp*EI digested DNA, 5.0 μL of the 10× End-Repair buffer, 5.0 μL of the dNTP mix, 5.0 μL of ATP, 1.0 μL of end-repair enzyme mix, and 6.0 μL of sterile dH_2O in a total reaction volume of 50.0 μL.
17. Incubate the reaction at room temperature for 45 min.
18. Stop the reaction by heating at 70°C for 10 min.
19. To purify the blunt-ended *Bsp*EI digested plasmids, add 100 μL of Buffer PBI from the QIAquick purification kit to the 50 μL blunt-ened *Bsp*EI digestion reaction. Pipette to mix.
20. Repeat steps 5 through 13.
21. To digest the blunt-ended *Bsp*EI digested pDCQ392W and pDCQ392 plasmid DNA with *Spe*I, set up the following digestion reactions: 30 μL of the blunt-ended *Bsp*EI digested DNA, 3.5 μL of 10× buffer, 0.3 μL of BSA, 1.0 μL of SpeI (10 U/μL), and 0.2 μL of sterile dH_2O in a total reaction volume of 35.0 μL.
22. Incubate the reaction at 37°C in a water bath for 2 h.
23. Stop the digestion reaction by heating at 65°C for 10 min.
24. To separate the blunt-end *Bsp*EI digested and *Spe*I digested DNA fragments of pDQ392W and pDQC392 via gel electrophoresis, place the preparative 0.7% TBE agarose gel into the gel box.
25. Fill the gel box with TBE buffer until the gel is submerged.
26. Load lambda *Hin*dIII digested DNA ladder (size marker DNA) in lane one.
27. Add 2 μL of Blue Juice™ tracking dye to each of the DNA.
28. Load the pDQC392W digested DNA sample into lanes two and three.
29. Load the pDQC392 digested DNA sample into lanes four and five.
30. Replace the gel box top, which also attaches the leads. Attach the other end of the leads to the power supply.
31. Run the gel at 100 V for 90 min. Check to make sure that the DNA is moving in the correct direction.
32. To visualize the DNA fragments on a low UV transilluminator.
33. For the pDCQ392W and the pDCQ392 samples, excise the larger DNA fragments from the gel using a single edge razor (see Note 7).

34. To purify the pDCQ392W and pDCQ392 DNA fragments using the Zymoclean™ Gel DNA Recovery Kit, add 3 volumes of Agarose Dissolving Buffer. Process the DNA sample using a 1.5-mL microcentrifuge tube.
35. Incubate the sample at 55°C for 10 min or until the agarose gel slice dissolves.
36. Transfer the DNA sample to a Zymo-Spin™ Column in a Collection Tube.
37. Centrifuge the spin column at 10,000 × *g* in a microcentrifuge for 30 s.
38. Discard the flowthrough.
39. Add 200 μL of Wash Buffer to the spin column.
40. Centrifuge the spin column at 10,000 × *g* in a microcentrifuge for 30 s.
41. Repeat the wash step.
42. Add 6 μL of water to the column matrix.
43. Transfer the column to a new 1.5-mL microcentrifuge tube.
44. Elute the DNA by centrifugation at 10,000 × *g* for 30 s.
45. Determine the DNA concentration using NanoDrop 1000 spectrophotometer. The prepared carotenoid DNA fragment from pDCQ392W and pDCQ392 will be used in a ligation reaction with the prepared pUTmTn5 vector DNA fragment (Subheadings 3.6 and 3.8).

3.6. Cloning the Kanamycin Resistance Gene from the Epicentre EZ::TN <Kan-2> into pUTmTn5 Transposon Vector

PCR amplification of the kanamycin resistance gene from EZ-TN™ <Kan-2>

1. To amplify the kanamycin gene present in the EZ::TN <KAN2> transposon, use PCR primers KnBstBI.R and KanSpeI.F (Fig. 2b).
2. Assemble the following PCR mixture for each PCR reaction: 1× Reaction: 0.5 μL of EZ-TN™ <KAN2> DNA, 45.0 μL of Supermix HF, 1.0 μL of Primer KnBstBI.R (20 μM), 1.0 μL of KnSpeI.F (20 μM), and 2.5 μL of sterile dH_2O for a total reaction volume of 50 μL.
3. Distribute 20.0 μL of the batch PCR mix into seven thin-wall PCR tubes.
4. Cover the PCR tubes with PCR caps.
5. Place the PCR tubes in to the gradient PCR machine in the following columns: Column 4 at 51.8°C, Column 6 at 54.2°C, Column 8 at 56.9°C, Column 9 at 58.4°C, Column 10 at 59.4°C, Column 11 at 59.8°C, and Column 12 at 60.1°C.
6. To check the efficiency of the PCR reaction by gel electrophoresis, load 3 μL of each PCR reaction on a 0.8% TBE agarose gel.

7. Fill the gel box with TBE buffer until the gel is submerged.
8. Load lambda *Hin*dIII digested DNA ladder (size marker DNA) in lane one.
9. Add 2 μL of Blue Juice™ tracking dye to each of the DNA .
10. Load the pDQC392W digested DNA sample into lanes two and three.
11. Load the pDQC392 digested DNA sample into lanes four and five.
12. Replace the gel box top, which also attaches the leads. Attach the other end of the leads to the power supply.
13. Run the gel at 100 V for 60 min. Check to make sure that the DNA is moving in the correct direction.
14. To visualize the PCR products, place the agarose gel on a UV transilluminator.
15. To clone the desirable PCR products into a TOPO™ vector, set-up the following TOPO reaction: 1.0 μL of fresh PCR product, 1.0 μL of dilute salt solution, 1.0 μL of TOPO vector, and 3.0 μL of sterile dH_2O to a final reaction volume of 6.0 μL.
16. Mix gently and incubate the reaction components for 5 min at room temperature.
17. Place the TOPO cloning reaction on ice.
18. Prepare for the TOPO transformation reaction. Warm the SOC medium to room temperature. Warm the LB/Kan^{50} agar plates at 37°C for 30 min. Spread 40 μL of X-gal (40 mg/mL) onto each LB/Kan^{50} agar plate and incubate the plates at 37°C until time to plate the transformation mixture. Thaw on ice one vial of One Shot cells for each transformation.
19. Add 2 μL of the TOPO Cloning reaction into one vial of the One Shot chemically competent *E. coli* cells and mix gently.
20. Incubate the transformation on ice for 15 min.
21. Heat-shock the cells by placing in a water bath heated to 42°C for 30 s. Do not shake.
22. Immediately transfer the tubes to ice.
23. Add 250 μL of the room temperature SOC medium.
24. Cap the tube tightly and shake the tube horizontally at 37°C for 60 min.
25. Plate 10 μL and 50 μL of the transformation mixture onto LB/Kan^{50}/X-gal agar plate and incubate plates overnight at 37°C.
26. Streak five white colonies for plasmid isolation onto LB/Kan^{50} agar plates and incubate the plates overnight at 37°C (see Note 8).

27. Isolate the plasmid DNA using the QIAprep Spin Miniprep kit. Follow steps 2–13 as described in Subheading 3.3.
28. To analyze the candidate plasmids, digest the DNA with *Xho*I using the following digestion mixture: 5.0 μL of miniprep DNA, 2.0 μL of 10× buffer, 0.2 μL of BSA, 1.0 μL of *Xho*I (10U/μL), and 11.8 μL for a total reaction volume of 20.0 μL.
29. Incubate the reaction for 1.5 h at 37°C in a water bath.
30. Separate the DNA fragments on a 0.8% TBE agarose gel (see Note 9).
31. Name the TOPO vector having the correct insert DNA fragment in the reverse orientation $pCR2.1Kn^R$.
32. To prepare the kanamycin resistance gene for ligation into the pUTmTn5 transposon vector,digest $pCR2.1Kn^R$ with *Bst*BI and *Spe*I in the following digestion reaction: 1× reaction: 5.0 μL of the $pCR2.1Kn^R$ (50 ng/μL), 2.0 μL of 10× buffer, 0.2 μL BSA, 1.0 μL of *Bst*BI (10 U/μL), 1.0 μL *Spe*I (10 U/μL), and sterile dH_2O for a total reaction volume of 20.0 μL.
33. Incubate the reaction at 37°C in a water bath for 2 h.
34. To separate the *Spe*I and *Bst*BI digested DNA fragments of $pCR2.1Kn^R$ via gel electrophoresis, follow steps 35–40 below.
35. Fill the gel box with TBE buffer until the gel is submerged.
36. Load lambda *Hin*dIII digested DNA ladder (size marker DNA) in lane one.
37. Add 2 μL of Blue Juice™ tracking dye to 20 μL of the *Spe*I and *Bst*BI digested $pCR2.1Kn^R$ DNA sample.
38. Load the DNA sample onto a 0.8% TBE agarose gel.
39. Replace the gel box top, which also attaches the leads. Attach the other end of the leads to the power supply.
40. Run the gel at 100 V for 60 min. Check to make sure that the DNA is moving in the correct direction.
41. To visualize the PCR products, place the agarose gel on a UV transilluminator.
42. Excise the smaller DNA fragment from the gel using a single edge razor for the $pCR2.1Kn^R$ DNA sample. The $pCR2.1Kn^R$ DNA fragment containing the kanamycin resistance genes is ~1.0 kb.
43. To purify the $pCR2.1Kn^R$ Kanamycin resistance DNA fragment, use the Zymoclean™ Gel DNA Recovery Kit
44. Add 3 volumes of Agarose Dissolving Buffer to the excised *pCR2.*$1Kn^R$ DNA gel sample. Process the DNA sample using a 1.5-mL microcentrifuge tube.
45. Incubate the sample at 55°C for 10 min or until the agarose gel slice dissolves.

46. Transfer the DNA sample to a Zymo-Spin™ Column in a Collection Tube.
47. Centrifuge the spin column at 10,000 × *g* in a microcentrifuge for 30 s.
48. Discard the flowthrough.
49. Add 200 μL of Wash Buffer to the spin column.
50. Centrifuge the spin column at 10,000 × *g* in a microcentrifuge for 30 s.
51. Repeat the wash step.
52. Add 6 μL of water to the column matrix
53. Transfer the column to a new 1.5-mL microcentrifuge tube.
54. Elute the DNA by centrifugation at 10,000 × *g* for 30 s.
55. Determine the DNA concentration using NanoDrop 1000 spectrophotometer.
56. The prepared kanamycin resistance DNA fragment from *pCR2.1KnR* will be used in a ligation reaction with the prepared pUTmTn5 vector DNA fragment (Subheading 3.7) and the prepared pDCQ392W and pDCQ392 cartenoid DNA fragments (Subheading 3.5) in Subheading 3.8.

3.7. Prepare the Transposon Vector (pUTmTn5) for Digestion with the First Restriction Endonuclease (MluI)

The digestion of pUTmTn5 with restriction endonucleases *Bst*BI and *Mlu*I will be carried out in separate reactions because the ends of the *Mlu*I digested DNA fragment will be blunt-ended prior to digestion with the second restriction enzyme.

1. To digest pUTmTn5 with *Mlu*I, set-up the following digestion reaction: 5.0 μL of pUTmTn5 (50 ng/μL), 2.0 μL of 10× buffer, 0.2 μL of BSA, 1.0 μL of *Mlu*I, and 11.8 μL of sterile dH_2O for a final reaction volume of 20.0 μL.
2. Incubate the reaction at 37°C in a water bath for 2 h.
3. To blunt-end the *Mlu*I digested DNAs, use the End It™ DNA End Repair Kit.
4. To the 28.0 μL of purified *Mlu*I digested DNA, add the following digestion components: 5.0 μL of 10× End-Repair buffer, 5.0 μL of dNTP mix, 5.0 μL of ATP, 1.0 μL of End-Repair Enzyme mix and 6.0 μL of sterile dH_2O in a total reaction volume of 50.0 μL.
5. Incubate the reaction at room temperature for 45 min.
6. Stop the reaction by heating at 70°C for 10 min.
7. To purify the blunt-ended *Mlu*I digested pUTmTn5 vector DNA, add 100 μL of Buffer PBI from the QIAquick purification kit to the 50 μL blunt-ended *Mlu*I digestion reaction. Pipette to mix.

8. Check to see if the color of the mixture remains yellow. If the mixture is orange or violet, add 10 μL of 3 M sodium acetate (pH 5) to the sample. Pipette to mix.
9. Put the QIAquick spin column into a 2-mL collection tube that is provided in the kit.
10. Apply the sample to the column and centrifuge for 60 s. The DNA will bind to the QIAquick column.
11. Discard the flowthrough.
12. Place the column back into the 2-mL collection tube.
13. Add 750 μL of Buffer PE to the QIAquick column to wash the DNA.
14. Centrifuge the sample for 60 s.
15. Discard the flowthrough.
16. Place the column back into the 2-mL collection tube and centrifuge for an additional 60 s (see Note 6).
17. Place the QIAquick column into a clean 1.5-mL microcentrifuge tube.
18. Elute the DNA from column. Add 30 μL of TE buffer to the center of the QIAquick member and let the column stand for 60 s.
19. Centrifuge the column for 60 s.
20. To digest the blunt-ended *Mlu*I digested pUTmTn5 DNA with *Bst*BI, set-up the following digestion reaction: 30 μL of the blunt-ended *Bst*I digested DNA, 3.5 μL of 10× buffer, 0.3 μL of BSA, 1.0 μL of *Bst*BI (10 U/μL), and 0.2 μL of sterile dH_2O in a 35 μL total reaction volume.
21. Incubate the reaction at 37°C in a water bath for 2 h.
22. Stop the digestion reaction by heating at 65°C for 10 min.
23. To purify the DNA using the QIAquick Nucleotide Removal Kit, add 10 volumes of Buffer PN to 1 volume of the reaction sample.
24. Mix the solution by moving the microcentrifuge tube back and forth.
25. Place a QIAquick spin column in the 2-mL collection tube provide in the kit.
26. Apply the sample to the QIAquick column.
27. Centrifuge the column for 60 s at 6,000×*g* in a microcentrifuge.
28. Discard the flowthrough.
29. Put the QIAquick column back into the same collection tube.
30. Wash the column using 750 μL of Buffer PE.
31. Discard the flowthrough.

32. Place the QIAquick column back into the empty collection tube.
33. Centrifuge the column for an additional 60 s at 13,000 × *g* in a microcentrifuge.
34. Place the QIAquick column into a clean 1.5-mL microcentrifuge tube.
35. Add 50 μL of Buffer EB to the center of the QIAquick membrane to elute the DNA.
36. Let the column stand for 60 s.
37. Centrifuge the column for 60 s at 13,000 × *g* in a microcentrifuge.
38. Determine the DNA concentration using NanoDrop 1000 spectrophotometer. The prepared pUTmTn5 vector DNA will be used in a ligation reaction with the prepared carotenoid insert DNA from pDCQ3929W and pDCQ392 (Subheadings 3.5 and 3.8).

3.8. Ligation of the Linearized Transposon Vector, pUTmTn5, the Insert DNA Fragment Containing the Kanamycin Resistance Gene from pCR2.1Kn^R and the Insert DNA Fragments Containing the Promoterless Carotenoid Genes from pDCQ392W and pDCQ392

To ligate the pUTmTn5 vector DNA fragment to either the pDCQ392W or pDCQ392 DNA fragments, each of which contains a promoterless set of carotenoid genes and the kanamycin resistance gene, use the Epicentre FastLink™ DNA Ligation Kit. The carotenoid genes present on the pDCQ392W fragment are *crt*WE*idi*YIB and when expressed from a suitable promoter produces canthaxanthin. In contrast, the genes present on the pDCQ392 fragment are *crt*WZE*idi*YIB. When expressed the *crt*WZE*idi*YIB gene cluster synthesizes of astaxanthin. The ligation of pUTmTn5 with the *crt*WE*idi*YIB gene cluster will result in the generation a promoterless canthanxanthin transposon vector that should be named pUTmTn5::392W. A promoterless astaxanthin transposon will also be constructed through the ligation pUTmTn5 with the promoterless *crt*WZE*idi*YIB gene cluster. This plasmid should be named pUTmTn5::392 (Fig. 3).

1. To ligate [pUTmTn5 + pDCQ392W + Kan] and [pUTmTn5 + pDCQ392+ Kan], set-up the following ligation reaction: 1.5 μL of vector DNA (pUTmTn5), 1.5 μL of insert DNA #1 (Kan^R), 1.5 μL of insert DNA #2 (pDCQ392W or pDCQ392), 1.5 μL of 10 mM ATP, 1.5 μL of 10× Fast-Link Ligation buffer, 1.0 μL of Fast Link DNA ligase, and 5.0 μL of sterile dH_2O in a 15.0 μL total reaction volume (see Note 10).
2. Incubate the ligation reaction for 15 min at room temperature.
3. Inactivate the Fast-Link DNA ligase by heating the reaction to 70°C for 15 min in a water bath.
4. Centrifuge the ligase reaction in microcentrifuge to remove the condensation from the lid.

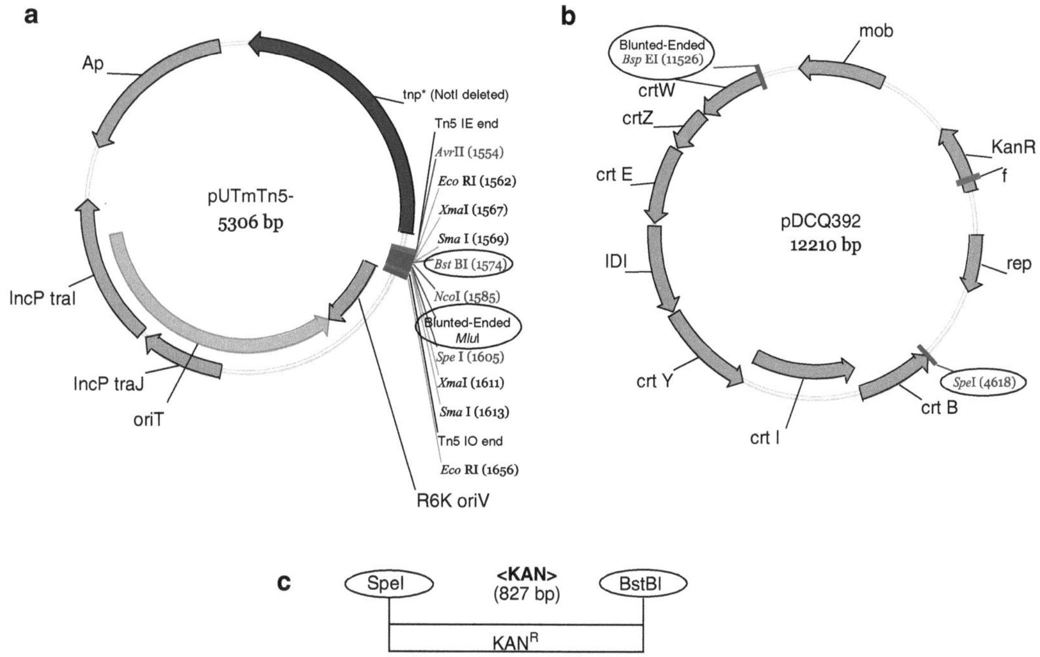

Fig. 3. Components of the ligation reaction in the formation of pUTmTn5::392. (**a**) The pUTmTn5– vector is prepared for the ligation reaction by digestion with *Bst*BI and *Mlu*I. The plasmid is digested with *Mlu*I first and blunt-ended prior to digestion with *Bst*I. (**b**) The carotenoid genes were prepared for the ligation reaction by digestion of pDCQ392 with *Spe*I and *Bsp*EI. Digestion of pDCQ392 occurs first with *Bsp*EI, which is blunt-ended prior to digestion with *Spe*I. (**c**) The kanamycin gene is prepared for the ligation reaction by the digestion of the kanamycin PCR product with *Spe*I and *Bst*BI.

5. Transform *E. coli* SY327 electrocompetent cells with 1.5 μL of the ligation reaction accord to the method outlined in Subheading 3.3.
6. Plate the transformation mixture on LB/Kan^{50} agar plates and incubate the plates at 37°C.
7. Inoculate individual colonies into 5 mL of LB/Kan^{50} broth and grow the cultures overnight with aeration (shaking at 250 rpm).
8. Harvest the cells by centrifugation at 4,000 × *g* for 10 min. Discard the supernatant.
9. To isolate the plasmid DNA using the QIAprep Spin Miniprep Kit, resuspend the pellet in 250 μL of Buffer P1. Vortex the cells well to ensure that no clumps of cells remain.
10. Add 250 μL of Buffer P2 and mix well by inverting the tubes 4–6 times. Mix by inverting the tube. Do not vortex at this step, which will result in shearing of the DNA.
11. Add 350 μL of Buffer N3. Mix well and immediately by inverting the tubes 4–6 times.
12. Centrifuge the solution at 13,000 × *g* in the microcentrifuge for 10 min.

13. Transfer the supernatants to a QIAprep spin column.
14. Centrifuge at 13,000 × *g* for 60 s. Discard the flow-though.
15. Add 500 μL of Buffer PE to the QIAprep spin column to wash the DNA on the column. Centrifuge the column for 60 s at 13,000 × *g*. Discard the flowthrough.
16. Centrifuge the column again for 60 s to remove any remaining wash buffer (see Note 6).
17. Transfer the QIAprep columns to clean 1.5 mL microcentrifuge tubes.
18. Add 50 μL of Buffer EB to the center of each QIAprep spin column.
19. Let the column stand for 60 s and then centrifuge the QIAprep column for 60 s to elute the plasmid DNA.
20. To confirm that the construction of plasmids pUTmTn5::392 W and pUTmTn5::392 are correct, digest the plasmid DNA with *Bst*BI and *Spe*I in the following digestion reaction: 1× reaction: 5.0 μL of the pCR2.1Kn^R (50 ng/μL), 2.0 μL of 10× buffer, 0.2 μL BSA, 1.0 μL of *Bst*BI (10 U/μL), 1.0 μL *Spe*I (10 U/μL), and sterile dH_2O for a total reaction volume of 20.0 μL.
21. Incubate the reaction at 37°C in a water bath for 2 h.
22. To separate the *Spe*I and *Bst*BI digested DNA fragments of pCR2.1Kn^R via gel electrophoresis, follow steps 26–29 below.
23. Fill the gel box with TBE buffer until the gel is submerged.
24. Load lambda *Hin*dIII digested DNA ladder (size marker DNA) in lane one.
25. Add 2 μL of Blue Juice™ tracking dye to 20 μL of the *Spe*I and *Bst*BI digested DNA sample.
26. Load the DNA sample onto a 0.8% TBE agarose gel.
27. Replace the gel box top, which also attaches the leads. Attach the other end of the leads to the power supply.
28. Run the gel at 100 V for 60 min. Check to make sure that the DNA is moving in the correct direction.
29. To visualize the DNA fragments, place the agarose gel on a UV transilluminator (see Note 11).

3.9. Transfer of pUTmTn5::392W and pUTmTn5::392 into E. coli Strain TOP10 via Conjugation (Triparental Mating)

The reporter plasmids (pUTmTn5::392W and pUTmTn5::392) prepared in Subheading 3.8 will be transferred from the donor *E. coli* strains into recipient *E. coli* strains containing the promoterless carotenoid genes through conjugation or specifically triparental mating. The donor plasmid, which harbors the promoterless carotenoid plasmid, has several important attributes. The donor plasmid contains the transfer functions necessary for mobilization from one strain to another. These include *tra*J, *tra*I, and *ori*T. The origin of replication (*ori*V) of the host plasmid is of the R6K type. The plasmid is able to replicate in lambda *pir* containing strains, such as SY327. However, it is incompatible with many commonly used strains. In these cases, pUTmTn5::392 W and pUTmTn5::392 will serve as a suicide vector. To follow the movement of the transposable element, the gene that confers resistance to kanamycin is positioned between the ends of transposable element. A key feature of the donor plasmid is the presence of a promoterless synthetic cluster of carotenoid genes. The carotenoid genes present on the pUTmTn5::392 W vector are *crt*B, *crt*I, *crt*Y, *idi*, *crt*E, and *crt*W. The expression of this synthetic operon allows for the production of canthaxanthin, which is an orangish pigment. The carotenoid genes present on the pUTmTn5::392 vector are *crt*B, *crt*I, *crt*Y, *idi*, *crt*E, *crt*Z, and *crt*W. The expression of this synthetic operon allow for the production of astaxanthin, which is reddish-orange in color. The intensity of the color is related to the level of expression of the carotenoid genes.

3.10. Growth of *E. coli* Donor, Helper, and Recipient Cells

1. Inoculate isolated colonies of the donor strain [*E. coli* SY327λpir (containing pUTmTn5::392 W or pUTmTn5::392)] into 5 mL of LB broth containing 50 μg/μL kanamycin in round bottom disposable test tubes.
2. Inoculate isolated colonies of the helper strain [*E. coli* DH5α (containing conjugative plasmid pRK2073)] into 5 mL of LB broth containing 25 μg/μL chloramphenicol in round bottom disposable test tubes.
3. Inoculate isolated colonies of the recipient strain [*E. coli* TOP10] into 5 mL of LB broth in round bottom disposable test tubes.
4. Grow the *E. coli* strains overnight (~12–16 h) at 30°C with aeration.
5. Harvest the cells from the *E. coli* donor and helper strains by centrifugation for 10 min (3,500 × *g*).
6. Discard the supernatant and resuspend both the *E. coli* donor and helper strains in equal volumes of LB broth (see Note 12).
7. Mix the washed *E. coli* donor/helper cells together and incubate the cells in stationary position at 30°C for ~2 h (see Note 13).
8. Harvest the *E. coli* donor and helper cells by centrifugation for 10 min (3,500 × *g*).
9. Remove as much of the liquid as possible, but do not allow the cells dry out.
10. Mix the damp *E. coli* donor/helper cell pellets with recipient cells during the triparental mating (Subheading 3.11).

3.11. Triparental Mating for E. coli

1. Grow the donor, helper, and recipient strains as stated in Subheading 3.10.
2. Resuspend the *E. coli* donor/helper cell pellets by adding 40 μL of the washed and concentrated recipient cells.
3. Mix the donor/helper/recipient cells together by moving the cells up and down in a 200 μL pipette tip.
4. Transfer the cell mixture to a single spot (the size of a nickel) onto a LB agar plates that do not contain antibiotics.
5. Incubate the LB plates containing the spots of *E. coli* cells overnight at 30°C.
6. Scrape the cells off the plate using an inoculating loop and resuspend the cells into 1.0 mL of LB broth.
7. Plate 10 μL, 50 μL and 100 μL of cells LB agar plates containing Kan^{50} μg/mL and Str^{50} μg/mL, which selects for recipient cells containing either the mTn5::392 W or the mTn5::392 transposon. Store the remainder of the cells at 4°C.
8. Incubate the plates at 30°C. Determine the appropriate amount of cells to plate on each plate (see Note 11).
9. Plate the reminder of the resuspended cell in the appropriate amounts onto fresh LB/Kan^{50}/Str^{50} agar plates.
10. Screen the colonies for those having the darkest hue (see Note 14).
11. Pick the colonies of interest and streak them twice to obtain will isolated colonies.
12. Determine the carotenoid levels produced (Subheading 3.12).

3.12. Analysis of Canthaxanthin and Astaxanthin in E. coli

To determine the levels of canthaxanthin or astaxanthin produced in strains having the darkest visual pigmentation, use the methods outlined in Subheading 3.5 of Chapter 7.

3.13. Determining the Chromosomal Location of Transposons mTn5::392 W and mTn5392 Within the Chromosome of E. coli Strain TOP10 via Direct Sequencing

1. To determine the chromosomal location of the carotenoid transposons, use the sequencing primers 392KnSeq.R and 392KnSeq.F (see Note 3) in the following DNA sequencing reaction: 3 μg/reaction of purified genomic DNA, 16 μL of BigDye v3.1 sequencing reagent, 3 μL of 10 μM Sequencing primer, 1 μL of Thermofidelase, and 12 μL of Molecular biology grade water (Fig. 4).
2. Perform sequencing reaction using the following thermal-cycling procedure: heat the sample for 3 min at 96°C; next 200 cycles [30 s at 95°C, 20 s at 55°C, and 2 min at 60°C]; store samples at 4°C.
3. To remove the unincorporated dideoxynucleoside triphosphates, use the QIAquick Nucleotide Removal kit prior to sequencing.

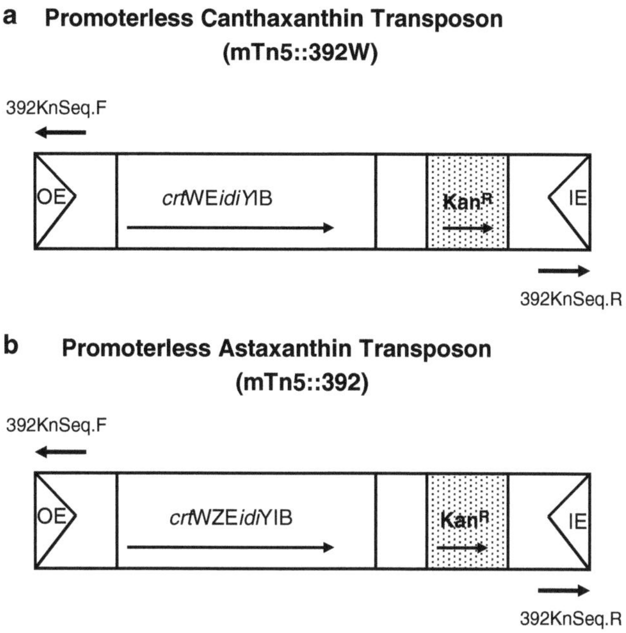

Fig. 4. Diagrams of the promoterless canthaxanthin (Tn5::392 W) and the promoterless astaxanthin (Tn5::392) transposons. (**a**) The Tn5::392 W transposon contains the carotenoid gene cluster (*crt*WE*idi*YIB) from pDCQ392W. (**b**) The Tn5::392 transposon contains the carotenoid gene cluster (*crt*WZE *idi* YIB) form pDCQ392. Both transposons have the following features. The triangles represent the mosaic ends of the Tn5 element. OE represents outside end and IE represents inside ends. Kan^R indicates the kanamycin resistance gene. The arrows at the ends of the transposons depict the DNA sequencing primers that are used to determine the location of the promoterless carotenoid transposons within the host's genome.

4. To purify the DNA sequencing reaction, add 10 volumes of Buffer PN to 1 volume of the reaction sample.
5. Mix the solution by moving the microcentrifuge tube back and forth.
6. Place a QIAquick spin column in the 2-mL collection tube provide in the kit.
7. Apply the sample to the QIAquick column.
8. Centrifuge the column for 60 s at 6,000 × *g* in a microcentrifuge.
9. Discard the flowthrough.
10. Put the QIAquick column back into the same collection tube.
11. Wash the column using 750 μL of Buffer PE.
12. Discard the flowthrough.
13. Place the QIAquick column back into the empty collection tube.
14. Centrifuge the column for an additional 60 s at 13,000 × *g* in a microcentrifuge.

15. Place the QIAquick column into a clean 1.5-mL microcentrifuge tube.
16. Add 50 μL of Buffer EB to the center of the QIAquick membrane to elute the DNA.
17. Let the column stand for 60 s.
18. Centrifuge the column for 60 s at 13,000×*g* in a microcentrifuge.
19. For each sequencing reaction mixture, transfer 40 mL to a well of a prespun 96-well cleanup plate.
20. Spin the plate for 5 min at 5,000×*g*.
21. Place the cleaned up reaction directly onto an Applied Biosystems 3700 DNA sequencer.
22. Sequence the DNA sample by automatic base calling.

4. Notes

1. Multiple Cloning Site (MCS). 58-nucleotide DNA primer and its reverse complement.
2. DNA PCR primers for amplification of the kanamycin gene.
3. DNA sequencing primers for direct chromosomal sequencing.
4. The digestion of pDCQ392W and pDCQ392 with restriction endonucleases *Bsp*EI and *Spe*I will be carried out in separate reactions because the buffering conditions needed for the two enzymes are incompatible.
5. Check to see if the color of the mixture remains yellow. If the mixture is orange or violet, add 10 μL of 3 M sodium acetate (pH 5) to the sample. Pipette to mix.
6. It is very important to remove all residual Buffer PE from the sample. The ethanol component of the buffer may interfere with subsequent reactions.
7. The pDCQ392W and the pDCQ392 DNA fragments containing the carotenoid genes should be approximately 6.4 kb and 6.9 kb, respectively.
8. Do not pick the dark blue colonies, which represent the TOPO cloning vectors that do not contain an insert DNA.
9. When the PCR fragment ligated in the reverse orientation in the TOPO vector, the expected size of the *Xho*I DNA fragments are ~4.0 kb (vector backbone) and ~0.94 kb (insert DNA).
10. Use a vector to insert molar ratio of 1:2 for each experimental reaction and set up a control reaction that does not contain

ligase (No Ligase Control) to detect the level of background colonies.

11. The expected sizes of the DNA fragments are 5.3 kb (Vector backbone), 0.8 kb (Kan gene) 6.5 kb (*crt*WE*idi*YIB genes), and 6.9 kb (*crt*WZE*idi*YIB genes).
12. This step serves to remove the antibiotics, which could interfere with the growth of the recipient cells.
13. During the liquid mating, some cells will acquire both the donor and helper plasmids.
14. Allow the plates to incubate until the colonies are pigmented. Some of the cells should appear orange or red due to the insertion of the promoterless carotenoid transposon behind a chromosomal promoter. The pigment produced is either canthaxanthin or astaxanthin.

References

1. Miura Y, Kondo K, Saito T, Shimada H, Fraser PD, Miswa N (1998) Production of the carotenoid lycopene, β-carotene, and astaxanthin in the food yeast *Candida utilis*. Appl Environ Microbiol 64:1226–1229
2. Sharpe P, DiCosimo D, Bosak M, Knoke K, Tao L, Cheng Q, Ye RW (2007) Use of transposon promoter-probe vectors in the metabolic engineering of the obligate methanotroph *Methylomonas* sp. Strain 16a for enhance C_{40} carotenoid synthesis. Appl Environ Microbiol 73:1721–1728
3. Tao L, Wilczek J, Odom J, Cheng Q (2006) Engineering a β-carotene ketolase for astaxanthin production. Metab Eng 8:523–531
4. Ye R, Yao H, Stead K, Wang T, Tao L, Cheng Q, Sharpe P, Suh W, Nagel E, Arcilla D, Dragotta D, Miller E (2007) Construction of the astaxanthin biosynthetic pathway in a methanotrophic bacterium *Methylomonas sp.* Strain 16a. J Ind Microbiol Biotechnol 34:289–299
5. Winso M, Simon S, Hill P, Sims C, Griesmayr G, Bycroft B, Williams P, Stewart G (1998) Engineering the *lux*CDABE genes from Photorhabdus luminescens to provide a bioluminescent reporter for constitutive and promoter probe plasmids and mini-Tn5 constructs. FEMS Microbiol Lett 163:193–202
6. Miller V, Mekalanos J (1988) A novel suicide vector and its use in construction of insertion mutations: osmoregulation of outer member proteins and virulence determinants in *Vibrio cholerae* requires *tox*R. J Bacteriol 170: 2575–2583
7. DeLorenzo V, Herrero M, Jakubzik U, Timmis K (1990) Mini-Tn5 transposon derivatives for insertion mutagenesis, promoter probing, and chromosomal insertion of cloned DNA in gram-negative Eubacteria. J Bacteriol 172:6568–6572
8. Herrero M, DeLorenzo V, Timmis K (1990) Transposon vectors containing nonantibiotic resistance selection markers for cloning and stable chromosomal insertion of foreign genes in gram-negative bacteria. J Bacteriol 172:6557–6567
9. Sedkova N, Tao L, Rouviere P, Cheng Q (2005) Diversity of carotenoid synthesis gene clusters from environmental Enterobacteriaceae strains. Appl Environ Microbiol 71:8141–8146
10. Tao L, Rouviere P, Cheng Q (2006) A carotenoid synthesis gene cluster from a non-marine *Brevundimonas* that synthesizes hydroxylated astaxanthin. Gene 379:101–108

Chapter 14

Directed Evolution of Carotenoid Synthases for the Production of Unnatural Carotenoids

Maiko Furubayashi and Daisuke Umeno

Abstract

Directed evolution is a well-established strategy to confer novel catalytic functions to the enzymes. Thanks to the relative ease of establishing color screening, carotenogenic enzymes can be rapidly evolved in the laboratory for novel functions. The combinatorial usages of the evolvants result in the creation of diverse set of novel, sometimes unnatural carotenoids. This chapter describes the directed evolution of diapophytoene (C_{30} carotenoid) synthase CrtM to function in the foreign C_{40} pathway, and the use of the CrtM variants thus obtained for the production of novel backbone structures.

Key words: Directed evolution, Library construction, Error-prone PCR, Color screening, Combinatorial biosynthesis

1. Introduction

Industrial value of carotenoids ranges from anti-oxidants, natural pigments, hormonal drugs, and micronutrients. Functions of these carotenoids are based on their molecular structures (molecular size, solubility, effective number of conjugated double bonds, and presence/absence of functional groups or cyclic ends), justifying the effort to construct biosynthetic pathways for various carotenoids on demand. Dozens of carotenogenic enzymes have been functionally expressed in noncarotenogenic hosts, such as *Escherichia coli*, and the combinatorial expression of these enzymes often yields the pathways for novel carotenoids (1). When the desired functions are unavailable (or do not exist in nature), natural carotenoid enzymes can be engineered to fulfill them (2). Phytoene desaturases (3, 4), cyclases (3, 5), hydroxylases (6), and carotene synthases (7) have been successfully engineered to have altered functions. These "mutant" enzymes are assembled into the

José-Luis Barredo (ed.), *Microbial Carotenoids from Bacteria and Microalgae: Methods and Protocols*, Methods in Molecular Biology, vol. 892, DOI 10.1007/978-1-61779-879-5_14,

new pathways (natural or unnatural) to the targeted carotenoids. Carotenoid biosynthetic enzymes cannot be rationally engineered due to the lack of structural information. Instead, directed evolution can be well adapted for the creation of novel carotenoid biosynthetic enzymes (2, 8).

Directed evolution is divided in two experimental steps: library creation and functional screening (Fig. 1). Starting from one or a set of parental genes, a large set of genetic variants is prepared. Thus, created "enzyme library" is then subjected to the screening to isolate the clones with novel/desired catalytic functions. Due to the pigmentation of the pathway products, high-throughput color screening can be developed to isolate the desired enzymatic functions along the path. For the enzymes, such as desaturases, cyclases, and ketolases, one can often see the emergence of new functions. This is because these enzymes directly alter the photochemical properties of carotenoids and thereby change in functions can be visualized. For evolving enzymes catalyzing the invisible steps in carotenoid synthesis, several enzymes are co-expressed to visualize their cellular functions. As a case study, this chapter describes the directed evolution of diapophytoene (C_{30} carotenoid) synthase CrtM to function in the foreign C_{40} pathway. CrtM variants thus obtained are used for the production of novel backbone structures, which can be further diversified by the addition of decoration enzymes.

Before the directed evolution experiment, screening system needs to be set up: one should construct a carotenoid pathway, including the target enzymes as its component. The pathway should be designed so that the change in the target enzyme properties results in the change in colony hue. *E. coli* has been an exclusive host for evolving carotenogenic enzymes due to the lack of its background color, rapid growth, and wealth in the genetics. In the case of screening for phytoene synthase activity, geranylgeranyl-diphosphate synthase (CrtE) is expressed to establish the precursor supply. Because the product of phytoene synthase is colorless, phytoene desaturase (CrtI) is additionally expressed to convert phytoene into red pigment, lycopene (see Note 1 and Fig. 2).

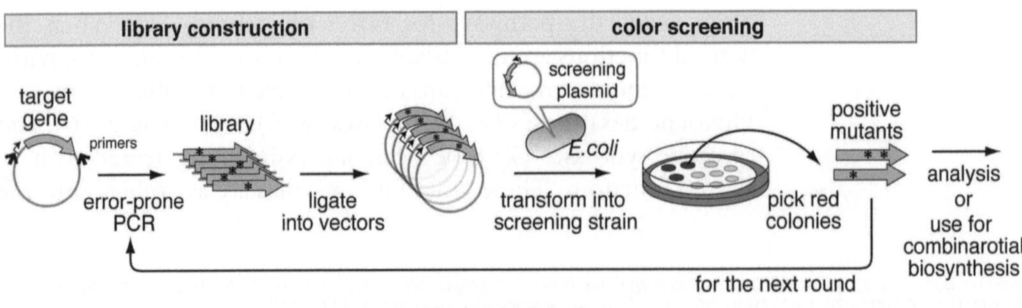

Fig. 1. General scheme of the directed evolution of carotenoid enzymes.

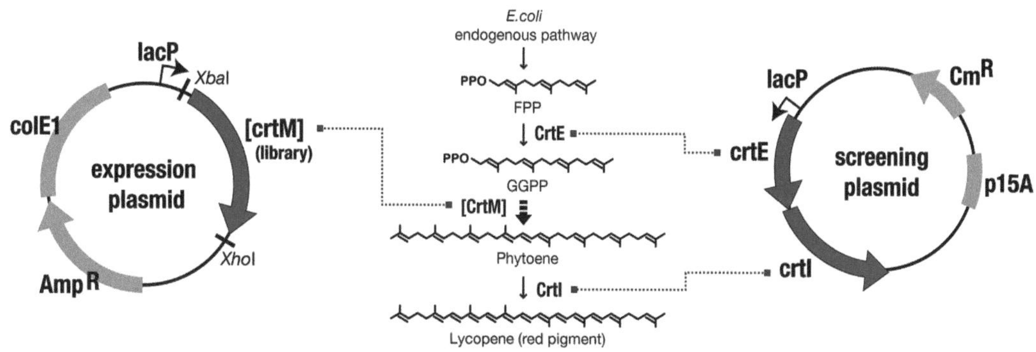

Fig. 2. Screening constructs for the CrtM mutants with phytoene synthase activity.

2. Materials

2.1. Library Creation (PCR Random Mutagenesis)

1. PCR primers forward and reverse (see Note 2).
2. pUC-crtM (see Note 3 and Fig. 2) (7).
3. *Taq* polymerase.
4. 10× *Taq* buffer.
5. 10× dNTP mixture: 2 mM each of dATP, dTTP, dCTP, dGTP. Prepare 50 μL aliquots of this mixture (to avoid excessive freeze/thaw cycles) and store at −20°C.
6. 10× $MnCl_2$ solution: 10 μM prepared in distilled water and stored at room temperature.
7. Nuclease-free water.
8. pUC-based vector (see Note 3) (7).
9. DNA clean and concentrator™—5 (Zymo Research Corporation, Orange, CA, USA).
10. Restriction enzymes *Xba*I and *Xho*I.
11. T4 DNA ligase.
12. TAE: 40 mM Tris–acetate, and 1 mM ethylene diamine tetraacetic acid (EDTA).
13. Ethidium bromide: 0.5 μg/mL in TAE.
14. Agarose gel: 0.7% LE agarose in TAE and ethidium bromide.
15. Zymoclean™ Gel DNA Recovery Kit (Zymo Research Corporation, Orange, CA, USA).

2.2. Color Screening

1. Screening plasmid (for details, see Note 4 and Fig. 2).
2. Z-Competent *E. coli* Transformation Kit (Zymo Research Corporation, Orange, CA, USA) (see Note 5).
3. *E. coli* XL1-Blue (Stratagene, La Jolla, CA, USA) (see Note 5).

4. Bio Trace NT Membrane, 82-mm discs (PALL Gelman Laboratory, Ann Arbor, MI, USA).
5. Sterilized toothpicks.
6. Carbenicillin 1,000× stock: 50 mg/mL (see Note 6).
7. Chloramphenicol 1,000× stock: 30 mg/mL.
8. LB (Luria–Bertani): Add 10 g of bacto-tryptone, 5 g of yeast extract, and 5 g of NaCl into 1 L of deionized water. Autoclave at 121°C for 15 min (9).
9. LB-agar: LB and 20 g/L agar.
10. SOC: Add 20 g of bacto-tryptone, 5 g of yeast extract, and 0.5 g of NaCl into 990 mL of deionized water. Add 10 mL of a 250 mM solution of KCl. Autoclave at 121°C for 15 min. Just before use, add 5 mL of a sterile solution of 2 M $MgCl_2$ and 20 mL of a sterile 1 M solution of glucose (9).

3. Methods

3.1. Library Creation

There are a variety of established methods for the construction of an enzyme library (10). Among them, whole gene error-prone PCR is probably the most popular method. Addition of the Mn^{2+} into the PCR sample lowers the fidelity of the polymerase whereby inserting random base substitutions into the target sequence (11).

1. For each PCR sample, add to tube: 2 fmol template DNA (see Notes 7 and 8), 5 μL 10× *Taq* buffer, 5 μL 10× dNTP mixture, 5 μL of 5 μM each primers (to be 25 pmol in final concentration), 1 μL *Taq* polymerase (5 units), 5 μL 10× $MnCl_2$ solution (to be 10–50 μM in final concentration, see Note 9), and distilled water to a final volume of 50 μL.
2. Mix the sample by pipetting, and confirm all the solution is on the bottom.
3. Place the tubes in a thermal cycler, and run the PCR program below: (1) 5 min initial denature at 94°C. (2) 30 s denature at 94°C, 30 s annealing at the temperature designed for primers, 1 min extension at 72°C. Repeat for 25 cycles. (3) 10 min at 72°C for final extension.
4. Run the product for gel electrophoresis to estimate the yield of the full-length gene (see Notes 7–9).
5. Clean the PCR product by using a Zymo DNA clean and concentrator Kit. In parallel, clean up the vector in the same way.
6. Digest both PCR product and the expression vector with appropriate enzymes (*Xba*I and *Xho*I).

7. Gel-purify the digested samples using Zymoclean™ Gel DNA Recovery Kit (see Note 10).
8. Treat the mixture of the purified insert (100 ng) and vector (100 ng) with T4 DNA ligase for 2 h (400 units/reaction to total volume of 10 μL) (see Note 11).
9. Optionally, after the reaction, concentrate the ligation product by using a Zymo DNA clean and concentrator kit (see Note 12).

3.2. Color Screening

1. To transform the ligation product into *E. coli* XL1-Blue Z-competent cells harboring the screening plasmid (see Note 5), add 5 μL of the ligation product into 100 μL competent cell on ice.
2. Incubate on ice for 10 min.
3. Add 400 μL SOC medium and shake for 1 h at 37°C.
4. Meanwhile, prepare the screening plates: mount the nitrocellulose membrane on each of 5–10 LB agar plates containing chloramphenicol and carbenicillin. Handle the nitrocellulose membrane with tweezers, not with bare hands. The nitrocellulose membrane is used to provide a white background to facilitate visualization of colony color.
5. Spread all the transformants on the screening plates (see Notes 13 and 14).
6. Incubate the plates at 37°C until the colonies become visible in size (see Note 15).
7. Leave the plates at room temperature for additional 12–36 h until the color develops (see Note 16).
8. When the color of colonies became distinguishable, determine which ones should be the positive variants (see Notes 17–19).
9. Pick all of the pink (positive) clones using sterile toothpicks and inoculate into a 2 mL LB liquid media containing chloramphenicol and carbenicillin. Shake the culture overnight at 37°C.
10. Miniprep the culture to collect the mutant plasmids.
11. At this point, plasmid contains both the mutant plasmid and the screening plasmid. To remove the latter, retransform the plasmid into *E. coli* strain (cloning strains without plasmids). Spread it on an LB agar plate containing carbenicillin. Pick the colony with no pigmentation, culture in LB liquid media, and miniprep to obtain the pure mutant plasmid.
12. For routine analysis of the mutants, determine the genotype of the mutant genes by DNA sequencing. Cellular function of the variants can be analyzed by the co-expression of various carotenogenic genes, followed by the pigment extraction and product analysis using HPLC-MS.

3.3. Toward the Novel Carotenoids

Sequence analysis of the above-obtained CrtM mutants revealed two amino acid residues (F26, W38) that control the size specificity of this enzyme (12, 13). In a different directed evolution experiment, farnesyl diphosphate synthase (FDS) from *Bacillus stearothermophilus* was also engineered into geranylfarnesyl diphosphate (C_{25}PP) synthase (14). Co-expression of this FDS mutant with size mutants of CrtMs resulted in the production of the novel phytoene-type carotenoid with C_{45} ($C_{25}+C_{20}$) and C_{50} ($C_{25}+C_{25}$) backbones (7). Further addition of carotenoid desaturases CrtI resulted in the production of diverse set of C_{45} and/or C_{50} carotenoids without protein engineering (15). Many of the downstream enzymes were shown to accept various different carotenoids. This way, structural variation of the carotenoids can be easily expanded by the addition of laboratory-evolved carotenogenic enzymes in a toolbox of the combinatorial biosynthesis.

4. Notes

1. For the convenience in mutant isolation and subsequent analysis, we prefer to place the target gene (the random library of *crtM*) on a pUC-based high copy expression vector (Fig. 2). The additional genes (*crtE* and *crtI*) are grouped into another pACYC-based vector. By physically separating the library (in expression vector) from other genes in the pathway (grouped in screening plasmid), one can significantly reduce the false positive clones by the accidental incorporation of mutations in nontarget genes during DNA handing. Note that this two-plasmid system has some drawbacks: first, one must prepare the competent cells harboring the screening plasmid by his/her own (for this purpose, we recommend Z-Competent *E. coli* Transformation Kit; see Note 5). Secondly, every time after the color screening, the screening plasmids must be removed from the miniprep of the positive colonies (see Subheading 3.2, step 11).
2. PCR primers are designed to share the same or similar melting temperatures. They are used at 5 μM and stored at −20°C. The primers for error-prone PCR are designed to anneal to the outside of the reading frame of the gene so that the entire reading frame are randomized. In this case, however, regulator sequence (promoter and ribosome binding site) is also under mutagenesis. In the situation where the mutations in the regulation sequence could lead to the false-positives, design the forward/reverse primer to anneal to the N/C terminus of the reading

frame, respectively. In this case, the priming sites of the target genes are virtually "masked" from the mutagenesis.

3. *crtM* gene is placed under *lac* promoter (see Note 20) followed by optimized ribosome binding site (AGGAGG). *crtM* library is to be ligated into the *Xba*I–*Xho*I site of the plasmid.
4. Two genes (*crtE*, *crtI*) are placed under *lac* promoter (see Note 20) followed by optimized ribosome binding site (AGGAGG).
5. Z-competent cells harboring screening plasmids. Z-Competent *E. coli* Transformation Kit is highly recommended to make competent cells of your own. Following the simple protocol, the transformation efficiency reliably exceeds 10^8 transformants/μg plasmid. With this efficiency, >10^4 colonies (variants) can be screened for single transformation using 1/5 of the ligation product.
6. The use of carbenicillin instead of ampicillin is important to prevent the formation of satellite colonies. Screening sometimes requires 2–3 days to fully develop the colony colors, and ampicillin does not stay effective that long.
7. In the given (fixed) concentration of Mn^{2+}, the average mutation rate of the library is proportional to the effective cycle numbers of the PCR. The most convenient way to control the cycle number is to change the amount of target DNA in the PCR. In theory, 10-fold decrease in template in error-prone PCR result in $\log_2 10$ ~ca. 3.3-fold elevation in mutation rate, given that final yield of PCR stay the same.
8. With a high amplification yield of error-prone PCR (>1,000-fold), one can keep the library free from contaminating the unamplified wild-type sequence.
9. Mutation rates of the library can be controlled by the concentration of Mn^{2+} in a PCR reaction. Typical concentration ranges for Mn^{2+} is from 0 to 100 μM. Higher concentration results in the higher mutation rate of the resultant library. In the high side (>200 μM), polymerase reaction is inhibited, resulting in the low amount or no product.
10. This step is important to remove the template plasmid. In addition to this size-fractionation, one can further eliminate the template plasmid by the treatment of PCR product with *Dpn*I, which selectively digests plasmid-borne methylated DNA, in prior to the gel purification.
11. Addition of 10 mM ATP in the reaction mixture would increase the ligation efficiency.
12. After purification, ligation can be stored at −20°C without losing the transforming efficiency (library size). In case where

the larger library is necessary, transform the (concentrated) ligation into commercial super-competent cells and propagate the transformant in a single mixed culture for 12 h. Miniprep this mixed culture to prepare plasmid library ready for transforming to screening strains.

13. Unlike agar surface, nitrocellulose membrane is too smooth to catch the cell well by using spreader. This way, most of the transformant are wiped out to the edge of the plate, forming the colonies only on the perimeter area of the membrane. Instead, directly spill the diluted transformant culture (1 mL/plate) on the screening plate. Make sure the media covers the entire surface of the plate. Wait until the surface gets dried and then put them into the incubator (37°C).
14. For the better color development, limit the colony density to be less than 500 colonies/plate by adjusting the culture to be spilled on each plate.
15. The time needed for colony formation depends both on the type of *E. coli* strain and the type of plasmids to be transformed. XL1-Blue cells harboring two plasmids in Fig. 2 take 20–24 h to form a colony at 37°C.
16. Two-step incubation is necessary because pigment production (color development) of C_{40} carotenoids is very low at 37°C. On the contrary, incubation/shaking at 37°C is desirable for growth and plasmid collection.
17. Count and record the number of the colonies formed on the plate. The total number of the colonies is referred to as "screening size."
18. In properly designed screening, clones of interest can be easily identified by naked eye. For the quantitative ranking and analysis of the performance of each of the entire library, consider the use of digital image analysis (16).
19. You might find nothing that appears positive. If that is the case, it could indicate (1) quality or size of the library was not high enough (failure in library creation), (2) screening was not good enough to visualize the activity (failure in screening design), or simply, (3) the enzyme was not capable of acquiring the activity by a couple of amino acid substitution (goal too ambitious).
20. In this system, no IPTG-induction is necessary; the leaky expression from *lac* promoter stably supports the expression needed for cellular pigmentation. Too much expression of carotenoid genes is sometimes toxic or even lethal. For instance, CrtE slows the cell growth, while CrtI is almost lethal upon full induction.

References

1. Lee PC, Momen AZ, Mijts BN, Schmidt-Dannert C (2003) Biosynthesis of structurally novel carotenoids in *Escherichia coli*. Chem Biol 10:453–462
2. Umeno D, Tobias AV, Arnold FH (2005) Diversifying carotenoid biosynthetic pathways by directed evolution. Microbiol Mol Biol Rev 69:51–78
3. Schmidt-Dannert C, Umeno D, Arnold FH (2000) Molecular breeding of carotenoid biosynthetic pathways. Nat Biotechnol 18: 750–753
4. Wang CW, Liao JC (2001) Alteration of product specificity of *Rhodobacter sphaeroides* phytoene desaturase by directed evolution. J Biol Chem 276:41161–41164
5. Cunningham FX Jr, Gantt E (2001) One ring or two? Determination of ring number in carotenoids by lycopene epsilon-cyclases. Proc Natl Acad Sci U S A 98:2905–2910
6. Sun Z, Gantt E, Cunningham FX Jr (1996) Cloning and functional analysis of the beta-carotene hydroxylase of *Arabidopsis thaliana*. J Biol Chem 271:24349–24352
7. Umeno D, Arnold FH (2004) Evolution of a pathway to novel long-chain carotenoids. J Bacteriol 186:1531–1536
8. Schmidt-Dannert C, Lee PC, Mijts BN (2006) Creating carotenoid diversity in *E. coli* cells using combinatorial and directed evolution strategies. Phytochem Rev 5:67–74
9. Sambrook J, Russell DW (2001) Molecular cloning: a laboratory manual, 3rd edn. Cold Spring Harbor Laboratory, Cold Spring Harbor
10. Arnold FH, Georgiou G (2003) Directed evolution library creation: methods and protocols. Humana, Totowa
11. Cirino PC, Mayer KM, Umeno D (2003) Generating mutant libraries using error-prone PCR. Methods Mol Biol 231:3–9
12. Umeno D, Tobias AV, Arnold FH (2002) Evolution of the C_{30} carotenoid synthase CrtM for function in a C_{40} pathway. J Bacteriol 184:6690–6699
13. Umeno D, Hiraga K, Arnold FH (2003) Method to protect a targeted amino acid residue during random mutagenesis. Nucleic Acids Res 31:e91
14. Ohnuma S, Nakazawa T, Hemmi H, Hallberg AM, Koyama T, Ogura K et al (1996) Conversion from farnesyl diphosphate synthase to geranylgeranyl diphosphate synthase by random chemical mutagenesis. J Biol Chem 271:10087–10095
15. Tobias AV, Arnold FH (2006) Biosynthesis of novel carotenoid families based on unnatural carbon backbones: a model for diversification of natural product pathways. Biochim Biophys Acta 1761:235–246
16. Tobias AV, Joern JM (2003) Solid-phase screening using digital image analysis. Methods Mol Biol 230:109–115

Chapter 15

A High-Throughput Screen for the Identification of Improved Catalytic Activity: β-Carotene Hydroxylase

Mark A. Scaife, Cynthia A. Ma, Phillip C. Wright, and Roberto E. Armenta

Abstract

Astaxanthin is a natural product of immense value. Its biosynthesis has been investigated extensively and typically requires the independent activity of two proteins, a β-carotene ketolase and β-carotene hydroxylase. Rational engineering of this pathway has produced limited success with respect to the biological production of astaxanthin. Random mutagenesis of the β-carotene ketolase has also been pursued. However, to date, no suitable method has been developed for the investigation of the β-carotene hydroxylase because β-carotene and zeaxanthin cannot be differentiated visually, unlike β-carotene and canthaxanthin. Thus, random mutagenesis and efficient selection of improved β-carotene hydroxylase clones is not feasible.

Presented here are the steps required for the efficient generation of a β-carotene hydroxylase random mutagenesis library in *Escherichia coli*. Subsequently presented is a novel high-throughput screening method for the rapid identification of clones with enhanced β-carotene hydroxylase activity. The validity of the presented method is confirmed by functional expression of the mutated proteins, combined with accurate quantification of produced carotenoids. The developed method has potential applications in the development of biological systems for improved carotenoid biosynthesis, as well as robust astaxanthin production.

Key words: Carotenoid, Astaxanthin, β-Carotene hydroxylase, Directed evolution, High-throughput screen

1. Introduction

Carotenoids compose a vast and diverse group of pigmented natural products that are produced ubiquitously by photosynthetic organisms. Colors vary from yellow through orange and on to red. This chromophore results from the absorption of specific wavelengths of light by the variable system of conjugated double bonds at the centre of some carotenoid compounds. Highly desaturated

José-Luis Barredo (ed.), *Microbial Carotenoids from Bacteria and Microalgae: Methods and Protocols*, Methods in Molecular Biology, vol. 892, DOI 10.1007/978-1-61779-879-5_15,

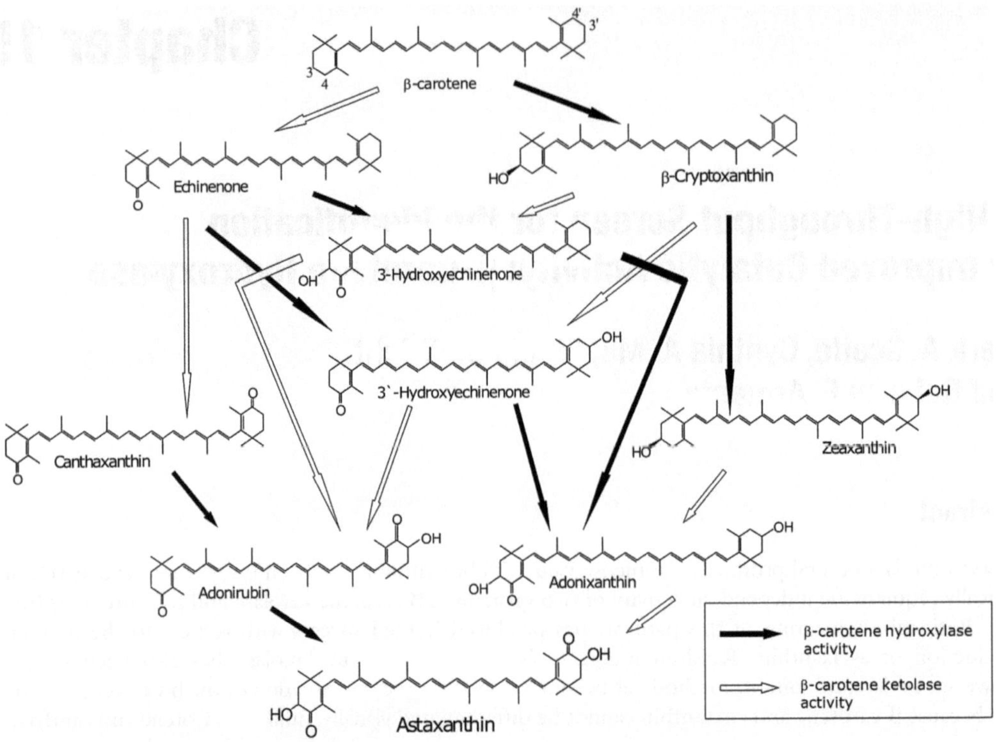

Fig. 1. Carotenoid biosynthetic pathway proceeding from β-carotene to astaxanthin.

carotenoids also function as strong antioxidants with the capacity to scavenge and quench cellular reactive oxygen species and free radicals. Their capacity as natural colorants and potent antioxidants leads to significant commercial value in nutraceutical, poultry, and aquaculture industries.

Astaxanthin is the ultimate product of carotenoid biosynthesis, being the most potent antioxidant, with the most associated health benefits (1). This results in astaxanthin claiming the highest commercial appeal and value. Due to this, astaxanthin is highly prized within the academic and industrial research arenas. However, currently no suitable technology for the efficient commercial production of astaxanthin for nutraceutical applications exists (2). This is because the final stages of astaxanthin biosynthesis require the combined, and potentially cooperative activity of two proteins, β-carotene hydroxylase and β-carotene ketolase (Fig. 1).

Previous research has demonstrated that efficient astaxanthin biosynthesis can be achieved via the application of carefully controlled expression systems (2). We formulated the hypothesis that increased β-carotene hydroxylase activity would increase astaxanthin yields from such a system. This hypothesis was based on observations made from both published reports and unpublished experimental observations. To achieve this, it was desirable to employ directed evolution methodologies. However, the use of directed evolution techniques is limited by the capacity of the research scientist to efficiently and accurately screen produced libraries. In the field of carotenoid research, visual screens to identify a desired outcome based on unique carotenoid chromospheres are common place. However, such screens are not universally applicable. For example, astaxanthin biosynthesis research, focusing on the manipulation of a β-carotene ketolase, responsible for the conversion of β-carotene to canthaxanthin, and potentially, zeaxanthin to astaxanthin, benefits from an associated color change in both reactions; yellow to orange, and potentially, yellow to orange/red. Thus, the rapid identification of desirable modifications to the target is possible. This feature has been exploited to study these reactions previously (3–5). However, visual screens are not applicable with respect to the second protein involved in the biosynthesis of astaxanthin, β-carotene hydroxylase. This problem arises because the conversion of β-carotene (yellow) to zeaxanthin (also yellow) does not produce a significant color change. Similarly, the putative conversion of canthaxanthin (orange) to astaxanthin (orange/red) would be difficult to differentiate via a visual screen.

To address this, and allow the rapid optimization of target β-carotene hydroxylase catalytic activity, we present a simple high-throughput screen for the rapid identification of β-carotene hydroxylase proteins with an enhanced capacity to synthesize zeaxanthin from β-carotene.

2. Materials

2.1. Competent Cell Preparation and Transformation

1. *E. coli* TOP10 (Invitrogen, Carlsbad, CA, USA) (see Note 1).
2. LB (Luria–Bertani) broth: add 10 g of bacto-tryptone, 5 g of yeast extract, and 10 g of sodium chloride (NaCl) into 1 L deionized water. Autoclave at 121°C for 15 min.
3. LB agar: LB broth, with the addition of 20 g/L agar.
4. Ice-cold sterile water.
5. Ice-cold sterile 10% glycerol solution.
6. 1 mm gap electroporation cuvette.
7. Carotenoid biosynthesis plasmids: pAC-Beta, pAC-Zeax, pAC-Zeax-IDI (kindly provided by Dr F.X. Cunningham, University of Maryland, MA, USA) (6, 7) (see Note 2).
8. S.O.C media: Add 20 g of tryptone, 5 g of yeast extract, 10 mM sodium chloride (NaCl), 2.5 mM potassium chloride (KCl), 10 mM magnesium chloride ($MgCl_2$), and 10 mM magnesium sulphate ($MgSO_4$) into 1 L deionized water. Autoclave at 121°C for 15 min. Cool to 60°C and add 20 mM glucose.

9. 100 mg/mL chloramphenicol (1,000× stock): to 0.1 g chloramphenicol, add 1 mL 70% ethanol.
10. 100 mg/mL ampicillin (1,000× stock): to 0.1 g ampicillin, add 1 mL sterile water, filter-sterilize via passage through a 0.22-μm syringe filter.

2.2. Cloning the Target β-Carotene Hydroxylase Gene

1. Bio-x-act short DNA polymerase system, including 10× Opti buffer, magnesium chloride ($MgCl_2$) (50 mM stock), and Bio-x-act short DNA polymerase (5U/μL) (Medicorp, Montreal, QE, Canada).
2. dNTP mix (10 mM) (Medicorp, Montreal, QE, Canada).
3. Nuclease-free water (Medicorp, Montreal, QE, Canada).
4. Eherb-CrtZ fwd (10 μM): 5′-ATGCTAGTAAATAGTTTAAT CGTCATC-3′.
5. Eherb-CrtZ rev (10 μM): 5′-TTATTCGGGCGAAGACGAC GAGG-3′.
6. Tris–acetate EDTA (TAE) buffer (50× stock): combine 242 g Tris base and 57.1 mL glacial acid in 750 mL deionized water. Add 100 mL of 0.5 M EDTA (pH 8.0), adjust final volume to 1 L. Prepare 0.5 M EDTA stock by dissolving 93.05 g EDTA in 400 mL deionized water, adjust pH to 8.0 with sodium hydroxide, top up to 500 mL final volume.
7. SYBR safe DNA gel stain (10,000×) (Invitrogen, Calrsbad, CA, USA).
8. 1% agarose gel: combine 1 g agarose (Medicorp, Montreal, QE, Canada) with 100 mL 1× TAE buffer, heat until agarose is completely melted.
9. 10× DNA loading buffer (Medicorp, Montreal, QE, Canada).
10. DNA molecular weight marker (Medicorp, Montreal, QE, Canada).
11. SureClean reagent (Medicorp, Montreal, QE, Canada).
12. pTrcHis2-TOPO TA cloning vector (Invitrogen, Carlsbad, CA, USA) (see Note 3).
13. Salt solution: 300 mM sodium chloride (NaCl), 15 mM magnesium chloride ($MgCl_2$) (Invitrogen, Carlsbad, CA, USA).
14. Plasmid miniprep kit (Zymo Research, Orange, CA, USA) (see Note 4).

2.3. Confirmation of pTrc-ZEH Function: Coexpression in *Escherichia coli*

1. Lyophilization equipment.
2. Carotenoid extraction solvent: methanol (HPLC grade).
3. Sonication water bath, chilled by addition of ice.
4. Argon gas.
5. Mobile phase solvents: methanol, ethyl acetate, water (HPLC grade).

6. HPLC system.
7. HPLC column: Luna C_{18} reverse phase (Phenomenex, Torrance, CA, USA).
8. C_{18} guard column (Phenomenex, Torrance, CA, USA).
9. Carotenoid standards: β-carotene and zeaxanthin (Sigma-Aldrich, Oakville, ON, Canada).

2.4. Reduced Target Gene Expression: Site-Directed Mutagenesis

1. Bio-x-act long DNA polymerase system, including 10× Opti buffer, magnesium chloride ($MgCl_2$) (50 mM stock), and Bio-x-act long DNA polymerase (5U/μL) (Medicorp, Montreal, QE, Canada) (see Note 5).
2. *Dpn*I restriction endonuclease and associated reaction buffer (New England Biolabs, Pickering, ON, Canada).
3. Commercial electrocompetent cells (transformation efficiency $\geq 1 \times 10^9$).
4. UV/Vis spectrophotometer compatible for 96-well plate.
5. MEP Fwd (10 μM): 5′-CTGAAATGAG<u>C</u>TGCTGACAATTA ATCATCCG-3′.
6. EP Rev (10 μM): 5′-GCGCTATTCAGATCCTCTTCTGA GATGAG-3′.

2.5. Directed Evolution of Target Gene: Error-Prone PCR

1. *Taq* DNA polymerase (New England Biolabs, Pickering, ON, Canada) (see Note 6).
2. Manganese chloride (1 M stock).

3. Methods

This method is targeted towards the identification of improved β-carotene hydroxylase genes within a biological system that uses these genes to catalyze the conversion of β-carotene to zeaxanthin. The described method may be applicable to other reactions within the carotenoid biosynthetic pathway. However, the application of this technique in such a manner would require prior validation. For this reason, we present details pertaining to the development and validation of this method.

Carotenoid pigmentation is the result of the absorption of specific wavelengths of light by the carotenoid molecule. This absorption is predominantly in the blue (450–495 nm) region of the spectrum; this concept is central to this method.

3.1. Competent Cell Preparation and Transformation

1. Grow an overnight culture of *E. coli* TOP10 at 37°C in LB.
2. Transfer 500 μL of overnight culture to 50 mL fresh LB media, within a 500-mL conical flask. Incubate at 37°C and 200 rpm, to OD_{600} 0.6.

3. Transfer the cells to 50-mL centrifuge tube and incubate on ice for 15 min. Complete all subsequent steps with minimal exposure to sources of heat.
4. Harvest the cells via centrifugation, 4,000 × *g*, at 4°C for 8 min.
5. Wash cell pellet three times in an equal volume ice-cold sterile water.
6. Wash the cells in an equal volume of ice-cold 10% glycerol.
7. Resuspend the cells in 200 μL ice-cold 10% glycerol and employ them immediately in transformation.
8. Transfer 60 μL electrocompetent cells to chilled electroporation cuvette, using a chilled pipette tip.
9. Add 100 ng plasmid DNA and electroporate at 2.2 kV.
10. Immediately add 700 μL room temperature S.O.C media and transfer to sterile 15-mL falcon tube, and incubate at 37°C, 250 rpm for 1 h.
11. Select for transformed cells on LB agar, supplemented with chloramphenicol (100 μg/mL), incubate at 37°C overnight.

3.2. Cloning of the Target β-Carotene Hydroxylase Gene

1. Amplify the β-carotene hydroxylase gene from *Pantoea agglomerans* (previously *Erwinia herbicola*) via PCR, using plasmid pAC-Zeax as the template DNA (7). Combine 5 μL 10× Opti buffer, 5 μL magnesium chloride, 1 μL dNTP mix, 2 μL Eherb-CrtZ fwd and Eherb-CrtZ rev, and 0.5 μL Bio-x-act short DNA polymerase with nuclease-free water in a 50 μL final volume.
2. Achieve amplification as follows: 94°C for 3 min, 30 cycles of 94°C for 30 s, 50°C for 30 s and 68°C for 30 s, followed by a final incubation at 68°C for 10 min.
3. Confirm successful isolation and amplification of target via agarose gel electrophoresis. Prepare a 1% agarose gel, cool agarose to 60°C, and add 5 μL SYBR safe DNA gel stain. Pour into assembled electrophoresis equipment and allow to solidify.
4. Flood electrophoresis chamber with 1× TAE buffer, combine 3 μL PCR product with 0.5 μL 10× loading buffer, and load gel.
5. Run electrophoresis and visualize the PCR product with a gel electrophoresis UV-transilluminator.
6. Purify the reminder of the PCR via precipitation. To the PCR sample, add an equal volume of SureClean reagent, incubate at room temperature for 10 min, and pellet DNA via centrifugation at 13,000 × *g*, for 10 min. Wash the precipitated DNA pellet in two volumes of 70% ethanol and resuspend in 10 μL nuclease-free water. Adjust the DNA concentration to 100 ng/μL.

7. Clone the purified PCR product into vector pTrcHis2-TOPO, to achieve this, combine 4 μL PCR product, 1 μL vector pTrcHis2-TOPO, and 1 μL salt solution and incubate at room temperature for 5–20 min.
8. Complete transformation as described in Subheading 3.1, except that 2 μL of the cloning reaction is employed.
9. Following transformation, perform selection at 37°C overnight, on LB agar plates supplemented with ampicillin (100 μg/mL) (see Note 7).
10. Select six colonies from the LB agar plate and use them to inoculate 3 mL of LB broth, supplemented with ampicillin (100 μg/mL), and incubate at 37°C, 200 rpm for 16 h.
11. Pellet biomass via centrifugation and extract plasmid DNA with a commercial kit (see Note 4).
12. Sequencing confirmed fidelity and orientation of the β-carotene hydroxylase gene, thereby confirming in the creation of pTrc-ZEH.

3.3. Confirmation of pTrc-ZEH Function: Coexpression in E. coli

1. Make *E. coli* TOP10 electrocompetent, as described previously and transform with plasmids pAC-Beta and pTrc-ZEH.
2. Select on LB agar plates, supplemented with ampicillin (100 μg/mL) and chloramphenicol (100 μg/mL), to ensure retention of both plasmids.
3. Pick three single colonies from the resultant plates and use them to inoculate 3 mL LB broth, supplemented with ampicillin (100 μg/mL) and chloramphenicol (100 μg/mL).
4. Achieve functional characterization via the transfer of 500 μL of the inoculum culture to 50 mL LB broth, similarly supplemented with ampicillin and chloramphenicol, within a 500-mL conical flask. Incubate the expression culture at 37°C to an OD_{600} 0.6, and subsequently 30°C for 24 h.
5. Transfer cultures to 50-mL centrifuge tubes and harvest biomass via centrifugation at 4,000 × *g*, 4°C for 10 min. Decant supernatant to waste and lyophilize cell pellets overnight.
6. Extract carotenoids from dry biomass via addition of 5 mL ice-cold methanol to the cell pellet, mechanical homogenization of the cell pellet, and sonication in a chilled water bath for 10 min (see Note 8).
7. Remove cell debris from the extract via centrifugation at 4,000 × *g*, at 4°C for 20 min.
8. Transfer supernatant to a 30-mL amber glass vial and evaporate the extraction solvent under a stream of argon.
9. Resuspend carotenoid extracts in 500 μL ice-cold methanol, transfer to 1.7-mL centrifuge tubes, centrifuge at 13,000 × *g* for 5 min at room temperature, and submit for HPLC analysis.

10. Employ 10–50 μL of carotenoid extract for analysis by HPLC.
11. Achieve separation with a Luna C_{18} reverse-phase column (4.6×250 mm, 5 μm) in combination with a C_{18} guard column. The employed mobile phase progressed from buffer A (methanol–ethyl acetate–water (88:10:2)) at 0.75 mL/min to buffer B (methanol–ethyl acetate–water (50:48:2)) at 1.5 mL/min over 30 min (see Note 9). Figure 2 presents a comparison of the resultant carotenoid profile.
12. Achieve carotenoid identification and quantification via retention time comparison to commercial standards, a standard curve generated with said standards allowed accurate quantification.

3.4. Reduced Target Gene Expression: Site-Directed Mutagenesis

1. Achieve reduced expression of the β-carotene hydroxylase gene, from plasmid pTrc-ZEH, by the introduction of a targeted mutation. Specifically, convert the first base of the pTrc-ZEH encoded −35 promoter sequence from a thymine residue (TTGACA) to cytosine (CTGACA) (4). Complete site-directed mutagenesis as follows: oligonucleotide primers designed to anneal to the vector pTrc-ZEH at positions 180–210 (MEP Fwd), and 454–483 (EP Rev) (also see Note 10). Complete PCR as described in Subheading 3.2. With an initial denaturing temperature of 94°C for 3 min, followed by 30 cycles of 94°C, 30 s, 60°C for 30 s, 68°C for 30 s.
2. Following a final incubation at 68°C for 5 min, add 5.5 μL 10× NEB buffer 4 and 1 μL *Dpn*I endonuclease to the reaction. Proceed with *Dpn*I digestion at 37°C for 4 h, removing the template DNA (pTrc-ZEH).
3. Purify the resultant sample via precipitation with SureClean reagent as described in Subheading 3.2.
4. Employ 1 μL of the resuspended DNA sample to transform commercially available electrocompetent *E. coli* strain TOP10. Perform selection on LB agar plates, supplemented with ampicillin (100 μg/mL).
5. Isolate plasmid DNA from resultant clones and sequence to confirm fidelity and the presence of the desired mutation.
6. Compare the plasmid SDM-Trc-ZEH, to the parental plasmid, pTrc-ZEH, in order to confirm the impact of the site-directed mutation on protein expression, by way of analysis of the resultant carotenoid profile.
7. Prepare electrocompetent *E. coli* strain TOP10 cells and transform with the resultant plasmid and pAC-Beta, as described in Subheading 3.1.
8. Culture modified *E. coli* strains under conditions conducive to carotenoid biosynthesis, described in Subheading 3.3.

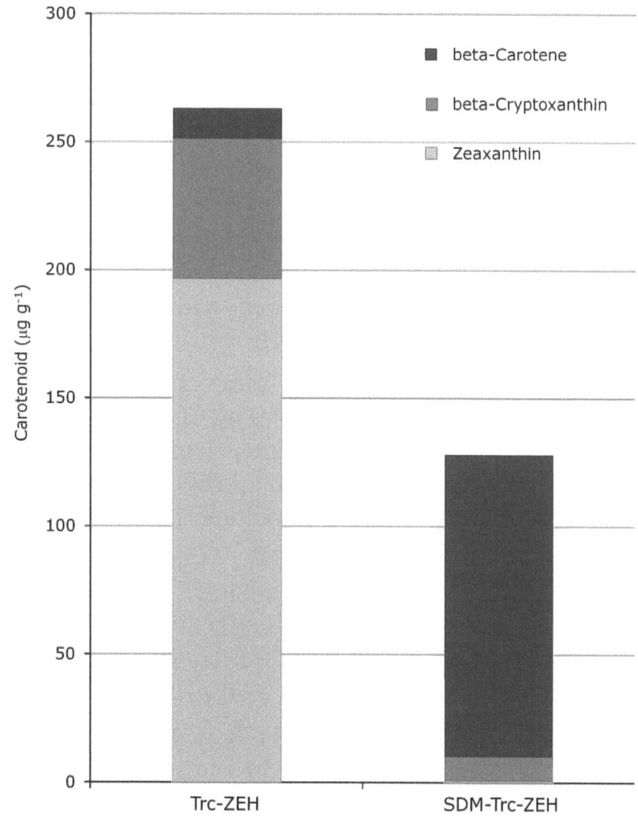

Fig. 2. Comparison of carotenoid profiles; *Escherichia coli*, transformed with pAC-Beta, plus plasmid pTrc-ZEH or SDM-Trc-ZEH. Demonstrates impact of site-directed mutation on carotenoid profile.

9. Harvest biomass, extract carotenoids, and analyze, as described in Subheading 3.3.
10. Figure 2 presents a comparison of the resultant carotenoid profiles. Employ plasmid SDM-Trc-ZEH as the template DNA in all subsequent directed evolution steps.

3.5. Directed Evolution of Target Gene: Error-Prone PCR

1. Generate a library of randomly mutated β-carotene hydroxylase genes via combined application of error-prone PCR and MEGAWHOMP cloning. To achieve this, exploit the natural mutation frequency of *Taq* DNA polymerase, increased via the addition of manganese chloride ions at 50–500 μM (see Note 11).
2. Set up a series of PCR reactions, employing plasmid SDM-Trc-ZEH as the template DNA. Briefly, combine 2.5 μL 10× standard *Taq* buffer, 0.5 μL dNTP mix (10 mM), 1 μL each of Eherb-CrtZ fwd and Eherb-CrtZ rev, 0.5 μL SDM-Trc-ZEH, and 0.25 μL Taq DNA polymerase with increasing manganese chloride concentrations (50–500 μM) in a final reaction volume of 25 μL (see Note 11).

3. Achieve amplification as follows, 94°C for 3 min, 30 cycles of 94°C for 30 s, 50°C for 30 s and 68°C for 30 s, with a final incubation at 68°C for 10 min.
4. Separate PCR products via agarose gel electrophoresis and visualize on a UV-transilluminator, as described in Subheading 3.2.
5. Precipitate PCR products subsequently via application of SureClean reagent, as described in Subheading 3.2, elute in nuclease-free water, and adjust the DNA concentration to 100 ng/μL.
6. Combine samples subsequently in equal concentrations and employ 100 ng of the resultant mixture as the template DNA, along with plasmid SDM-Trc-ZEH (isolated from *Dam*$^{+}$ *E. coli*), in a MEGAWHOMP PCR reaction (8) designed such that the randomly mutated β-carotene hydroxylase gene replaces the wild type version, within the plasmid SDM-Trc-ZEH.
7. Briefly, combine 5 μL 10× Opti buffer, 3 μL magnesium chloride, 1 μL dNTP mix, 100 ng error-prone PCR product, 100 ng SDM-Trc-ZEH, and 0.5 μL Bio-x-act long DNA polymerase in a 50-μL final volume.
8. Perform PCR at 94°C for 3 min, 30 cycles of 94°C for 30 s, 60°C for 30 s and 68°C for 6 min, followed by a final incubation at 68°C for 10 min.
9. Following amplification, remove template DNA (SDM-Trc-ZEH) via *Dpn*I digestion. For this, add 5.5 μL 10× NEB buffer 4, and 1 μL *Dpn*I endonuclease directly to the PCR, and incubate at 37°C for 4 h.
10. Subsequent precipitation of DNA with SureClean reagent removes contaminants (primers, $MgCl_2$, $MnCl_2$, etc.) and concentrates the sample.
11. Transform products of this MEGAWHOMP cloning reaction into electrocompetent *E. coli* cells, previously modified via transformation with the plasmid pAC-Beta, as described in Subheading 3.1.
12. The resultant library is screened via the presented method.

3.6. Method Development: Initial Wave Scan

Aim: To demonstrate a correlation between absorption at specific wavelengths and carotenoid content within different zeaxanthin producing *E. coli* cultures.

1. Culture *E. coli* strain TOP10, transformed separately with plasmid pAC-Zeax or pAC-Zeax-IDI under conditions conducive to carotenoid biosynthesis described in Subheading 3.3. Culture wild type *E. coli* strain TOP10 in an identical fashion.
2. Transfer 200 μL of each culture, to a 96-well plate, in triplicate, and employ in an absorption scan over a wavelength range of 200–900 nm, using a 96-well plate UV/Vis spectrophotometer.

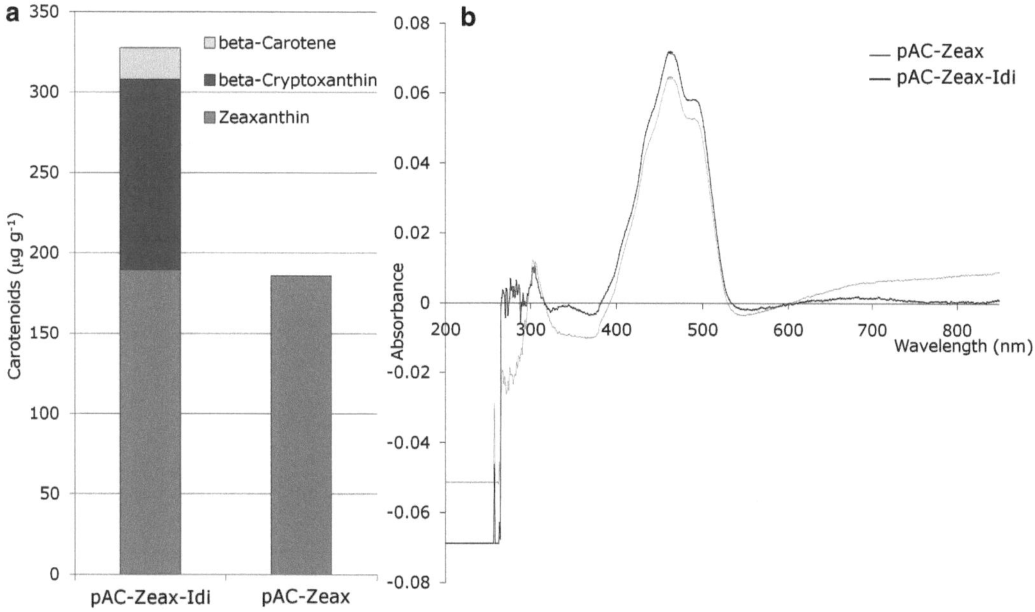

Fig. 3. (**a**) Accurate quantification of *Escherichia coli* strains engineered to biosynthesize different quantities of zeaxanthin. (**b**) Comparative absorbance values for the same *E. coli* strains, generated by a 96-well plate UV/Vis spectrophotometer wave scan.

3. Harvest remaining culture volume by centrifugation, lyophilize, extract, and quantify total carotenoids via HPLC, using the previously described method, Subheading 3.3. Figure 3a presents obtained results.
4. Process absorbance data to generate the average of the replicates and account for variations in cell density (see Note 12). Figure 3b presents a graphical representation of this data. Use wild type *E. coli* to generate baseline absorbance, and data for the two zeaxanthin producing cultures are based upon this.
5. A significant difference in absorption levels between carotenoid producing and wild type *E. coli* strains is apparent, as is a substantial difference between the two carotenoid producing strains, within the 450–495 nm wavelength range. This correlates with the blue region of the visible light spectrum. Therefore, absorbance in the 450–500 nm range may be employed to detect changes in total carotenoid content for *E. coli*.

3.7. Method Validation via Small-Scale Application

1. To confirm the validity of the developing method, randomly select 330 clones and analyze from the created mutant library.
2. Select single colonies randomly from LB agar plates and use to inoculate triplicate wells containing 200 µL LB broth, supplemented with ampicillin (100 µg/mL) and chloramphenicol (100 µg/mL), within a 96-well plate.

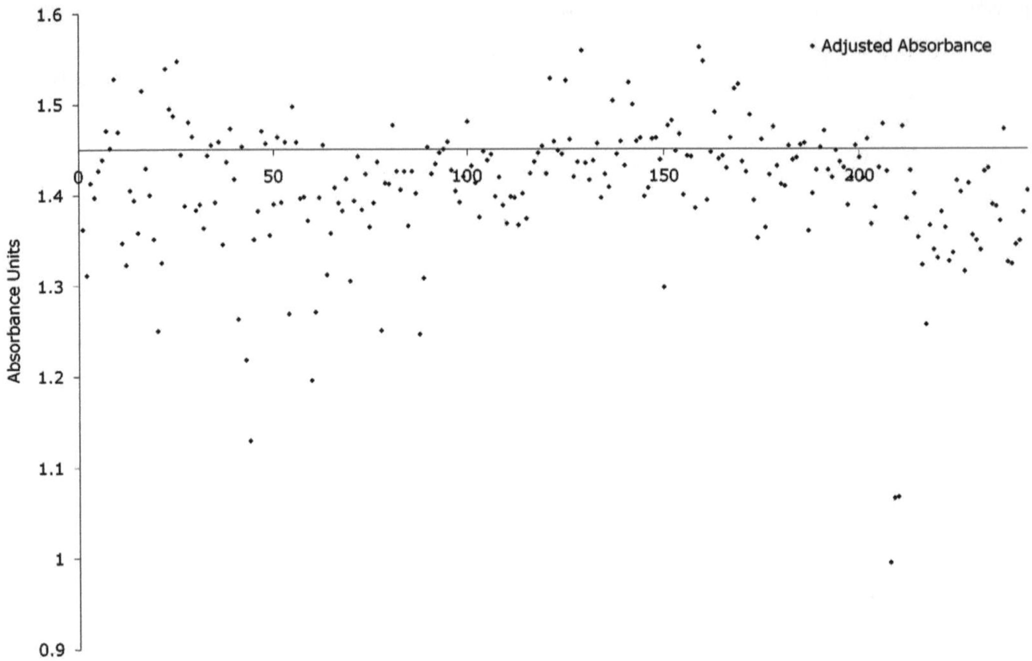

Fig. 4. Summary of high-throughput screen absorbance data. *X*-axis represents average absorbance of SDM-Trc-ZEH (wild type) strain, for this assay. Data points above the *x*-axis therefore represent putatively improved clones.

3. Incubate assay plates at 30°C, 250 rpm for 24 h, in the dark, and take end point absorption readings at 595 nm and 450 nm.
4. Normalize data for absorbance at OD_{595} and apply a cell density cutoff ($OD_{595} > 0.6$) to remove artifacts (see Note 13). A summary of the produced data is presented in Fig. 4.
5. To validate this screen as an efficient method, grow clones with predicted improvements in carotenoid biosynthesis in 50 mL culture volumes under conditions conducive to carotenoid biosynthesis, described in Subheading 3.3.
6. Collect biomass and submit to total carotenoid extraction and HPLC analysis. Refer to Subheading 3.3.
7. Analysis demonstrated that, of 330 colonies screened, 84 were eliminated by the OD_{595} cutoff, while 63 were identified as putatively improved.
8. HPLC analysis confirms increased carotenoid biosynthesis for 50 clones, with increases in total carotenoid biosynthesis up to fourfold being obtained (see Note 14).

4. Notes

1. *E. coli* strain for expression of carotenoid biosynthesis genes and molecular biology work must be *Dam/Dcm*$^+$; no other specific genotype is required.
2. DNA plasmids containing genes for efficient biosynthesis of specific carotenoids within *E. coli.* e.g. pAC-Beta (β-carotene biosynthesis), pAC-Zeax, or pAC-Zeax-Idi (zeaxanthin biosynthesis) (6, 7), others could be used.
3. Alternate DNA plasmids suitable for coexpression of a target β-carotene hydroxylase gene could be used. Criteria for selection are: origin of replication (*Ori*) and selective marker compatibility with employed carotenoid biosynthesis plasmid, as well as capacity to manipulate target gene sequence and expression level (e.g., encode a suitable and well characterized promoter system).
4. Any commercial kit would be suitable for this application. However, yield and ease of use combined with speed and simplicity of protocols lead the authors to use the Zyppy miniprep kit.
5. Any high-fidelity long-read PCR reagents, including supplier provided DNA polymerase specific reaction buffer, would be suitable for this application.
6. Any low-fidelity *Taq* DNA polymerase (non-proofreading), including DNA polymerase specific reaction buffer, would be suitable for this application.
7. Differential antibiotic selection ensures that only the desired construct is present in the transformed *E. coli* strains. pAC-Zeax encodes chloramphenicol resistance while pTrcHis2 employs ampicillin resistance. Therefore, the use of ampicillin for selection eliminated residual pAC-Zeax plasmid from the strain.
8. Due to the oxidation properties of carotenoids, extraction should be performed with minimal sample exposure to sources of heat and light.
9. Mobile phase is isocratic, buffer A, at 0–10 min, and isocratic, buffer B, at 20–30 min. Thus, the gradient proceeding from buffer A to buffer B, and a flow rate of 0.75 mL/min to 1.5 mL/min, is imposed at 10–20 min. Achieve carotenoid identification and quantification via retention time comparison to commercial standards; a standard curve generated with said standards allows for accurate quantification.
10. The underlined base within primer MEP Fwd represents the intentionally introduced single base mismatch. Primers designed such that the mutation is introduced and entire β-carotene hydroxylase open reading frame is amplified, including pTrc promoter.

11. Increase manganese chloride concentrations at 50 µM intervals, giving six discrete concentrations within six different PCR tubes. This allows for variable error rates and PCR efficiencies to be attained, and combine in subsequent steps.
12. Adjusting the data such that OD_{595} for each sample is equal to 1 eliminates cell density as an influential factor, thereby makes trends more apparent.
13. For this assay, we demonstrate that reduced OD_{595} values, below 0.6, produce artificially elevated values for carotenoid content. Therefore, a cutoff of $OD_{595} > 0.6$ is employed. This eliminates the bias, significantly increasing the accuracy of the screen.
14. Reference to Fig. 2 demonstrates that the mutation introduced within the −35 region of the pTrc promoter, in the creation of SDM-Trc-ZEH, significantly reduces carotenoid biosynthesis. Increases observed in carotenoid profile are within the SDM-Trc-ZEH construct. The actual impact on total carotenoid biosynthesis, when compared to pTrc-ZEH or other β-carotene hydroxylase expression studies, should not be inferred from this data.

Acknowledgments

This work was supported in part by the National Research Council Canada Industrial Research Assistance Program (NRC-IRAP), project number 674299. We also thank the United Kingdom's EPSRC under the ChELSI Initiative (EP/E036252/1).

References

1. Higuera-Ciapara I et al (2006) Astaxanthin: a review of its chemistry and applications. Crit Rev Food Sci Nutr 46:185–196
2. Scaife MA et al (2009) Characterization of cyanobacterial beta-carotene ketolase and hydroxylase genes in *Escherichia coli*, and their application for astaxanthin biosynthesis. Biotechnol Bioeng 103:944–955
3. Tang XS et al (2007) Improvement of a crtO-type of beta-carotene ketolase for canthaxanthin production in *Methylomonas* sp. Metab Eng 9:348–354
4. Tao L et al (2006) Engineering a beta-carotene ketolase for astaxanthin production. Metab Eng 8:523–531
5. Ye RW et al (2006) Mutational and functional analysis of the beta-carotene ketolase involved in the production of canthaxanthin and astaxanthin. Appl Environ Microbiol 72: 5829–5837
6. Cunningham FX Jr et al (1996) Functional analysis of the beta and epsilon lycopene cyclase enzymes of arabidopsis reveals a mechanism for control of cyclic carotenoid formation. Plant Cell 8:1613–1626
7. Sun Z et al (1996) Cloning and functional analysis of the beta-carotene hydroxylase of *Arabidopsis thaliana*. J Biol Chem 271: 24349–24352
8. Miyazaki K (2003) Creating random mutagenesis libraries by megaprimer PCR of whole plasmid (MEGAWHOMP). In: Arnold FH and Georgiou G (eds) Meth Mol Biol, vol. 231. Humana Press, New Jersey, pp 23–28

Chapter 16

DNA Fingerprinting Intron-Sizing Method to Accomplish a Specific, Rapid, and Sensitive Identification of Carotenogenic *Dunaliella* Species

Jorge Olmos-Soto, J. Paniagua-Michel, Rosalía Contreras, and Leonel Ochoa

Abstract

Dunaliella salina has become the most important microorganism for the production of β-carotene around the world. Natural carotenoids are a source of active metabolites utilized in different areas of food nutrition and pharmaceuticals, both in humans and also in animals. Identification of *Dunaliella* species from natural environments or certified culture collections is not precise and it is time consuming. However, accurate identification is extremely important because a slight difference in *Dunaliella* species generates great differences in carotenoids production. Here, we describe an intron-sizing method to make a rapid and precise identification for each of the most important carotenogenic species, showing that each hyper-producer species has an exclusive 18S rDNA fingerprint profile.

Key words: *Dunaliella*, Molecular identification, Specific oligonucleotides, *18S rDNA*, Fingerprint

1. Introduction

Dunaliella was originally described by Teodoresco in 1905. Since then, taxonomic studies among *Dunaliella* have identified several new species. However, even today, differentiation among halophilic and carotenogenic *Dunaliella* species in both green and red stages is difficult and time consuming. In addition, some strains and species in culture collections are misnamed and have given rise to unnecessary strains and species names (1). This confusion of strains and species names makes comparison of results by different authors difficult (2). *Dunaliella salina* and *D. salina/bardawil* are the only reported β-carotene hyperproducer species of the genus that grows in salt-saturated lagoons (3). However, a controversy still exists about identification of *D. salina/bardawil* as a different

José-Luis Barredo (ed.), *Microbial Carotenoids from Bacteria and Microalgae: Methods and Protocols*, Methods in Molecular Biology, vol. 892, DOI 10.1007/978-1-61779-879-5_16, © Springer Science+Business Media, LLC 2012

species of *D. salina* (2). Molecular identification provides a useful tool to distinguish between inter- and intraspecific morphologically similar species (4, 5) and mixed populations (6, 7). Intraspecies identification of community members without cultivation avoids some selective biases associated with pure culture methods (8). Species-specific oligonucleotides could be useful to identify either species from culture collections or from natural environments (9, 10). Slight phylogenetic and taxonomic differences in *Dunaliella* species can conceal profound differences in their potential for production of metabolites such as carotenoids (4). To study these organisms one must discriminate them from their presumed non-carotenogenic congeners and from among species with similar morphological features (11).

In this work, using the intron-sizing method, we demonstrated that each hyperproducer species has an exclusive *18S rDNA* fingerprint profile. Non-β-carotene hyperproducer species of *Dunaliella* were easily discriminated with our methodology, even in the green stage, avoiding long cleaning, purification, and growing process. The intron-sizing method provides a novel and very powerful DNA fingerprinting technique to accomplish a specific, rapid, and sensitive identification of carotenogenic *Dunaliella* species (12).

2. Materials

Prepare all solutions using ultrapure water molecular biology grade, analytical/biotech-grade reagents, and HPLC/Unisolv-grade solvents. Prepare and store at room temperature unless indicated otherwise. Carefully follow all waste-disposal regulations when disposing waste materials.

2.1. Isolation of Dunaliella Species from Environmental Samples

1. Sterile tubes (1.5 and 15 mL) and bottles (200 mL).
2. Pasteur pipettes.
3. Microscope slides and cover glasses.
4. Mesh of 40 μm pore size membrane (BD Falcon, Bedford, MA, USA).
5. Refrigerated centrifuge for tubes and bottles.
6. Single-channel pipettes with adjustable volume.
7. Pipette tips of 10, 200, and 1,000 μL.
8. Acidic Lugol's iodine solution: Dissolve 20 g of KI into 200 mL of water and add 10 g crystal I_2. Then add 20 mL glacial acetic acid and mix. Filter before storing and mix gently before using (13).
9. Phase contrast microscope.
10. Hemacytometer with 0.1-mm deep chamber.

11. TE buffer pH 8.0 (10 mM Tris–HCl pH 8 and 1 mM EDTA): Weigh 0.1211 g of Tris, transfer to a glass, and add 70 mL of water. Mix with a magnetic bar and adjust pH with 6 N HCl. Weigh 0.0372 g of EDTA and add to the solution, dissolve and bring to 100 mL with water. Sterilize at 121°C and 15 lb/in.^{2} for 15 min. Cool to room temperature and pass through 0.2-µm membrane disc filters. Store aliquots at -20°C (14).
12. Antibiotic/antifungal solution: Dissolve 100 mg of penicillin G (sodium or potassium salt), 25 mg of streptomycin sulfate, 10 mg of chloramphenicol, and 0.5 mg of Amphotericin B, all in 10 mL of water. Sterilize by 0.2-µm membrane filtration, prepare aliquots of 1 mL, and keep frozen until used (15).
13. Environmental chamber at 20°C and 10/14 h light–dark period.
14. Filters of 0.45 µm (Millipore, Bedford, MA, USA).
15. f/2 medium: This is a common and widely used general enriched medium designed for growing coastal marine algae. Into 950 mL natural seawater filtered through 0.45 µm, add the components shown in Table 1. Bring the final volume to 1 L with filtered natural seawater. If silicate is not required, omit to reduce precipitation. Autoclave at 121°C for 20 min (16, 17).

2.2. Chromosomal DNA Purification

1. TE Buffer pH 8 (see Subheading 2.1).
2. 10% SDS: 10% (w/v) sodium dodecyl sulfate (SDS) in water.
3. 10 mg/mL lysozyme: Weigh 0.010 g of lysozyme and add 1 mL of water, mix gently, and store in small single-use aliquots at -20°C. Place on ice to use.
4. RNase A (Sigma Chemical Company, St. Louis, MO, USA).
5. RNase A buffer: Weigh 0.8203 g sodium acetate (0.1 M) and 0.0111 g EDTA (0.3 mM) and transfer to a glass beaker. Add 50 mL of water, mix with a magnetic bar, and adjust to pH 4.8 with acetic acid. Bring to 100 mL (14).
6. 10 mg/mL ribonuclease A: Weigh 0.010 g of RNase A and add 1 mL RNase buffer, mix gently, and store at -20°C (see Note 1).
7. 20 mg/mL proteinase K: Weigh 0.020 g of proteinase k and add 1 mL of water, mix gently, and store at -20°C (see Note 1).
8. 5 M NaCl: Weigh 29.22 g of NaCl and add water to 100 mL.
9. CTAB/NaCl solution (10% CTAB/0.7 M NaCl): Dissolve 4.1 g of NaCl in 80 mL of water and slowly add 10 g of CTAB (hexadecyltrimethyl ammonium bromide) while heating and stirring. If necessary, heat to 65°C to dissolve. Adjust final volume to 100 mL.
10. 24:1 Chloroform/isoamyl alcohol (see Note 2).
11. Phenol solution equilibrated with 10 mM Tris–HCl, pH 8, 1 mM EDTA (see Note 2).

Table 1
Composition of f/2 medium

Component	Stock solution (g/L dH_2O)	Quantity used	Concentration in final medium (M)
(1) $NaNO_3$	75	1 mL	8.82×10^{-4}
(2) $NaH_2PO_4{\cdot}H_2O$	5	1 mL	3.62×10^{-5}
(3) $Na_2SiO_3{\cdot}9H_2O$	30	1 mL	1.06×10^{-4}
(4) Trace metals solution	(See following recipe)	1 mL	
Into 950 mL of dH_2O, dissolve the EDTA and other components. Bring the final volume to 1 L with dH_2O			
$FeCl_3{\cdot}6H_2O$	–	3.15 g	1.17×10^{-5}
$Na_2EDTA{\cdot}2H_2O$	–	4.36 g	1.17×10^{-5}
$MnCl_2{\cdot}4H_2O$	180.0	1 mL	9.10×10^{-7}
$ZnSO_4{\cdot}7H_2O$	22.0	1 mL	7.65×10^{-8}
$CoCl_2{\cdot}6H_2O$	10.0	1 mL	4.20×10^{-8}
$CuSO_4{\cdot}5H_2O$	9.8	1 mL	3.93×10^{-8}
$Na_2MoO_4{\cdot}2H_2O$	6.3	1 mL	2.60×10^{-8}
(5) Vitamins solution	(See following recipe)	0.5 mL	
Into 950 mL of dH_2O, dissolve the thiamine·HCl, and add 1 mL of biotin and cyanocobalamin stock solutions. Bring final volume to 1 L with dH_2O. Sterilize by filtering and store frozen aliquots			
Thiamine·HCl (vitamin B_1)	–	200 mg	2.96×10^{-7}
Biotin (vitamin H)	1	1 mL	2.05×10^{-9}
Cyanocobalamin (vitamin B_{12})	1	1 mL	3.69×10^{-10}
(6) $MgSO_4{\cdot}7H_2O$[a]	246.4	5 mL	0.005
(7) KNO_3[a]	0.5	5 mL	2.5×10^{-5}
(8) NaCl[b]	–	23.4 g	1
	–	257.2 g	5

[a]Added for better growth
[b]Added to adjust the salinity of the culture medium

12. 25:24:1 Phenol/chloroform/isoamyl alcohol (see Note 2).
13. Isopropanol (see Note 3).
14. 70% Ethanol (v/v) water solution (see Note 3).

2.3. Gel Preparation and Chromosomal DNA Analysis

1. Horizontal gel electrophoresis equipment, gel casting platform 8 × 10 cm, gel combs (slot formers), and DC power supply.
2. 5× TBE (5× electrophoresis TBE buffer stock solution pH 8.3): Weigh 43.2 g of Tris, 22 g boric acid, and 3.68 g of EDTA. Add water and dissolve. Bring to 800 mL with water. For 1× working solution, dilute stock solution 1:5 with sterilized water.
3. 1.2% Agarose gel: 1.2 g of agarose in 100 mL of 1× TBE buffer.

4. 10 mg/mL ethidium bromide stock solution: Weigh 10 mg of ethidium bromide and dissolve in 1 mL of water (see Note 4).
5. 10 μg/mL ethidium bromide working solution: Dilute the stock solution 1:1,000 using sterilized water (see Note 4).
6. 10× Loading buffer: 0.25% (w/v) bromophenol blue, 0.25% (w/v) xylene cyanol, and 30% (v/v) glycerol, in water (14).
7. DNA molecular weight markers.
8. UV transilluminator.

2.4. Chromosomal DNA Quantification

1. 1 cm quartz cell.
2. UV spectrophotometer.
3. Single-channel pipettes with adjustable volume.
4. Pipette tips.
5. Sterilized water.

2.5. PCR Amplification

1. Single-channel pipettes with adjustable volume.
2. Aerosol-resistant barrier pipette tips.
3. Sterile 0.2-mL microcentrifuge tubes.
4. Ice flakes.
5. *D. salina* CCAP 19/18 (Culture Collection Algae and Protozoa, www.ccap.ac.uk).
6. *Dunaliella* chromosomal DNA on TE buffer pH 8 (18).
7. 200 ng/μL 5′ and 3′ stock primers (see Table 2).

Table 2
Combination of primers used to amplify the full sequence of *18S rDNA* gene and to identify species of *Dunaliella* (5, 11, 12)

Strain	Primers	bp
D. bardawil	MA1-MA2	~2,500
	MA1-MA3	~2,100
	DBs-MA2	~1,000
D. parva	MA1-MA2	~2,500
	MA1-MA3	~2,100
	DPs-MA2	~1,000
D. salina	MA1-MA2	~2,100
	MA1-MA3	~2,100
	DSs-MA2	~750
D. tertiolecta	MA1-MA2	~1,700
	MA1-MA3	~1,700

8. 100 mM dNTP's stock solution (Invitrogen, Carlsbad, CA, USA).
9. dNTP's working solution: Dilute 1:10 with sterilized water the 10 mM dNTP's stock solution.
10. 10× PCR buffer: 200 mM Tris–HCl pH 8.4, 500 mM KCl (Invitrogen, Carlsbad, CA, USA).
11. Sterilized distilled and deionized water for molecular biology reactions.
12. *Taq* DNA Polymerase 5 U/μL (Invitrogen, Carlsbad, CA, USA).
13. Thermal cycler.

3. Methods

Carry out all procedures at room temperature unless otherwise specified.

3.1. Isolation and Culture of Dunaliella Species

Dunaliella occurs in a wide variety of marine habitats such as oceans, brine lakes, salt marshes, and salt water ditches near the sea, predominantly in water bodies containing more than 2 M of NaCl and high levels of magnesium. The phenomenon of orange-red algal bloom in such hypersaline environments is usually related to combined sequential growth of *Dunaliella*, brine shrimps, and halophilic bacteria (18). This phenomenon occurs in our region, in these places, we have a high seawater evaporation level and crystals formation (11).

A generalized schematic method of isolation (15), purification, culture, and identification of *Dunaliella* species is shown in Fig. 1.

1. Collect 200 mL of pink colored water samples and take to the laboratory for further analysis.
2. Analyze 15 μL of each sample in a phase contrast microscope at 10×, 40×, and 100× objectives. Once confirmed the presence of red colored and flagellated *Dunaliella* species, look forward for potential contamination of bacteria, fungi, or ciliates.
3. Filter 50 mL of each field sample through 40-μm membrane mesh.
4. Transfer 15 mL of each filtered sample to sterile conic tubes, centrifuge at 750 ×*g* for 10 min at 18°C, and discard the supernatant (see Note 5).
5. Resuspend the pellets in 10 mL of f/2 5 M NaCl culture medium, centrifuge at 750 ×*g* for 5 min at 18°C, and discard the supernatant. Repeat this step three times. Make a duplicate.

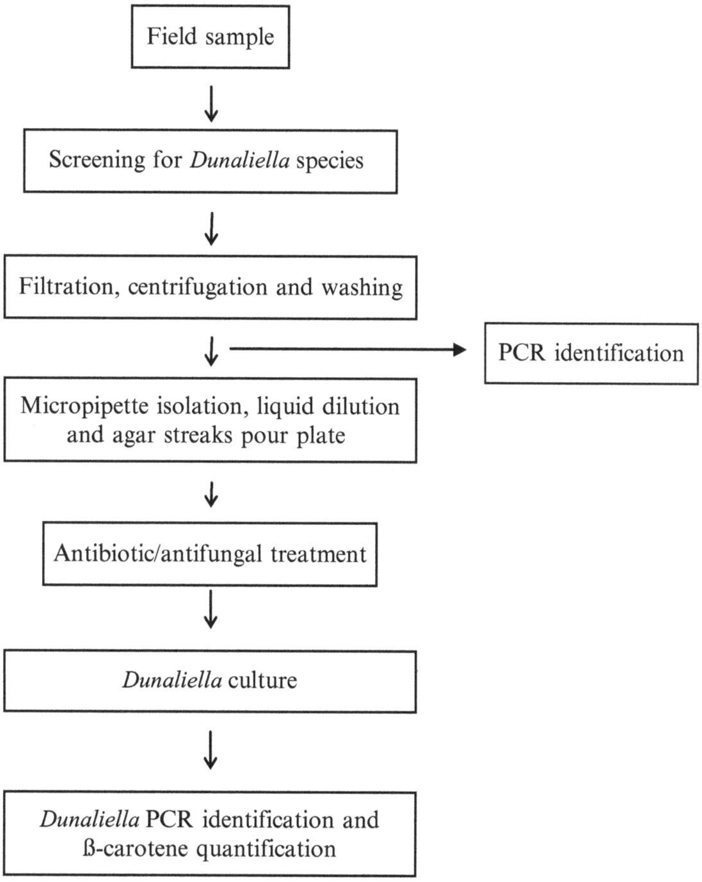

Fig. 1. Schematic method of isolation, purification, culture, and identification of *Dunaliella* species.

6. Resuspend pellets in 3 mL of f/2 5 M NaCl culture medium.
7. To select stronger cells, cover the tubes with aluminum foil and let 1 cm space foil free (without cover) on the upper part of the liquid.
8. Place the tubes 30 cm away from a cold white light source with 50 $\mu mol/m^2.s$ intensity, for 24 h at 25°C.
9. Take 1 mL from the top of the tube of each sample and transfer to new tubes.
10. Check microscopically 15 μL of each sample searching for motility and abundance of *Dunaliella* species.
11. Pick up single cells under a phase contrast microscope and place without damage into a sterile droplet (see Note 6). Repeat this process until a single algal cell can be confidently placed into 1 mL of sterile medium (19).
12. Place 1 mL of each concentrated sample into a test tube containing 9 mL sterile medium. Mix, remove 1 mL, and add to a second tube with 9 mL of medium (see Note 7).

13. Repeat this step six times (six 1:10 dilutions) (see Note 8).
14. At the same time, spread 100 μL of each dilution on 1.5% agar plates with f/2 1 M or 5 M of NaCl media.
15. Incubate at 20°C with 50 μmol/m^2/s light intensity until colony formation occurs.
16. Pick up the selected isolated algal colonies with the help of a micropipette tip.
17. Discharge the cells into 1 mL of f/2 liquid medium (19).
18. Illuminate the cultures with 50 μmol/m^2/s using a 10:14 h light–dark cycle at 25°C, letting cells to growth for at least 4–6 weeks (see Note 9).
19. Add 5 μL of acidic Lugol's iodine solution to a 500 μL culture sample and place 10 μL of the mixture in a hemacytometer (see Note 10) (13).
20. Analyze 15 μL of each fresh culture in a phase contrast microscope to confirm the identity and mobility of *Dunaliella* species, and also to check some possible contamination with other microorganisms.
21. If microbial contamination is detected, add 0.5 mL of antibiotic/antifungal solution to 50 mL algal culture (see Note 11), and incubate 24 h at 25°C (15).
22. Centrifuge at 750 × *g* for 10 min at 18°C and discard supernatant.
23. Add fresh medium free of antibiotics/antifungal, and incubate at 18°C for 48 h.
24. Add 1 mL of culture to 1.5-mL tubes, centrifuge by 5 min at 16,000 × *g*, and discard supernatant. Cultures are concentrated by five times.
25. Resuspend the pellets in 1 mL of TE buffer (14) and centrifuge for 5 min at 16,000 × *g* at room temperature.
26. Discard supernatant and repeat the process two times to eliminate salt excess before extracting chromosomal DNA.

3.2. Chromosomal DNA Purification (20)

1. Resuspend the cell pellet obtained in 3.1 in 567 μL of TE buffer by vortex.
2. Add 10 μL of 10 mg/mL lysozyme, mix gently, and incubate for 15 min at 37°C.
3. Add 10 μL of 10 mg/mL RNase A, mix gently, and incubate 20 min at 37°C.
4. Add 30 μL of 10% SDS, mix, add 3 μL of 20 mg/mL proteinase K, mix gently, and incubate 1 h at 37°C. Mix gently every 15 min.
5. Add 100 μL of 5 M NaCl and mix thoroughly.

6. Add 80 μL CTAB/NaCl solution, mix, incubate 10 min at 65°C. Cool for 5 min at room temperature.
7. Add 700 μL of chloroform/isoamyl alcohol, mix, centrifuge for 5 min at 16,000 × *g*. Transfer 700 μL supernatant to a fresh tube.
8. Add equal volume phenol/chloroform/isoamyl alcohol, mix, and centrifuge for 5 min at 16,000 × *g*. Transfer 600 μL of supernatant to a fresh tube.
9. Add 360 μL of isopropanol and mix gently until DNA precipitates. Centrifuge for 10 min at 16,000 × *g*. Discard supernatant by drawing off liquid taking care of the DNA pellet.
10. Wash pellet using 1 mL of 70% ethanol, centrifuge for 5 min at 16,000 × *g*. Remove supernatant and dry pellet under vacuum at 30°C.
11. Resuspend the pellets in 100 μL of TE buffer, and store at -20°C.

3.3. Gel Preparation and Chromosomal DNA Resolution

1. Dissolve agarose powder in 1× TBE buffer using a microwave oven by 1–2 min with 50% power, until agarose is completely dissolved and the solution looks clear (be sure all agarose particles are completely melted), avoiding boiling.
2. Cool liquid agarose to reach 45°C and transfer it to a horizontal electrophoresis chamber previously prepared. Place gel comb with enough wells for your samples.
3. Let gel solidify for 25 min and add 1× TBE buffer until gel is covered 5 mm up. Wait for 5 min to remove the comb. Place the chamber lid and connect the negative and positive chamber wires to a power supply (the DNA will migrate to the positive).
4. Turn on the power source and pre-warm the electrophoresis chamber for 15 min at 10 mA. Turn off the chamber; disconnect it completely before loading the DNA samples.
5. Use 1 μL of 10× loading buffer per each 5 μL of *Dunaliella* chromosomal DNA sample.
6. After loading 6 μL of each sample in each well, let stand for 5 min. Cover the chamber and turn on the power source. Run for 20 min at 10 mA or until the samples enter the gel and the xylene cyanol and the bromophenol blue dyes are separated (light blue and dark blue respectively), then increase to 20 mA and wait for 30 min more to samples spread better. Run until the first dye gets 2 cm before to the gel ending limit. Turn off the chamber and disconnect it completely.
7. Take out the gel with a spatula and place it in a plastic container. Add ethidium bromide working solution (10 μg/mL), incubate for about 5 s, and carefully return the solution to the bottle. Add distilled water, gently wash for 10 min, and discard the remaining solution to a plastic container (see Note 12).

8. Put the gel inside a DNA analyzer containing a UV transilluminator. You will see bright bands where you got DNA because the ethidium bromide binds to DNA.
9. Get the image on the screen and print the picture in special paper or send the picture to the computer.
10. To discard the gel, place it in a hermetic plastic bag and transfer to the container (see Note 12).

3.4. Chromosomal DNA Quantification

1. To quantify the chromosomal DNA concentration, measure at both OD_{260} and OD_{280} in 1 cm quartz well. In a standard base, 50 μg/mL of dsDNA has a lecture of 1.0 at 260 nm (21).
2. Estimate chromosomal DNA concentration using the formula: $(X\mu g/mL)/(OD_{260}) = (50\ \mu g/mL)/(1.0\ OD)$ $X = \mu g/mL$. To obtain the final concentration multiply X by the dilution factor of the sample.
3. Estimate DNA purity with the OD_{260}/OD_{280} ratio. A pure sample should have from 1.65 to 1.9 (see Note 13).

3.5. PCR Amplification

PCR reactions should be done in a DNA-free environment. Use of "clean" exclusive automatic pipettes and aerosol-resistant barrier tips is recommended.

1. Add the following components to a sterile 0.2-mL microcentrifuge tube placed on ice: 2 μL (50 ng) of chromosomal DNA in TE buffer pH 8, 1 μL of each 5′ and 3′ primer (see Table 2), 2 μL dNTP mixture, 3 μL $MgCl_2$, 10 μL of 10× PCR buffer, and 80.75 μL of autoclaved distilled water. Mix well, add 0.25 μL of *Taq* DNA polymerase (1 U), and mix gently (22). A PCR reaction is performed in a total volume of 100 μL (see Note 14).
2. Incubate tubes in a thermal cycler at 95°C for 5 min to completely denature the DNA template.
3. Perform 25 cycles of PCR amplification as follows: denature at 95°C for 1 min, anneal at 52°C for 1 min, and extend at 72°C for 2 min.
4. Incubate for an additional 10 min at 72°C and then cool to 4°C. The samples can be stored at −20°C until use.

4. Notes

1. Store in small single-use aliquots and place on ice before to use to avoid enzyme inactivation.
2. Store in amber glass bottles at 4°C. Use cold.

3. Store at −20°C. Use cold.
4. *Caution*: Ethidium bromide is mutagenic and must be handled with care. Store the stock solution at −20°C and the working solution at 25°C; protected from light in both cases.
5. You need to see a small pellet; if not, repeat the step three times in the same tube.
6. The micropipette isolation is usually performed with a Pasteur pipette heated in a flame, extended, and broken.
7. The dilution technique has been used for many years and it is effective for organisms that are rather abundant in a sample. The goal of the dilution method is to place only one cell into a test tube, thereby establishing a single-cell isolate.
8. Six 1:10 dilutions theoretically place a single cell into the final tube if the original sample had 10^6 cell/mL (19).
9. Daily shake the tubes gently. In tubes with clear growth, it is important to double culture medium volume, until density reaches 10^5–10^6 cells/mL.
10. For counting cultured cells, acidic Lugol's iodine solution is the best for preserving flagella. To DNA extraction and β-carotene quantification, culture's density must be around 10^5–10^6 cells/mL.
11. It yields 100 mg/L of penicillin, 25 mg/L of streptomycin, 10 mg/L of chloramphenicol, and 0.5 mg/L of amphotericin B.
12. Ethidium bromide is classified as a mutagenic compound. It must be discarded according to waste-disposal regulations.
13. Proteins have the maximal OD at 280 nm, because of the tryptophan, measurements at this wavelength show the presence of protein contamination. The most common method to purify DNA is making a phenol extraction to remove protein contamination (21). However, maximal phenol wavelength absorption is OD_{270}. DNA with phenol contamination can give false positive measurements at OD_{260}, raising DNA concentration values.
14. Prepare a master mix for multiple reactions to minimize reagent loss and errors. Conserved primers MA1, MA2, and MA3 (see Table 3) are designed from conserved sequences of the 5′ and 3′ edges of the *18S rDNA* gene and used to amplify the full sequence of all strains. Specific primers DBs, DPs, and DSs (see Table 3) are designed from specific sequences (introns) of the *18S rDNA* gene and are used to identify strains *D. bardawil*, *D. parva*, and *D. salina*, respectively (see Fig. 2).

Table 3
Primers used to amplify the *18S rDNA* gene of *Dunaliella* species (5, 11, 12)

	Primers	Sequence
Conserved	MA1 (f)	5′-CGGGATCCGTAGTCATATGCTTGTCTC-3′
	MA2 (r)	5′-CGGAATTCCTTCTGCAGGTTCACC-3′
	MA3 (r)	5′-GGAATTCCGGAAACCTTGTTACGAC-3′
Specific	DBs (f)	5′-GGGAGTCTTTTTCCACCT-3′
	DPs (f)	5′-GTAGAGGGTAGGAGAAGT-3′
	DSs (f)	5′-GCAGGAGAGCTAATAGGA-3′

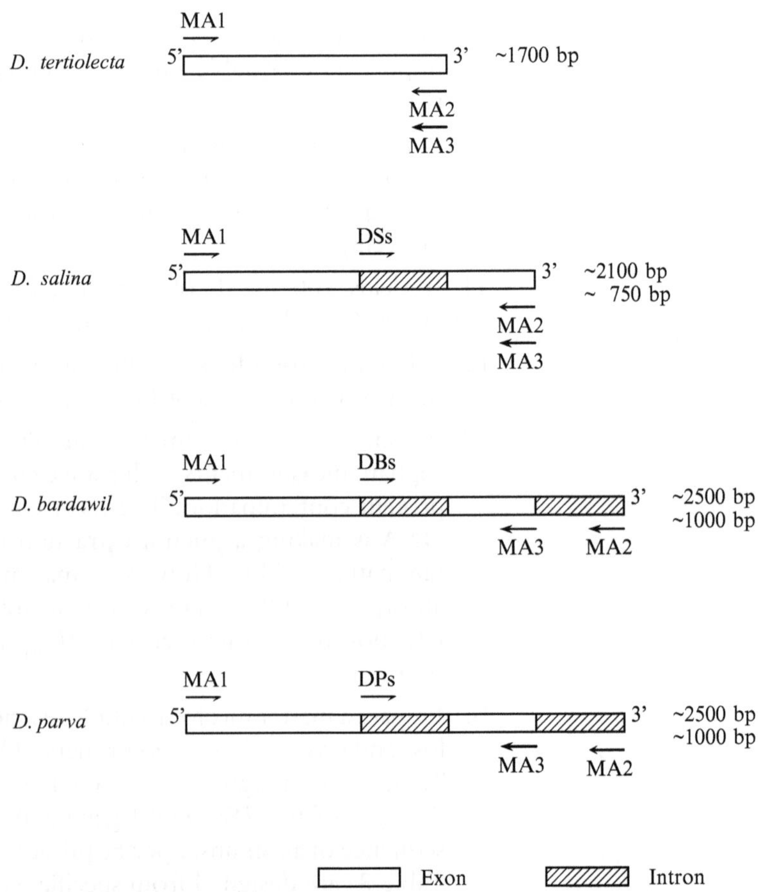

Fig. 2. Graphical representation of *18S rDNA* genes in different species of *Dunaliella*; the *arrows* indicate the binding sites of conserved and specific primers.

References

1. Oren A (2005) A hundred years of Dunaliella research: 1905–2005. Saline Syst 1:2
2. Borowitzka MA, Siva CJ (2007) The taxonomy of the genus *Dunaliella* (*Chlorophyta*, *Dunaliellales*) with emphasis on the marine and halophilic species. J Appl Phycol 19: 567–590
3. Ben-Amotz A, Asís A, Avron M (1991) The biotechnology of cultivating *Dunaliella* for production of β-carotene rich algae. Bioresour Technol 38:233–235
4. Gómez PI, González MA (2004) Genetic variation among sever strains of *Dunaliella salina* (*Chlorophyta*) with industrial potencial, based on RAPD bandung patterns and on nuclear ITS rDNA sequences. Aquaculture 233:149–162
5. Olmos SJ, Paniagua MJ, Contreras FR (2000) Molecular identification of *Dunaliella* sp. utilizing the 18S rDNA gene. Lett Appl Microbiol 30:80–84
6. García TA, Olmos SJ (2007) Quantification by fluorescent in situ hybridization of bacteria associated with *Litopenaeus vannamei* larvae in Mexican shrimp hatchery. Aquaculture 262:211–218
7. Hernández ZG, Olmos SJ (2006) Identification of bacterial diversity in the oyster *Crassostrea gigas* by fluorescent in situ hybridization and polymerase chain reaction. J Appl Microbiol 100:664–672
8. DeLong EF, Wickham GS, Pace NR (1989) Phylogenetic stains: ribosomal RNA-based probes for the identification of single cells. Science 243:1360–1363
9. Ki SJ, Han MS (2005) Sequence-based diagnostics and phylogenetic approach of uncultured freshwater dinoglagellate *Peridinium* (*Dinophyceae*) species, based on single-cell sequencing of rDNA. J Appl Phycol 17:147–153
10. Ki JS, Han MS (2007) Rapid molecular identification of the harmful freshwater dinoglagellate *Peridinium* in varios life stages using genus-specific single-cell PCR. J Appl Phycol 19:467–470
11. Olmos SJ, Paniagua MJ, Contreras FR, Trujillo L (2002) Molecular identification of β-carotene hyper-producing strain of *Dunaliella* from saline environments using specie-specific oligonucleotides. Biotechnol Lett 24:365–369
12. Olmos SJ, Ochoa SJL, Paniagua-Michel JJ, Contreras FR (2009) DNA fingerprinting differentiation between β-carotene hyperproducer strain of *Dunaliella* from around the world. Saline Syst 5:5
13. Guillard RRL, Sieracki MS (2005) Counting cells in cultures with the light microscope. In: Andersen RA (ed) Algal culturing techniques. Elsevier Academic Press, Burlington, p 239
14. Moore DD (1992) Reagent and solutions. Appendix 1. In: Ausubel FM, Brent R, Kingston RE, Moore DD, Seidman JG, Smith JA, Struhl K (eds) Short protocols in molecular biology. Wiley, New York, pp A1–57
15. Guillard RRL (2005) Purification methods for microalgae. In: Andersen RA (ed) Algal culturing techniques. Elsevier Academic Press, Burlington, p 117
16. Harrison PJ, Berges JA (2005) Marine culture media. In: Andersen RA (ed) Algal culturing techniques. Elsevier Academic Press, Burlington, p 21
17. Andersen RA, Berges JA, Harrison PJ, Watanabe MM (2005) Recipes for freshwater and seawater media. Appendix A. In: Andersen RA (ed) Algal culturing techniques. Elsevier Academic Press, Burlington, p 507
18. Ben-Amotz A (2004) Industrial production of microalgal cell-mass and secondary products—major industrial species *Dunaliella*. In: Richmond A (ed) Handbook of microalgae culture: biotechnology and applied phycology. Blackwell Science Ltd, Oxford, UK, p 273
19. Andersen RA, Kawachi M (2005) Traditional microalgae isolation techniques. In: Andersen RA (ed) Algal culturing techniques. Elsevier Academic Press, Burlington, p 83
20. Moore DD (1992) Preparation and analysis of DNA. In: Ausubel FM, Brent R, Kingston RE, Moore DD, Seidman JG, Smith JA, Struhl K (eds) Short protocols in molecular biology. Wiley, New York, pp 2–1
21. Sambrook J, Fritsch EF, Maniatis T (1989) Molecular cloning: a laboratory manual. Cold Spring Harbor Laboratory Press, NY (vols. 1, 2, 3)
22. Innis MA, Gelfand DH (1990) Optimization of PCRs. In: Innis MA, Gelfand DH, Sninsky JJ, White TJ (eds) PCR protocols: a guide to methods and applications. Academic, San Diego, p 3

Chapter 17

Ketocarotenoid Biosynthesis in Transgenic Microalgae Expressing a Foreign β-C-4-carotene Oxygenase Gene

Marta Vila, Aurora Galván, Emilio Fernández, and Rosa León

Abstract

Ketocarotenoids are obtained by the action of the β-carotene ketolase, which catalyses the addition of a keto-group at the C4 position of carotenoids β-ion-rings. Most microalgae and higher plants do not posses the carotene ketolase activity and consequently do not synthesize ketocarotenoids, which are highly demanded as feed supplements and as nutraceutical for human nutrition. Here we propose the use of the unicellular microalga *Chlamydomonas reinhardtii* to express the *Bkt* (β-carotene ketolase) gene from *Haematococcus pluvialis* and synthesize a new ketocarotenoid that the vegetative cells of *Chlamydomonas* do not synthesize in the natural way. The methodology needed to successfully achieve metabolic engineering of ketocarotenoids synthesis in *Chlamydomonas* is described in this chapter, including the construction of a vector containing the *Bkt* gene, transformation of *Chlamydomonas*, selection of transformants, and carotenoids analysis.

Key words: *Chlamydomonas reinhardtii*, *Haematococcus pluvialis*, β-Carotene ketolase, Ketocarotenoids, Transgenic microalgae

1. Introduction

Ketocarotenoids are xanthophylls obtained by oxygenation of carotenes or their hydroxylated derivatives by the action of a group of β-carotene ketolases, which catalyse the addition of a keto-group in the C4 position of carotenoids β-ione-rings (see Fig. 1). There is a huge worldwide market for ketocarotenoids, such as astaxanthin and cantaxanthin, which are highly demanded as feed supplements for fish aquaculture (1) and as nutraceutical for human nutrition (2). Astaxanthin is an essential feed additive to provide their characteristic pigmentation in aquaculture-growth salmon, trout, or shrimp (1).

Several classes of carotenoid ketolases have been reported. Those encoded by *CrtW* genes, found in bacteria (3, 4), which are

José-Luis Barredo (ed.), *Microbial Carotenoids from Bacteria and Microalgae: Methods and Protocols*, Methods in Molecular Biology, vol. 892, DOI 10.1007/978-1-61779-879-5_17,

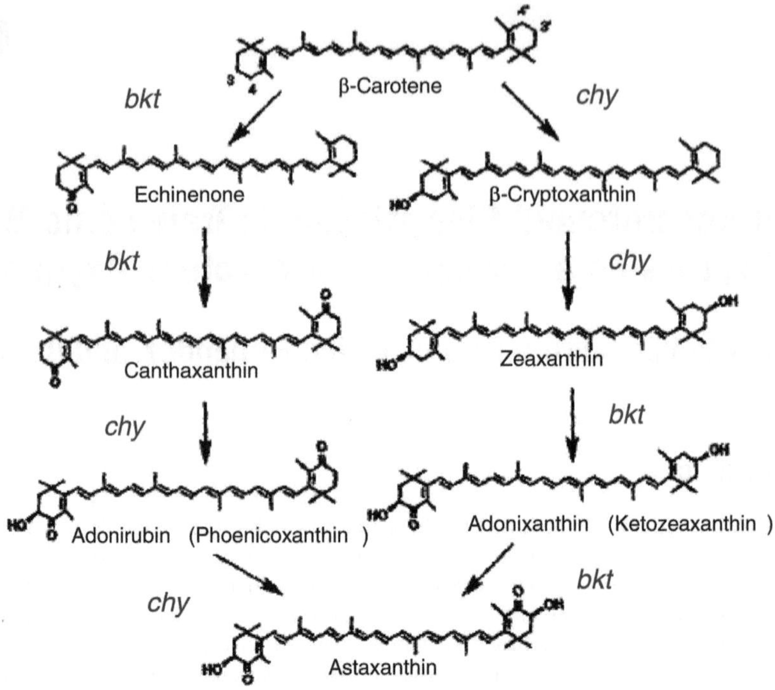

Fig. 1. Metabolic pathways for the synthesis of cantaxanthin and astaxanthin in plant cells. *Chy* β-carotene hydroxylase, *bkt* β-carotene ketolase.

very similar in functionality and sequence to the *Bkt* genes found in some algae and plants, and catalyze the oxygenation of both ione-rings of β-carotene. And the *CrtO* type, found in cyanobacteria (5), which asymmetrically add one keto-group to β-carotene to form echinenone, that seems to have arisen independently in evolution as they share very little sequence similarity with the *CrtW* type. *CrtW* ketolases are related to bacterial (*CrtZ*) and plant (*Bhy*) beta-ring hydroxylases; all members of a large class of membrane-integral, di-iron oxygenase enzymes (6).

Most microalgae and higher plants do no posses the carotene oxygenase activity and consequently do not synthesize ketocarotenoids. One important exception is the green microalga *Haematococcus pluvialis*, which is able to accumulate around 4% of its dry weight of astaxanthin under stressing conditions and has become the first biotechnological source of natural ketocarotenoids (1), but its culture at industrial scale is not free of problems. In vegetative cells of *Chlamydomonas reinhardtii* no ketocarotenoids have been found, although recent studies have demonstrated the existence of a putative *Bkt* gene in its genome and the presence of ketocarotenoids in mature zygospores of this microalga (7).

In vivo production of astaxanthin and other ketocarotenoids by organisms that do not synthesize them in the natural way has been achieved by metabolic engineering in *E. coli* (8, 9), cyanobacteria (10), and several higher plants (11–15). Here we propose the use of the unicellular microalga *C. reinhardtii* to express the *Bkt* gene from *H. pluvialis*, inserted in an adequate construction, to synthesize a new ketocarotenoid that the vegetative cells of *Chlamydomonas* do not synthesize in the natural way. The methodology needed to successfully achieve metabolic engineering of ketocarotenoids synthesis in *Chlamydomonas* is developed in the next subheadings of this chapter and includes: (1) total RNA extraction from *H. pluvialis.* (2) Synthesis of *Bkt* cDNA from *H. pluvialis.* (3) Integration of the obtained *Bkt* gene in pS104-PLK-tp plasmid. (4) Transformation of *C. reinhardtii* with the obtained construction. (5) Selection of positive transformants.

2. Materials

2.1. Isolation of RNA

1. *H. pluvialis* SAG 192-80 (16).
2. Hutner traces solution: Solve 10 g of EDTA (free acid) in 250 mL of H_2O. In a different recipient, solve in 500 mL of water at 100°C and in the following order 2.28 g of BO_3H_3, 4.40 g of $ZnSO_4{\cdot}7H_2O$, 1.02 g of $MnCl_2{\cdot}4H_2O$, 1 g of $FeSO_4{\cdot}7H_2O$, 0.32 g of $CoCl_2{\cdot}6H_2O$, 0.32 g of $CuSO_4{\cdot}5H_2O$, and 0.22 g of $(NH_4)_6MoO_{24}{\cdot}4H_2O$. Mix both solutions and heat the mixture again to 100°C. Let the solution to cool until 80–90°C and adjust to pH 6.8 with 20% KOH. Complete the final solution to 1 L total volume and store it in the dark for 2 days before use (17).
3. Liquid medium for *H. pluvialis*: 0.72 g/L of KPO_4H_2, 1.44 g/L of K_2PO_4H, 0.25 mM $MgSO_4{\cdot}7H_2O$, 0.13 mM $CaCl_2{\cdot}2H_2O$, 0.5 g/L of NH_4Cl, and 5 mL/L of Hutner traces solution. For nitrogen-free medium no ammonium was added.
4. Lysis RNA solution: 50 mM Tris–HCl, pH 8.0, 0.3 M NaCl, 5 mM EDTA, and 2% SDS, prepared in DEPC treated water (see Note 1).
5. PCI: phenol–chloroform–isoamyl alcohol 25:24:1, saturated with 10 mM Tris buffer pH 8.0 (see Note 2).
6. Chloroform saturated with 10 mM Tris buffer pH 8.
7. 0.15 M sodium acetate.

2.2. Synthesis of Bkt cDNA from H. pluvialis

1. Oligo dT (Invitrogen, Carlsbad, CA, USA).
2. SuperScript II (Invitrogen, Carlsbad, CA, USA).
3. RNaseH (Invitrogen, Carlsbad, CA, USA).

4. dNTP mix 10 mM: For 2 mL add 200 μL of 100 mM dATP, 200 μL of 100 mM dCTP, 200 μL of 100 mM dGTP, 200 μL of 100 mM dTTP, and 1,200 μL of milli-Q water.
5. Forward primer A: 5′-TGCCGCTCGAGAGCCTCAAATAA-3′. XhoI site is underlined.
6. Reverse primer B: 5′-CACTCCTGCAGACGCAAGACATC-3′. PstI site is underlined.
7. Agarose (BMA, Rockland, ME, USA).
8. TAE buffer stock 50× solution: 242 g of Tris, 57.1 mL of glacial acetic acid, and 100 mL of 0.5 M EDTA pH 8.
9. Ethidium bromide 5 mg/mL in water (see Note 3).
10. UV-transilluminator.
11. Quiaquick gel extraction kit (Qiagen, Hilden, Germany).
12. pGemT plasmid (Promega, Madison, WI, USA).

2.3. Integration of the Obtained Bkt Gene in the Constructed Plasmid

1. pSI103 (18) (see Note 4).
2. pSI104-tpPLK (19) (see Note 5).
3. LB (Luria Bertani): Add 10 g of bacto-tryptone, 5 g of yeast extract, and 10 g of NaCl into 1 L of deionized water. Autoclave at 121°C for 15 min.
4. LB-agar: LB and 20 g/L agar. Autoclave at 121°C for 15 min.
5. TCM: 10 mM Tris, 10 mM $CaCl_2$, and 10 mM $MgCl_2$, pH 7.
6. Solution I: 50 mM Glucose, 25 mM Tris–HCl pH 8, and 10 mM EDTA.
7. Solution II: 0.2 M NaOH and 1% SDS.
8. Solution III: 60 mL of 5 M potassium acetate, 11.5 mL of glacial acetic acid, and 28.5 mL of water.
9. T4 DNA ligase.
10. Ligase buffer 10×: 300 mM Tris–HCl (pH 7.8 at 25°C), 100 mM $MgCl_2$, 100 mM DTT, and 10 mM ATP.
11. *E. coli* strain DH5α (20).
12. 50 mM $CaCl_2$.
13. Ampicillin (Sigma, St. Louis, MO, USA).

2.4. Transformation of Chlamydomonas reinhardtii with the Obtained Construction

1. *Chlamydomonas reinhardtii* 704 strain (Cw15, Arg7, mt +) (21).
2. Solution Tris: Tris 121 g/L.
3. Solution A: 5 g/L of $CaCl_2 \cdot 2H_2O$, 10 g/L of $MgSO_4 \cdot 7H_2O$, and 40 g/L of NH_4Cl.
4. Solution B: 115 g/L of K_2HPO_4, and 6 g/L of KH_2PO_4.
5. TAP: For 1 L of TAP medium add 20 mL of solution Tris 1 M, 10 mL of solution A, 1 mL of solution B, 1 mL of Hutner

traces solution, and 0.95 mL of glacial acetic acid, and adjust to pH 7. Autoclave at 121°C for 15 min.

6. TAP-agar: TAP and 20 g/L agar. Autoclave at 121°C for 15 min.
7. Microalgal transformation mix: 0.3 g of sterile glass beads (0.4–0.6 mm ∅) and 0.2 mL of 20% polyethylene glycol 8000.
8. Paromomycin (Sigma, St. Louis, MO, USA).

2.5. Selection of Positive Transformants

1. Lysis buffer for genomic DNA: 10 mM Tris–HCl, pH 8.0, 1 mM EDTA, and 3% SDS.
2. TE buffer: 10 mM Tris–HCl pH 8.0, and 1 mM EDTA.
3. TE buffer for genomic DNA: 10 mM Tris–HCl, pH 8 and 0.1 mM EDTA.
4. 5 mM Tris–HCl, pH 8.
5. Forward primer C: 5′-AGCGGTGCCCTCCTGATAAAC-3′.
6. Reverse primer D: 5′-TTCCGGTAAGCTGCTCCAACAT-3′.
7. Nylon syringe filters (Fisher Scientific, Hudson, NH, USA).
8. 3 M sodium acetate, pH 5.2.
9. HPLC apparatus equipped with UV-Vis detector and RP-18 column.
10. Mobile phase: Solvent A (ethyl acetate) and solvent B (acetonitrile/water) 9:1 (v/v).

3. Methods

3.1. Total RNA Extraction from H. pluvialis

1. Transfer a well-grown *H. pluvialis* culture to high light *conditions* at 25°C for 3 days (see Note 6).
2. Harvest 200 mL of the *H. pluvialis* cells by centrifugation at 4,000 × *g* for 5 min.
3. Resuspend the pellet in 3 mL of the buffer lysis solution.
4. Vortex for 20 min and incubate on ice for 5 min (see Note 7).
5. Add an equal volume of PCI, mix vigorously, and centrifuge at 4,000 × *g* for 5 min at room temperature.
6. Transfer the upper aqueous phase to another tube and repeat the extraction with PCI until the interface is clear.
7. Add the same volume of chloroform, mix, and centrifuge at 4,000 × *g* for 10 min at room temperature. Take the aqueous phase.
8. Add 2.4 volumes of ethanol absolute and incubate at -20°C for at least 4 h.
9. Centrifuge at 4°C and 4,000 × *g* during 10 min.

10. Discard flow-through, resuspend the pellet in 0.5 mL of water-DEPC, and add the same volume of LiCl 8 M.
11. Incubate a 0–4°C overnight.
12. Centrifuge at 4,000×*g* for 30 min at 4°C.
13. Wash the pellet with ethanol 70% two times.
14. Dry the pellet and resuspend it in water-DEPC.
15. Add 2.5 volumes of 0.15 M sodium acetate and incubate a –20°C overnight.
16. Centrifuge and wash the pellet with ethanol 70%.
17. Dry the pellet and resuspend in water-DEPC.
18. Store at –80°C.

3.2. Synthesis of Bkt cDNA from H. pluvialis

1. To synthesize single-strand *Bkt* cDNA, add the following components to a nuclease-free microcentrifuge tube: 1 μL of Oligo dT (500 μg/mL), 1 μL of the total RNA previously obtained, 1 μL dNTP mix (10 mM each), and 9 μL of sterile distilled water.
2. Heat the mixture to 65°C for 5 min and quickly chill on ice.
3. Add 4 μL of the 5× first strand buffer supplied with the SuperScript II RNaseH-reverse transcriptase, mix contents of the tube gently, and incubate at 42°C for 2 min.
4. Add 1 μL of the reverse transcriptase (SuperScript II RNaseH 200 units) and mix very gently again.
5. Incubate at 42°C for 50 min.
6. Inactivate the reaction by heating at 70°C for 15 min.
7. Add 1 μL of RNAase H and incubate at 37°C for 20 min to remove the RNA complementary to the cDNA.
8. Using 2 μL of the retrotranscriptase reaction mixture, and the specific forward primer A and reverse primer B for *HpBkt*, carry out the PCR reaction.
9. To carry out the PCR amplification add in a total volume of 50 μL: 2 μL of the RT reaction mixture, 20 pmol of each primer, 0.2 mM dNTPs, 1 U pfu *Taq* DNA polymerase, 5 μL of specific 10× buffer (containing 2.5 mM $MgCl_2$), and 1% dimethylsulfoxide (DMSO). Run the following PCR program: 0.5 min at 96°C, 0.5 min at 42°C, and 1.5 min at 72°C for 30 cycles.
10. To purify the PCR product, run an electrophoresis in a 1% agarose gel.
11. The 1% agarose gel is prepared by solving 1 g of agarose in 100 mL of TAE buffer 1× and melting it in a microwave. Add 5 mL of ethidium bromide solution and gently pour the gel in the tray (try to avoid air bubbles) and place a separation comb

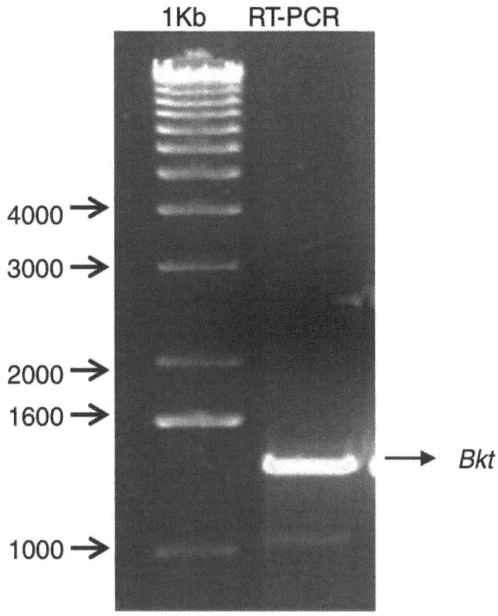

Fig. 2. DNA fragment corresponding to the *HpBkt* cDNA obtained by RT-PCR from total RNA and visualized in an EtBr-stained agarose 1% gel. The obtained band has a size of 1.4 kb.

into the tray. Allow the gel to completely solidify. The gel is ready to be loaded and run.

12. Visualize the band on a UV-transilluminator by staining with ethidium bromide (see Note 3). The expected band has a size of 1.4 kb (see Fig. 2).
13. Isolate the DNA containing the *HpBkt* cDNA fragment from the agarose band using the Quiaquick gel extraction kit (see Note 8).
14. Store frozen the obtained cDNA until its utilization or clone it in pGemT plasmid (see Note 9).

3.3. Integration of the Obtained Bkt Gene in the pSI104-tpPLK Plasmid

1. Inoculate the bacteria which contain plasmid pSI104-tpPLK and plasmid pGemT-*HpBkt* in different tubes with 3 mL of LB medium. Incubate overnight at 37°C with agitation (200 rpm) and isolate the plasmid following the alkaline lysis protocol (22) as described below.
2. Harvest 1.5 mL of the culture by centrifugation at 12,000 × *g* in a bench microcentrifuge for 30 s. Remove the medium, leaving the bacterial pellet as dry as possible.
3. Resuspend the pellet in 0.2 mL of cold solution I by vigorous vortexing.
4. Add 0.2 mL of cold freshly prepared solution II, and mix by inverting the tube several times. Do not vortex. Leave the tube in ice for 5 min.

5. Add 0.2 mL of cold solution III (60 mL of potassium acetate 5 M, 11.5 mL of glacial acetic acid and 28.5 mL of water) and mix gently by inverting the tube 6 or 8 times. Leave the tube in ice between 5 and 10 min.
6. Centrifuge at 12,000 × *g* for 5 min in a microfuge. Transfer the supernatant to a fresh tube.
7. Add 600 mL of PCI (see Note 2), mix by vortexing, and centrifuge at 12,000 × *g* for 3 min.
8. Transfer the aqueous upper phase to a fresh tube and repeat PCI extraction until the interphase is clear.
9. Precipitate the plasmidic DNA with 2 volumes of ethanol at −20°C for 10–20 min. Centrifuge at maximum speed for 5 min at 4°C in a microfuge and remove the supernatant.
10. Rinse the pellet with 1 mL of 70% ethanol, remove the supernatant, and allow the plasmidic pellet to dry in the air for 10 min.
11. Digest pSI104-tpPLK and pGemT-*HpBkt* with the restriction enzymes *XhoI* and *PstI* (see Note 10).
12. Run electrophoresis in 1% agarose gel (see Subheading 2.2).
13. Cut the bands corresponding to the digested pSI104PLK (about 3.9 kb) and *HpBkt* cDNA (about 1.3 kb).
14. Purify the DNA using the Quiaquick gel extraction kit (see Note 8).
15. To ligate both fragments with T4 ligase, add the following components to a nuclease-free microcentrifuge tube: vector DNA, insert DNA, ligase 10× buffer, T4 DNA ligase, and nuclease-free water to final volume of 10 μL. Use a 1:3 molar ratio of vector:insert DNA when cloning a fragment into a plasmid vector. Incubate the reaction at 16°C overnight.
16. Transform *E. coli* competent cells with the obtained pSI104-tpPLK-*Bkt* plasmid (see Notes 11 and 12).
17. Select 2 or 3 of the colonies that have appeared and grow them in liquid LB with ampicillin (see Note 13).
18. Isolate the plasmid from the chosen colonies by alkaline lysis method (see steps 2–10) and check by digestion or sequencing that the isolated plasmids are pSI104PLK-tp-*HpBkt*.
19. Use the new plasmid pSI104PLK-tp-*HpBkt* to transform *C. reinhardtii* cells.

3.4. Transformation of *C. reinhardtii* with pSI104PLK-tp-HpBkt Plasmid

Transformation is carried out according to the glass-beads method of Kindle (23) with minor modifications.

1. Grow a culture of *C. reinhardtii* in TAP medium at 25°C to a cell density of about 10^7 cells/mL (see Note 14).

2. Harvest 600 mL of *C. reinhardtii* cells and resuspend them in 0.6 mL fresh TAP medium to obtain a 100-fold concentrated cell suspension.
3. Add the concentrated cell suspension (0.6 mL) to a conical tube containing the transformation mix and about 500 ng of the plasmid pSI104PLK-tp-*HpBbkt*, obtained as described in Subheading 3.3, and 500 ng of plasmid pSI103 which carries resistance to the antibiotic paromomycin.
4. Vortex for 8 s, add 5 mL of TAP medium, centrifuge 3 min at 6,000 × *g*, and discard supernatant.
5. Resuspend the cellular pellet in a small volume of TAP medium and transfer to a new 50-mL tube. Complete with fresh sterile TAP until 50 mL and incubate at 25°C and standard light overnight.
6. After this incubation in the absence of antibiotic, pellet the cells and spread them onto TAP Petri dishes with 30 μg/mL paromomycin (see Note 13).
7. Incubate at 25°C. Transformed colonies will be visible after 4 or 5 days.

3.5. Selection of Positive Transformants

The next step is to check which of the paromomycin-resistant transformants of *C. reinhardtii* has the plasmid inserted in their genome.

1. To isolate the genomic DNA from the transformants, scrap one loop of each *Chlamydomonas* transformant from the plate, resuspend it in 10 μL of lysis buffer and incubate at room temperature for 15 min.
2. Add 500 μL of TE buffer and 60 μL of 3 M sodium acetate, pH 5.2.
3. Extract the DNA with PCI and precipitate with isopropanol.
4. Wash the pellet with 70% ethanol, dry, and resuspend in 5 mM Tris–HCl, pH 8 (see Note 15).
5. Use 1 μL of the genomic DNA solution to carry out a PCR reaction using the forward primer C and the reverse primer D, which anneal with *RbcS2* promoter and *HpBkt* gene amplifying a 547-bp fragment. It is expected that about 30–50% of the transformants obtained with two plasmids acquired both of them.
6. Take several of the positive transformants in which insertion of the *HpBkt* gene has been shown to analyse their pigment composition.
7. Grow a culture (3 mL) of each transformant in TAP medium at 25°C.
8. Take 1 mL of a well grown culture, heat at 100°C for 5 min, and add 4 mL of acetone.

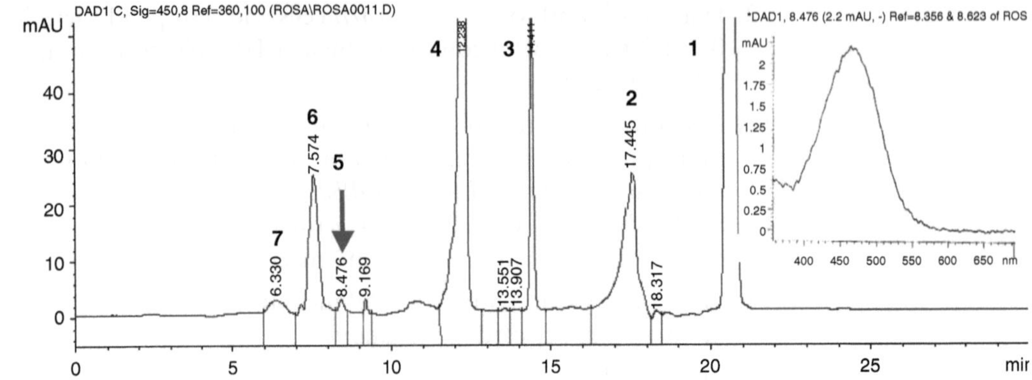

Fig. 3. HPLC profile of carotenoids present in acetone extracts from transgenic *Chlamydomonas* cells containing the *H. pluvialis Bkt* gen. The *arrow* indicates the new carotenoid present in the transformed cells. The absorlion spectra corresponding to the new peak is also shown. Peaks identification: 1-β-carotene, 2-chlorophyll a, 3-chlorophyll b, 4-lutein, 5-keto-lutein, 6-violaxanthin, 7-neoxanthin.

9. Vortex vigorously and centrifuge at 6,000 × *g* for 10 min to eliminate the cell debris.
10. Filter the supernatant through nylon syringe filters and inject a fraction of the filtered supernatant in an HPLC equipped with UV-Vis detector and RP-18 column.
11. Apply the mobile phase at a flow rate of 1 mL/min with the following gradient programme: 0–16 min 0–60% A; 16–30 min 60% A; 30–35 min 100% A. Injection volume 100 μL. Carry out pigments detection at 450 nm (24). In Fig. 3 it is shown an example of the chromatographic pigment analysis of one of the transgenic *Chlamydomonas* strains expressing the *Bkt* gene from *H. pluvialis* showing the presence of the new ketocarotenoid, keto-lutein.

4. Notes

1. Diethylpyrocarbonate (DEPC) is used to inactivate the RNase enzymes from water and other laboratory utensils. Solve 0.1% (v/v) DEPC in the water or solution to be treated and stir with a magnetic bar for at least 1 h at room temperature. Then autoclave to transform traces of DEPC, into CO_2, H_2O, and EtOH that will evaporate. DEPC cannot be used with Tris buffer or HEPES since their amino groups will inactivate DEPC by reacting with it.
2. To make PCI 25:24:1. Add to 50 mL of phenol, 50 μL of dihydroxiquinoline dihydrate and 50 mL of Tris 1 M and mix with a magnetic bar. Add 450 mL of water and mix again. Wait until two phases appear. Control the pH of the upper aqueous phase. It must be between 7.8 and 8. Discard most of the aqueous

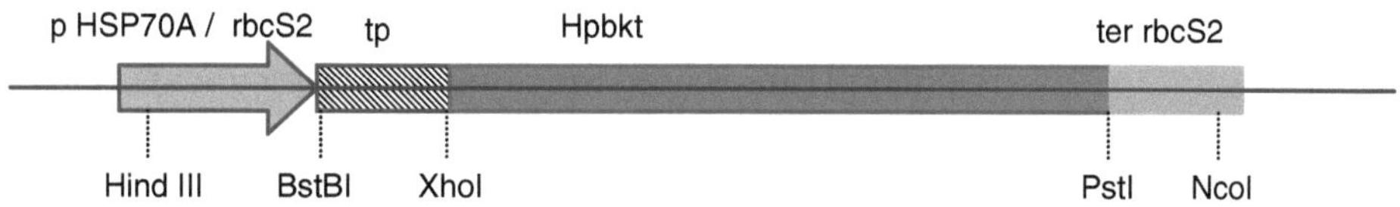

Fig. 4. Schematic representation of the construction used to express the *HpBkt* gene from *H. pluvialis* in *Chlamydomonas*. *prbcS2* ribulose 1,5 bisphosphate carboxylase small subunit promoter; *ter rbcS2* ribulose 1,5 bisphosphate carboxylase small subunit terminator region; *pHSP70A* heat shock protein 70A promoter; *tp* transit peptide; *Bkt* β-carotene ketolase. The whole construction has 2,550 bp.

phase (leave only 1 or 2 cm) and add 24 parts of chloroform and 1 part of isoamyl alcohol. Keep the solution at 4°C preserved from the light. All the process should be carried out with gloves and in a fume hood due to high toxicity of phenol.

3. Ethidium bromide (EtBr) is a very dangerous mutagenic agent. It must be handled with gloves and with extreme caution. Solutions containing EtBr and solids contaminated with EtBr should be discarded as toxic chemical waste.

4. Plasmid pSI103 (18) contains the *AphVIII* gene from *Streptomyces rimosus*, coding for an aminoglycoside 3′ phosphotransferase that confers resistance to the antibiotic paromomycin, under the control of the strong constitutive promoters *RbcS2* and *Hsp70A* and terminated by the 3′ untranslated region of *RbcS2*.

5. Plasmid pSI104-tpPLK contains a polylinker region under the control of the strong constitutive promoters *RbcS2* and *Hsp70A* and terminated by the 3′ untranslated region of *RbcS2*. It also has the RBCS2 transit peptide encoding sequence immediately after the promoter and before the polylinker to ensure transport to the chloroplast of the desired protein (see Fig. 4). This plasmid has been obtained in our lab after modification of plasmid pSI103 (19).

6. Cultures are grown in mineral liquid medium at 25°C and bubbled with air containing 3% (v/v) CO_2. Use a light intensity of 500 $\mu E/m^2/s^1$.

7. All tips, tubes, and solutions used for manipulation of DNA should be treated in autoclave and handled with gloves to avoid degradation of DNA by nucleases.

8. Gel extraction kits are commercialized by many companies, such as Quiagen or Eppendorf. These kits allow the purification of DNA fragments up to 10 kb from agarose gels based on silica-membrane columns. Buffers for binding, washing, and elution of DNA are provided with the kit.

9. Direct digestion of PCR products can be difficult, especially when restriction sites are near the ends of the fragment. Cloning the fragment into a plasmid allows easy conservation and propagation of the obtained DNA and easy digestion and

isolation of the desired fragment. pGEMT is one of the most popular plasmids used for cloning PCR fragments. This vector is provided linearized and with 3′ terminal thymidine overhangs (3′-T) which prevent recircularization of the vector and a compatible overhang for ligation of PCR products generated by certain thermostable polymerases which often add a single deoxyadenosine to the 3′ ends of amplified fragments. The plasmid is provided in a kit with 2× rapid ligation buffer which allows reactions to be completed in 1 h at room temperature.

10. To carry out plasmid digestion incubate the DNA (up to 1 μg) with 1 unit of the desired restriction enzyme and the corresponding buffer at 37°C for 2 or 3 h.

11. Competent bacterial cells are bacteria that have gained the ability to take up extracellular DNA and can be obtained by different methods. The method that we use in our lab consists on:

    (a) Inoculate 3 mL of LB medium with *E. coli* DH5α and incubate overnight at 37°C with shaking at 200 rpm. Transfer 60-μL aliquots of the culture into new tubes with 3 mL of LB media and incubate for 2–3 h at 37°C with shaking at 200 rpm until DO_{580} near to 0.5 (it is important that the optical density is not higher than 0.5).

    (b) Transfer the 6 mL (two tubes) of the cells to a sterile 15-mL polypropylene centrifuge tube, and collect the cells by centrifugation at 6,000 × *g* for 5 min. After centrifugation, decant the supernatant and resuspend gently the cell pellet in one-half of the original volume (3 mL) of cold, sterile 50 mM $CaCl_2$. Incubate in ice for 20–30 min, and centrifuge at 4°C at 6,000 × *g* for 5 min.

    (c) Decant the supernatant and gently resuspend the cell pellet in one-tenth the original volume (0.6 μL) of cold, sterile 50 mM $CaCl_2$. Keep the competents cells obtained at −80°C until their use.

12. For the transformation of competent cells, transfer 200 μL of competent cell to a nuclease-free microcentrifuge tube, and add 100 μL TCM buffer and DNA (no more than 50 ng in a volume of 10 μL). Mix the contents and incubate in ice for 20–30 min. Incubate at 43°C for 90 s. Do not shake the tubes. Add 0.7 mL of LB medium and incubate at 37°C for 30–40 min with shaking at 200 rpm. Transfer the transformed competent cells (1 mL) onto agar LB medium plates containing 100 μg/mL ampicillin. Incubate overnight at 37°C. All the process should be carried out in sterile conditions.

13. Antibiotics (ampicillin and paromomycin) are solved in water, sterilized by filtration through 0.2 μm filters, and stored at −20°C.

14. Standard cultures are grown in mineral liquid medium at 25°C, bubbled with air containing 3% (v/v) CO_2, and continuously

illuminated with cool white and daylight from fluorescent lamps (100 μE/m^2/s^1).

15. Additional information can be seen at http://www.chlamy.org/methods/dna.html.

References

1. Lorenz R, Cysewski GR (2000) Commercial potential for *Haematococcus* microalgae as a natural source of astaxanthin. Trends Biotechnol 18:160–167
2. Guerin M, Huntley ME, Olaizola M (2003) *Haematococcus* astaxanthin applications for human health and nutrition. Trends Biotechnol 21:210–216
3. Misawa N, Nakagawa M, Kobayashi K et al (1990) Elucidation of the *Erwinia uredovora* carotenoid biosynthetic pathway by functional analysis of gene products expressed in *Escherichia coli*. J Bacteriol 172:6704–6712
4. Misawa N, Satomi Y, Kondo K et al (1995) Structure and functional analysis of a marine bacterial carotenoid biosynthesis gene cluster and astaxanthin biosynthetic pathway proposed at the gene level. J Bacteriol 177:6575–6584
5. Fernández-González B, Sandmann G, Vioque A (1997) A new type of asymmetrically acting β-carotene ketolase is required for the synthesis of echinenone in the cyanobacterium *Synechocystis* sp. J Biol Chem 272:9728–9733
6. Cunningham FX, Gantt E (1998) Genes and Enzymes of carotenoid biosynthesis in plants. Ann Rev Plant Physiol Plant Mol Biol 49: 557–583
7. Lohr M, Chung-Soon I, Grossman AR (2005) Genome-based examination of chlorophyll and carotenoid biosynthesis in *Chlamydomonas reinhardtii*. Plant Physiol 138:490–515
8. Breitenbach J, Misawa N, Kajiwara S et al (1996) Expression in *Escherichia coli* and properties of the carotene ketolase from *Haematococcus pluvialis*. FEMS Microbiol Lett 140:241–246
9. Lotan T, Hirschberg J (1995) Cloning and expression in *E. coli* of the gene encoding β-C-4 oxygenase, that converts β-carotene to the ketocarotenoid canthaxanthin in *Haematococcus pluvialis*. FEBS Lett 364:125–128
10. Harker M, Hirschberg J (1997) Biosynthesis of ketocarotenoids in transgenic cyanobacteria expressing the algal gene for β-C-4-oxygenase, *crtO*. FEBS Lett 404:129–134
11. Ralley L, Fraser PD (2004) Metabolic engineering of ketocarotenoid formation in higher plants. Plant J 39:477–486
12. Stalberg K, Lindgren O, Ek B et al (2003) Synthesis of ketocarotenoids in the seed of *Arabidopsis thaliana*. Plant J 36:771–779
13. Mann V, Harker M, Pecker I et al (2000) Metabolic engineering of astaxanthin production in tobacco flowers. Nat Biotechnol 18:888–892
14. Gerjets T, Sandmann G (2006) Ketocarotenoid formation in transgenic potato. J Exp Bot 57: 3639–3645
15. Morris WL, Ducreux LJM, Fraser PD et al (2006) Engineering ketocarotenoid biosynthesis in potato tubers. Metab Eng 8:253–263
16. Pringsheim EG (1966) Nutritional requirements of *Haematococcus pluvialis* and related species. J Phycol 2:1–7
17. Hutner SH, Provosoli L, Schatz A, Haskins CP (1950) Some approaches to the study of the role of metals in the metabolism of microorganisms. Proc Am Philos Soc 94:152–170
18. Sizova I, Fuhrmann M, Hegemann P (2001) A *Streptomyces rimosus AphVIII* gene coding for a new type phosphotransferase provides stable antibiotic resistance to *Chlamydomonas reinhardtii*. Gene 277:221–229
19. León R, Fernández E, Couso I (2007) Metabolic engineering of ketocarotenoids biosynthesis in the unicelullar microalga *Chlamydomonas reinhardtii*. J Biotechnol 130: 143–152
20. Hanaha D (1983) Studies on transformation of *Escherichia coli*. Eur J Biochem 179:185–194
21. Loppes R, Radoux M, Ohresser MC, Matagne RF (1999) Transcriptional regulation of Nia1 gene encoding nitrate reductasa in *Chlamydomonas reinhardtii*. Plant Mol Biol 41:701–711
22. Sambrook J, Russell DW (2001) Molecular cloning. A laboratory manual, vol 3. Cold Spring Harbour Laboratory Press, Cold Spring Harbour, New York
23. Kindle KL (1990) High-frequency nuclear transformation of *Chlamydomonas reinhardtii*. Proc Natl Acad Sci U S A 87:1228–1232
24. Young A, Orset S, Tsavalos A (1997) Methods for carotenoids analysis. In: Pessarakli M (ed) Handbook of photosynthesis. Marcel Dekker, New York

Chapter 18

Characterization of Carotenogenesis Genes in the Cyanobacterium *Anabaena* sp. PCC 7120

Hajime Masukawa, Mari Mochimaru, and Shinichi Takaichi

Abstract

Cyanobacteria produce many kinds of carotenoids for light harvesting and light protection in photosynthesis. To elucidate the biosynthetic pathways of carotenoids in *Anabaena* sp. PCC 7120 (also known as *Nostoc* sp. PCC 7120), we have produced gene-disruption mutants lacking selected proposed carotenoid biosynthetic enzymes. Here we describe the construction of mutants by triparental mating. A cargo plasmid, bearing a target gene interrupted by an antibiotic-resistant cassette, is transformed to *E. coli* donor containing a helper plasmid, and is introduced into *Anabaena* cells by conjugation. Double-reciprocal recombination replaces the target genes in *Anabaena* genome with mutated ones on the plasmid. Carotenoids in the selected double recombinants are identified using high-performance liquid chromatography.

Key words: *Anabaena* sp. PCC 7120, Carotenogenesis, Carotenoid, Cyanobacterium, Gene disruption, *sacB*, Triparental mating

1. Introduction

We have previously identified the molecular structures of carotenoids in some *Anabaena* and *Nostoc* strains (1–3). In these genera, the major carotenoids are β-carotene and echinenone, and the polar carotenoids are myxol and 4-ketomyxol glycosides, whose presence is limited to cyanobacteria. In *Anabaena* sp. PCC 7120 (also known as *Nostoc* sp. PCC 7120), the myxol and 4-ketomyxol glycosides are (3*R*,2′*S*)-myxol 2′-fucoside and (3*S*,2′*S*)-4-ketomyxol 2′-fucoside, respectively (1). Their glycoside moiety is α-L-fucose, although it was previously believed that rhamnose was the common glycosylated sugar in cyanobacteria (4, 5).

Elucidation of the cyanobacterial carotenogenesis pathways is essential for better understanding the diversity of carotenoids. We have proposed the candidates of the carotenogenesis genes in

José-Luis Barredo (ed.), *Microbial Carotenoids from Bacteria and Microalgae: Methods and Protocols*, Methods in Molecular Biology, vol. 892, DOI 10.1007/978-1-61779-879-5_18,

Anabaena sp. PCC 7120 by sequence similarity based on amino acid sequences from the functionally confirmed genes (1, 5), because its genome sequencing has already been completed (6). To identify the function of these candidates, the construction of the gene disruption mutants and comparison of the carotenoids from these mutants with those from the wild type are considered effective. *Anabaena* sp. PCC 7120 is one of the most suitable materials, because their cells can be manipulated by genetical techniques based on conjugative transformation (7–9).

Here, we describe our protocols, which we have used for functional identification of two distinct β-carotene ketolases, CrtW and CrtO, from *Anabaena* sp. PCC 7120 (10). Gene disruption mutants of these genes were constructed by interrupting the target genes with neomycin resistance (Nm^r) gene, *npt*, as a selectable marker. DNA fragments containing the target gene interrupted by *npt* gene were cloned to pRL271, a cargo plasmid for triparental mating, containing the *sacB* gene (11, 12). The resulting plasmids were transformed to competent cells of HB101 (*E. coli* donor strain) containing pRL623, a helper plasmid. The cargo plasmid carrying the *npt*-interrupted target gene was introduced into *Anabaena* cells by conjugation that depends on conjugative plasmid RP4 in *E. coli* strain J53 and finally integrated into *Anabaena* chromosome via homologous recombination by a single-crossover event. Double-reciprocal recombination replaced the target genes with mutated ones, and double-crossover recombinants could be selected by the conditionally lethal gene, *sacB*, in cargo plasmid pRL271. The isolated double recombinants were grown photoautotrophically, and the organic solvent-soluble pigments were extracted from the cells. The extracted pigments were analyzed by high-performance liquid chromatography (HPLC).

2. Materials

Sequence information for the genomic DNA from *Anabaena* sp. PCC 7120 is available in the CyanoBase database (http://genome.kazusa.or.jp/cyanobase/Anabaena/) (6).

2.1. Reagents for DNA Manipulation

1. DNA polymerase: High-fidelity enzymes.
2. Restriction enzymes: BamHI, XbaI, SmaI, SacI, XhoI, and SalI.
3. Antibiotic stock solutions: Ampicillin (Ap), neomycin (Nm), and kanamycin (Km) are dissolved in distilled water. Chloramphenicol (Cm) and erythromycin (Em) are dissolved in ethanol. Each at final concentration of 25 mg/mL and these solutions are filter-sterilized (pore size 0.22 μm). Store at −20°C.
4. Nitrocellulose filters HATF 08250 (Millipore, Billerica, MA, USA).

2.2. Anabaena sp. PCC 7120 and E. coli Strains

1. *Anabaena* sp. PCC 7120 (Pasteur Culture Collection of Cyanobacteria, Paris, France).
2. *E. coli* strain XL1-Blue MRF' (Agilent Technologies, La Jolla, CA, USA).
3. *E. coli* strain HB101 bearing pRL623 (Dr. C.P. Wolk, Michigan State University, MI, USA) (12).
4. *E. coli* strain J53 bearing RP4 (Dr. C.P. Wolk, Michigan State University, MI, USA) (13).

2.3. Media for Culture

1. LB: Add 10 g of bacto-tryptone, 5 g of yeast extract, and 10 g of NaCl into 1 L of deionized water. Autoclave at 121°C for 20 min.
2. LB-antibiotic(s): Final concentrations used for growth of *E. coli* recombinants are as follows: 50–100 μg/mL Ap, 50 μg/mL Km, 50–100 μg/mL Cm, and 50–100 μg/mL Em.
3. LB agar: LB and 20 g/L agar.
4. LB-antibiotics agar plates: Autoclave LB-agar, and when the solution cools down to 60°C, add the antibiotic(s) to the concentrations showed above. Pour 20–25 mL into each Petri dish.
5. BG11: Add 1.5 g of $NaNO_3$, 40 mg of $K_2HPO_4 \cdot 3H_2O$, 75 mg of $MgSO_4 \cdot 7H_2O$, 36 mg of $CaCl_2 \cdot 2H_2O$, 20 mg of Na_2CO_3, 1 mg of disodium EDTA, 6 mg of citric acid, 6 mg of ferric ammonium citrate, 20 mL of 1 M TES-KOH (pH 8.2), and 1 mL of A6 solution described below into 1 L of deionized water. Autoclave at 121°C for 20 min (14).
6. A6 solution: Add 2.86 g of H_3BO_3, 1.81 g of $MnCl_2 \cdot 4H_2O$, 0.22 g of $ZnSO_4 \cdot 7H_2O$, 0.39 g of $Na_2MoO_4 \cdot 2H_2O$, 79 mg of $CuSO_4 \cdot 5H_2O$, and 49 mg of $Co(NO_3)_2 \cdot 6H_2O$ into 1 L of deionized water. Store at 4°C.
7. BG11-antibiotic(s): Final concentrations used for growth of *Anabaena* recombinants are as follows: 30 μg/mL Nm, 5–10 μg/mL Em, and 5% sucrose (see Note 1).
8. BG11 agar: BG11 and 20 g/L agar.
9. BG11-antibiotics agar plates: Add the antibiotic(s) to the concentrations showed above onto BG11 agar and spread them with bacteria spreader.

2.4. Plasmids

1. pBluescript II SK(+) (Agilent Technologies, La Jolla, CA, USA).
2. pRL648: Containing the *npt* gene (Km^r/Nm^r) (11).
3. pRL271: Cargo plasmid for triparental mating, containing Cm^r, Em^r, and *sacB* genes (see Note 2) (15).
4. pRL623: Helper plasmid for triparental mating, containing genes encoding methylase that protect against activities of

AvaI, AvaII, and AvaIII (see Note 3) (12). Maintained in *E. coli* strain HB101.

5. RP4: Conjugal plasmid for triparental mating, maintained in *E. coli* strain J53 (see Note 2).

2.5. HPLC

1. Centrifuge Himac CR21 (Hitachi, Tokyo, Japan).
2. Centrifuge MR-15A (Tommy Seiko, Tokyo, Japan).
3. Acetone/methanol (7:2, v/v) for extraction. Store at room temperature.
4. Ultra-sonicator (VP-5S) (Taitec, Tokyo, Japan).
5. Chloroform/methanol (3:1, v/v) for solvent. Store at room temperature.
6. μBondapak C_{18} column (8×100 mm, RCM-type) (Waters, Milford, MA, USA) (see Note 4) (1, 10, 16).
7. Methanol/water (9:1, v/v) for HPLC eluent. Store at room temperature.
8. MCPD-3600 photodiode array detector (200–800 nm, 1.4 nm interval) (Otsuka Electronics, Osaka, Japan).

3. Methods

3.1. Cloning to a Cargo Plasmid

1. Construct a plasmid containing a target gene interrupted by an *npt* gene. It should contain two approximately 1-kb PCR fragments of a target gene, which sandwich an *npt* gene (see Note 5 and Fig. 1, Step 1). A 1.1-kb fragment of the *npt* gene can be cut out from pRL648 by digestion with BamHI, XbaI, or SmaI (see Notes 6 and 7).
2. Ligate the SacI–XhoI (or SacI–SalI) fragment containing the target gene and *npt* of the plasmid between SacI–XhoI sites of pRL271 (see Fig. 1, Step 2) (15).
3. Transform the ligated DNA into competent cells of *E. coli* (XL1-Blue MRF'), select transformants on LB agar medium containing Cm and Km, and purify the resulting cargo plasmid DNA.
4. Transform the purified cargo plasmid into competent cells of HB101 bearing helper plasmid pRL623 (see Fig. 1, Step 3) (12), and select transformants on LB agar medium containing Cm, Em, and Km.

3.2. Conjugal Transformation

1. For the preparation of *E. coli* donor cells, grow the selected HB101 (created in Subheading 3.1) carrying the cargo and helper plasmids and conjugative strain J53 (RP4) overnight at 37°C in 3 mL LB liquid medium with Cm, Em and Km, and with Ap and Km, respectively.

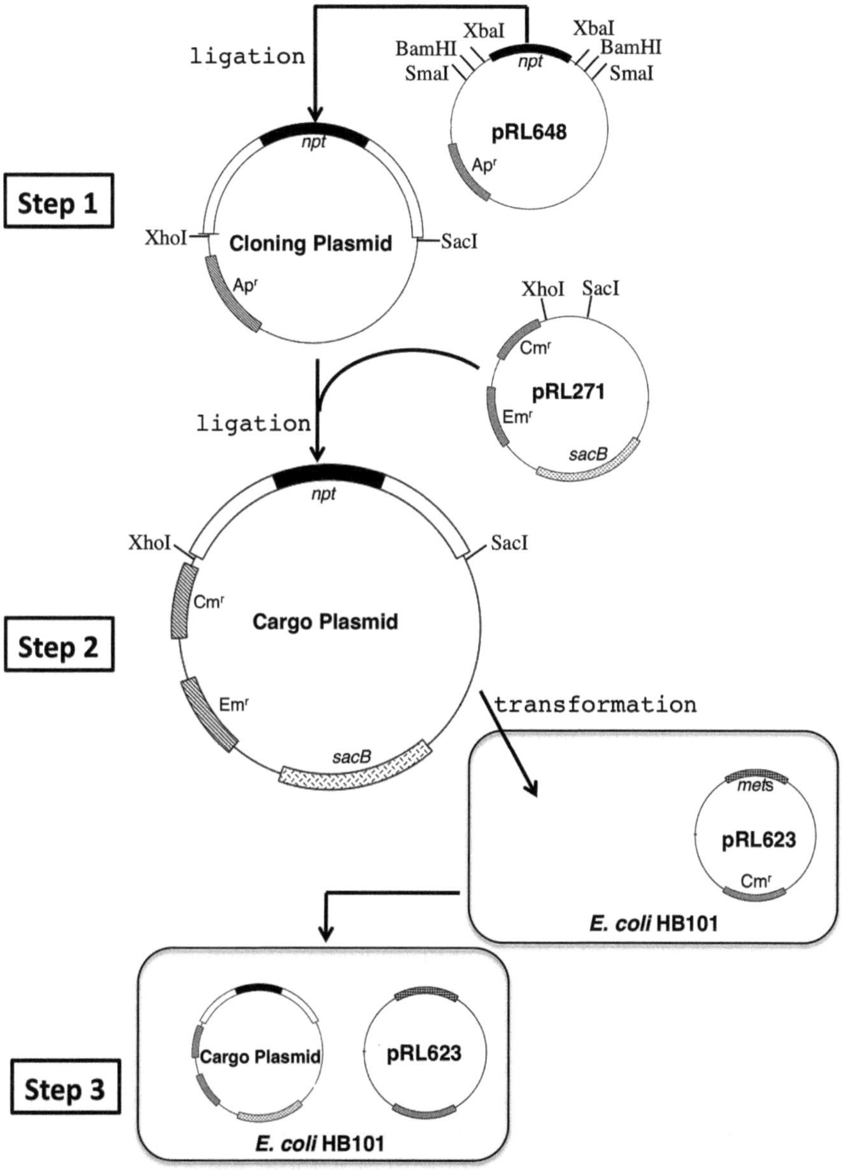

Fig. 1. Preparation of *E. coli* host for triparental mating. Step 1: Insertion of *npt* (neomycin phosphotransferase) gene conferring resistance to kanamycin and neomycin (shown as a *black box*) between PCR fragments of target gene (shown as *white boxes*) and construction of cloning plasmid. Step 2: Construction of cargo plasmid. Step 3: Transformation of the cargo plasmid to *E. coli* HB101 bearing helper plasmid, pRL623. Other antibiotic-resistance genes are shown as Ap^r, Cm^r, and Em^r. *sacB* encoding levansucrase is used for selection of double-crossover recombinants (17) and *mets* indicates three methylase genes that protect the cargo plasmid against restriction by the restriction enzymes of *Anabaena* sp. PCC 7120 (12).

2. Dilute the two overnight cultures about 40 times in fresh LB liquid medium with appropriate antibiotics (as in step 1) and grow at 37°C for about 3 h up to $OD_{610}=0.5$.
3. Transfer 1.5 mL of the two cultures to separate microcentrifuge tubes, centrifuge for 3 min at ~3,000 × *g* at room temperature, and discard medium.

4. Wash cell pellets by adding 300 μL of LB liquid medium without antibiotics, centrifuge as in step 3, and discard medium.
5. Resuspend cell pellets by gently pipetting in 100–150 μL of LB liquid medium without antibiotics.
6. Mix the two cell suspensions gently in the same tube, centrifuge as in step 3, and discard medium.
7. Resuspend as in step 5.
8. For the preparation of *Anabaena* sp. PCC 7120 recipient cells, collect 1-week-grown *Anabaena* cultures by centrifugation at ~3,000 × *g*, and discard medium.
9. Suspend cells with fresh BG11 liquid medium at a final concentration of about 10 μg chlorophyll *a* (Chl)/mL (see Note 8).
10. For the triparental spot mating, autoclave and dry at room temperature the nitrocellulose filters (HATF 08250), and place the filters onto BG11 agar medium without antibiotics.
11. For spot mating, apply 5–10 μL of cell suspension of *Anabaena* onto nitrocellulose filters.
12. Apply 2–5 μL of a two-strain mixed suspension of *E. coli* onto each *Anabaena* spot.
13. Incubate the filters under dim light conditions at 26–30°C for about 1 day.
14. Transfer the filters to fresh BG11 agar medium containing either Em or Nm (see Note 9) and incubate at 26–30°C under normal light conditions.
15. After about 1 week of incubation, suspend some mating spots with 200 μL of BG11 liquid medium, spread the cell suspension onto BG11 agar medium containing the same antibiotics as used in step 14, and incubate as in step 14.
16. When exconjugants grow to form visible colonies (after further about 2 weeks of incubation), pick up some of them (several to a couple of tens of colonies) and transfer to BG11 agar medium containing Nm with or without Em.

3.3. Selection of Double Recombinants

1. Grow a couple of isolates in 20–30 mL BG11 liquid medium with Nm at 26–30°C for about 1 week (see Note 10).
2. Dilute the 1-week-old cultures 50–100 times in 20–30 mL fresh BG11 liquid medium with Nm and incubate for 3–4 more days at 26–30°C.
3. Take 1–2 mL of the cell suspension, spread it onto BG11 agar medium containing Nm and 5% sucrose (see Note 11), and incubate under normal light conditions at 26–30°C.
4. Pick up the colonies (tens of colonies, if possible) that are probable double-crossover recombinants and maintain them on Nm-containing BG11 agar medium.

5. For confirmation of the (Nm^r Em^s sucrose^r) phenotype that true double-crossover recombinants exhibit, transfer very small amounts of the recombinant cells onto BG11 agar medium containing Nm with or without 5% sucrose and separately onto BG11 agar medium containing Em (18).
6. Select the recombinants resistant to Nm and sucrose, and sensitive to Em.
7. Verify the complete segregation of the target gene of the wild type and replacement of it by the mutated version in the recombinants by PCR and/or Southern blotting.

3.4. Extraction and Analysis of Pigments

1. The wild and mutant cells of *Anabaena* sp. PCC 7120 are cultured in 200–500 mL BG11 medium (1, 10, 16), and around 0.5 g cells in the stationary state are collected by centrifugation and stored at −30°C.
2. Carotenoids and Chl are extracted from the cells with acetone/methanol (7:2, v/v) using an ultra-sonicator and are collected by centrifugation.
3. After evaporation to dryness, the pigments are dissolved in a small volume of chloroform/methanol (3:1, v/v) (see Note 12).
4. The pigments are injected to HPLC and eluted with methanol/water (9:1, v/v) from the first 20 min and then with 100% methanol (2.0 mL/min) (1, 10, 16). Carotenoids and Chl are monitored at 475 and 664 nm, respectively, as indicated in Fig. 2.

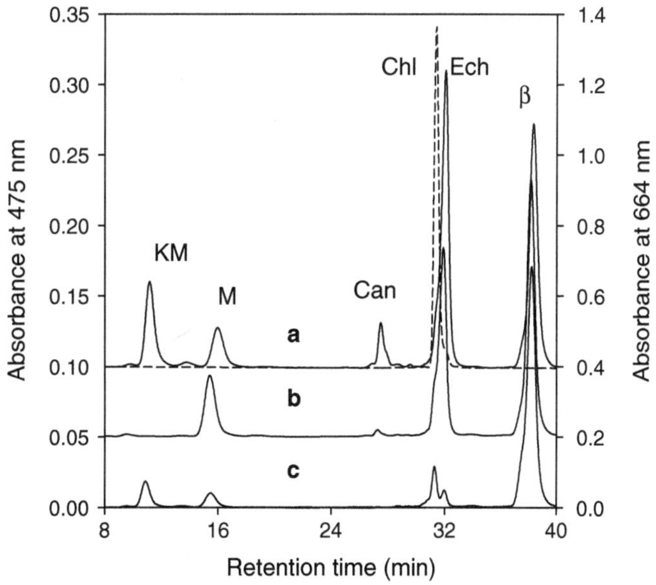

Fig. 2. HPLC elution profiles of pigments extracted from wild-type (**a**), and *crtW* (*alr3189*) (**b**), and *crtO* (*all3744*) (**c**) deleted mutants of *Anabaena* sp. PCC 7120 (see Note 13). *KM* 4-ketomyxol 2′-fucoside, *M* myxol 2′-fucoside, *Can* canthaxanthin, *Ech* echinenone, β β-carotene, *Chl* chlorophyll *a* (see Note 14) (10).

4. Notes

1. Filter-sterilized sucrose should be added to autoclaved BG11 agar medium, before the agar solidified, to prevent sucrose degradation during autoclave treatment.
2. The pRL271 plasmid harbors Cm- and Em-resistance genes and *sacB* gene encoding levansucrase, whose expression in the presence of sucrose kills some cyanobacteria. It is mobilized by the broad-host-range conjugal plasmid RP4 (Ap^r, Km^r, tetracycline^r) from the *E. coli* donor to the *Anabaena* host in triparental mating and can replicate in *E. coli* but not in *Anabaena* (non-replicating vector).
3. The pRL623 plasmid with methylase genes modifies AvaI, AvaII, and AvaIII sites in the cargo plasmid in the same *E. coli* cell to protect the cargo plasmid from restriction by AvaI, AvaII, or AvaIII in *Anabaena* host.
4. Since characteristics of each C_{18} column are different, use the same column and eluent, or alternatively modify suitable eluent conditions for the C_{18} column.
5. Fragments of the target gene are amplified by PCR. Genomic DNA isolated from *Anabaena* is used as a template for PCR. Efficient homologous recombination events in *Anabaena* cells require approximately 1-kb DNA regions flanking an antibiotic-resistance cassette inserted in the target gene on both sides.
6. The 1.1-kb fragment containing *npt* of pRL648, which is derived from Tn5, can be cut out with BamHI, XbaI, or SmaI. For isolation of the recombinants of *Anabaena*, another Tn903-derived *npt* gene from pUC4K (Pharmacia, Uppsala, Sweden) and a streptomycin/spectinomycin-resistance (Sm^r/Sp^r) gene cassette with transcriptional and translational terminators on either side from pHP45Ω (19) can be also used.
7. Since the *npt* gene of Tn5 from pRL648 was fused to the *psbA* promoter from chloroplasts of *Amaranthus hybridus*, which is quite strong in *Anabaena* sp. PCC 7120 (11), it may have a polar effect on the expression of downstream and upstream genes of the *npt*-interrupted target gene.
8. Growth medium BG11 can be replaced by an eightfold dilution of Allen and Arnon (AA/8) liquid medium or non-diluted AA agar (20) with added nitrate.
9. Single-crossover recombination retains the integrated plasmid into the *Anabaena* chromosome. Since the cargo plasmid (see Subheading 2.4) carries Em^r and Nm^r genes and *sacB* gene as well, either 5–10 μg/mL Em or 30 μg/mL Nm (not their combination) can be used for selection of single-crossover exconjugants.

10. Second homologous recombination allows for replacement of the wild-type version of the target gene by the *npt*-interrupted version of the gene, resulting in loss of the plasmid and Em^r and *sacB* genes from the chromosome. Therefore, Nm (but not Em) is used for selection of double-crossover recombinants.
11. The expression of *sacB* gene is lethal for many Gram-negative bacteria including some cyanobacteria in the presence of sucrose, and thus *sacB* facilitates isolation of double-crossover recombinants (17).
12. Chloroform is very toxic, but solubility of carotenoids in chloroform is high.
13. 4-Ketomyxol 2′-fucoside was absent in the *crtW*-deleted mutant (b), and the amount of echinenone was decreased in the *crtO*-deleted one (c). Therefore, two distinct β-carotene ketolases, CrtW and CrtO, catalyze in two distinct biosynthetic pathways due to different substrate specificities (10).
14. Pigments were identified based on the retention times on HPLC, absorption spectra in HPLC eluent, and relative molecular masses after purification. The glycoside moiety was identified using ^1H NMR (1).

Acknowledgments

We thank C.P. Wolk of Michigan State University for the gift of the pRL plasmids and conjugative plasmid RP4 required for triparental mating and H. Sakurai of Waseda University for his help with construction of the *Anabaena* sp. PCC 7120 mutants. This work was supported in part by Grants-in-Aid for Scientific Research from JSPS to S.T. (16570038) and H.M. (16.9494) and by the Great Lakes Bioenergy Research Center (DOE BER Office of Science DE-FC02-07ER64494) to R.P. Hausinger of Michigan State University (for H.M.).

References

1. Takaichi S, Mochimaru M, Maoka T, Katoh H (2005) Myxol and 4-ketomyxol 2′-fucosides, not rhamnosides, from *Anabaena* sp. PCC 7120 and *Nostoc punctiforme* PCC 73102, and proposal for the biosynthetic pathway of carotenoids. Plant Cell Physiol 46:497–504
2. Takaichi S, Mochimaru M, Maoka T (2006) Presence of free myxol and 4-hydroxymyxol and absence of myxol glycosides in *Anabaena variabilis* ATCC 29413, and proposal of a biosynthetic pathway of carotenoids. Plant Cell Physiol 47:211–216
3. Takaichi S, Maoka T, Mochimaru M (2009) Unique Carotenoids in the terrestrial cyanobacterium *Nostoc commune* NIES-24: 2-hydroxymyxol 2′-fucoside, nostoxanthin and canthaxanthin. Curr Microbiol 59:413–419
4. Takaichi S, Maoka T, Masamoto K (2001) Myxoxanthophyll in *Synechocystis* sp. PCC 6803 is myxol 2′-dimethylfucoside, (3*R*,2′*S*)-myxol

2′-(2,4-di-*O*-methyl-α-L-fucoside), not rhamnoside. Plant Cell Physiol 42:756–762

5. Takaichi S, Mochimaru M (2007) Carotenoids and carotenogenesis in cyanobacteria: unique ketocarotenoids and carotenoid glycosides. Cell Mol Life Sci 64:2607–2619
6. Kaneko T, Nakamura Y, Wolk CP, Kuritz T, Sasamoto S, Watanabe A, Iriguchi M, Ishikawa A, Kawashima K, Kimura T, Kishida Y, Kohara M, Matsumoto M, Matsuno A, Muraki A, Nakazaki N, Shimpo S, Sugimoto M, Takazawa M, Yamada M, Yasuda M, Tabata S (2001) Complete genomic sequence of the filamentous nitrogen-fixing cyanobacterium *Anabaena* sp. strain PCC 7120. DNA Res 8:205–213
7. Wolk CP, Vonshak A, Kehoe P, Elhai J (1984) Construction of shuttle vectors capable of conjugative transfer from *Escherichia coli* to nitrogen-fixing filamentous cyanobacteria. Proc Natl Acad Sci USA 81:1561–1565
8. Wolk CP, Ernst A, Elhai J (1994) Heterocyst metabolism and development. In: Bryant DA (ed) The molecular biology of cyanobacteria. Kluwer, Dordrecht, pp. 769–823
9. Cohen MF, Meeks JC, Cai YA, Wolk CP (1998) Transposon mutagenesis of heterocyst-forming filamentous cyanobacteria. Methods Enzymol 297:1–17
10. Mochimaru M, Masukawa H, Takaichi S (2005) The cyanobacterium *Anabaena* sp. PCC 7120 has two distinct β-carotene ketolases: CrtO for echinenone and CrtW for ketomyxol synthesis. FEBS Lett 579: 6111–6114
11. Elhai J, Wolk CP (1988) A versatile class of positive-selection vectors based on the nonviability of palindrome-containing plasmids that allows cloning into long polylinkers. Gene 68:119–138
12. Elhai J, Vepritskiy A, Muro-Pastor AM, Flores E, Wolk CP (1997) Reduction of conjugal transfer efficiency by three restriction activities of *Anabaena* sp. strain PCC 7120. J Bacteriol 179:1998–2005
13. Thomas MC, Smith CA (1987) Incompatibility of group P plasmids: genetics, evolution, and use in genetic manipulation. Annu Rev Microbiol 41:77–101
14. Rippka R (1988) Isolation and purification of cyanobacteria. Methods Enzymol 167:3–27
15. Black TA, Cai Y, Wolk CP (1993) Spatial expression and autoregulation of *hetR*, a gene involved in the control of heterocyst development in *Anabaena*. Mol Microbiol 9:77–84
16. Mochimaru M, Masukawa H, Maoka T, Mohamed HE, Vermaas WFJ, Takaichi S (2008) Substrate specificities and availability of fucosyltransferase and β-carotene hydroxylase for myxol 2′-fucoside synthesis in *Anabaena* sp. strain PCC 7120 compared with *Synechocystis* sp. strain PCC 6803. J Bacteriol 190:6726–6733
17. Cai Y, Wolk CP (1990) Use of a conditionally lethal gene in *Anabaena* sp. strain PCC 7120 to select for double recombinants and to entrap insertion sequences. J Bacteriol 172:3138–3145
18. Masukawa H, Mochimaru M, Sakurai H (2002) Disruption of the uptake hydrogenase gene, but not of the bidirectional hydrogenase gene, leads to enhanced photobiological hydrogen production by the nitrogen-fixing cyanobacterium *Anabaena* sp. PCC 7120. Appl Microbiol Biotechnol 58:618–624
19. Prentki P, Krisch HM (1984) In vitro insertional mutagenesis with a selectable DNA fragment. Gene 29:303–313
20. Allen MB, Arnon DI (1955) Studies on nitrogen-fixing blue-green algae. I. Growth and nitrogen fixation by *Anabaena cylindrica* Lemm. Plant Physiol 30:366–372

Chapter 19

Obtaining Lutein-Rich Extract from Microalgal Biomass at Preparative Scale

José M. Fernández-Sevilla, F. Gabriel Acién Fernández, and Emilio Molina Grima

Abstract

Lutein extracts are in increasing demand due to their alleged role in the prevention of degenerative disorders such as age-related macular degeneration (AMD). Lutein extracts are currently obtained from plant sources, but microalgae have been demonstrated to be a competitive source likely to become an alternative. The extraction of lutein from microalgae posesses specific problems that arise from the different structure and composition of the source biomass. Here is presented a method for the recovery of lutein-rich carotenoid extracts from microalgal biomass in the kilogram scale.

Key words: Lutein, Microalgae, Extraction

1. Introduction

The carotenoid lutein and its extracts have a potential role in preventing retinal degeneration (1) and a protective effect on other diseases, such as cardiovascular and cancer (2), arising from its antioxidant properties. Lutein is present in an ample variety of foods, particularly in green leafy vegetables, but the dietary intake does not guarantee the recommended daily dose. Besides, the bioavailability of the lutein in those foods is low (3) because of its high content in fiber. Therefore, the use of lutein-rich extracts is on the rise and is widely recommended as a means of ensuring the recommended intake of 5 mg/day. Lutein extracts are currently obtained from dried marigold petals (*Calendula* sp.), which has the advantage of having a relatively simple composition and a weak tissue structure that makes unnecessary a disintegration step. On the other hand, marigold petals have low lutein content (4), are seasonal, likely to change in composition, and are, as any other land crop,

José-Luis Barredo (ed.), *Microbial Carotenoids from Bacteria and Microalgae: Methods and Protocols*, Methods in Molecular Biology, vol. 892, DOI 10.1007/978-1-61779-879-5_19, © Springer Science+Business Media, LLC 2012

exposed to uncontrolled contamination and environmental conditions. Microalgae have recently been suggested as a potential alternative source of lutein (5) because its biomass has a relatively high content of lutein (1% dry wt.) (6, 7) and its productivity is much higher, particularly when it is considered in a land-occupation basis.

Since microalgae are a novel source of lutein because of their high content in the carotenoid, a very high productivity, and the potential to supply year-round a uniform, homogeneous biomass, cultured under controlled conditions. Microalgae being a novel source of lutein, there is little information on the appropriate extraction procedures that allow the recovering of carotenoids. Compared to marigold petals, microalgae biomass presents significant differences, being the most important the presence of a thick, resistant cell wall and a more complex composition, particularly in pigments, that makes more difficult the separation of the carotenoid fraction. This chapter describes a procedure for the recovery of the carotenoid fraction contained in microalgal biomass that has been developed for species with a hard cell wall, such as *Scenedesmus almeriensis* or *Murielopsis* sp.

2. Materials

Solvents and chemicals are reactive grade. Solutions are to be prepared using deionized water. Dry nitrogen and argon are used as sweeping gas or inert atmosphere. Solutions are to be prepared fresh as short in advance as possible. The use of inert atmosphere by sweeping with nitrogen and the use of airtight flasks are advised throughout. Storage under argon atmosphere is recommended. Material size and reactive amounts are given considering a 100-g batch of dry biomass. The method has been tested for up to 500-g biomass.

2.1. Cell Breakage

1. Bench-top ball mill of 5 L working volume (18 cm inner diameter 15 cm depth approx.) filled with 50 ceramic balls 27 mm diameter (2,700 kg/m^3 density).
2. 100 g alumina (aluminum oxide, type A-5) (Sigma–Aldrich Inc., St. Louis, MO, USA).

2.2. Alkaline Treatment

1. KOH solution: 40 g/L in deionized water (see Note 1).
2. Glass, cylindrical jacketed reaction tank 1 L working volume (340 mm height, 100 mm diameter) provided with internal baffles, an airtight 3-neck lid and an outlet at the bottom.
3. Agitator Heidolph RZR 2020 (Heidolph Instruments GmbH & Co., Schwabach, Germany).

4. Stirrer Teflon Rushton-type (6-blades, 50 mm diameter for a base tank working volume of 1 L).
5. Thermostatic bath.

2.3. Extraction

1. Separating funnel (2 L).
2. Buchner funnel (90 mm diameter).
3. Kitasato flask.
4. Filter paper.
5. Nylon net filter (10 μm) (Millipore, Billerica, MA, USA).
6. Vacuum pump.

2.4. Solvent Removal, Purification, Resuspension in Oil, and One-Step Alkaline Extraction

1. Rotary evaporator.
2. Alkaline Water–Ethanol–Hexane (WEH) extracting solution: Mix distilled water, ethanol (96%, v/v, reagent grade), and hexane (reagent grade) in a proportion of 10:84:6 (v/v/v), and add 40 g/L of high grade KOH.

3. Methods

Microalgal biomass should be processed right after freeze-drying or must be stored under −18°C for up to 2 weeks as longer delay induces significant degradation of the carotenoid fraction. Solutions are to be prepared fresh as short in advance as possible. Timing is critical in the extraction procedures. The use of inert atmosphere (nitrogen or preferably argon) is advised throughout and requested in specific steps. All the procedures described here are to be carried out at room temperature (20–24°C) unless otherwise stated. The procedure described here is for an amount of 100 g biomass dry weight, but it can be scaled up or down by keeping the proportions described. A scheme of the overall process is shown in Fig. 1a.

3.1. Cell Breakage

1. Dry microalgal biomass, typically freeze-dried, to be used for the extraction procedure. After the dying process the biomass is likely to form lumps of irregular size that can hamper the procedure and must therefore be broken to a particle size less than 500 μm. The use of a sieve is advised to ensure a proper granulometry.
2. Weigh the same alumina as the sample weight. Add the alumina and mix thoroughly by hand or in a tumbler, until both solids are undistinguishable to the eye.
3. Place in a ball mill and process for 10 min. The drum is filled to half its working volume with 4-cm diameter ceramic balls (see Note 2).

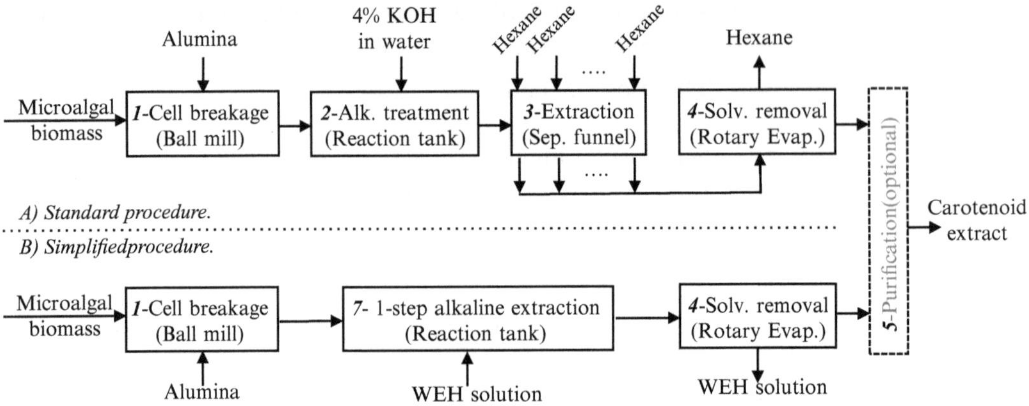

Fig. 1. Processing scheme for the production of lutein-rich carotenoid extracts from microalgal biomass at preparative scale. Two alternative extractions are proposed. (**a**) The first one, depicted in the *upper part,* combines an alkaline treatment and a multiple extraction with hexane, while (**b**) the second alternative, shown in the *lower part*, is based on the use of a specially formulated extraction mixture which greatly reduces the requirements of solvent and the size of the equipment.

4. Transfer the milled biomass and alumina to a 1-L working volume jacketed glass tank thermostated at 40°C. Process without delay.

3.2. Alkaline Treatment

1. With the milled biomass in the tank, add enough 4% KOH aqueous solution at 40°C to complete 1 L.
2. Close the tank with the stirrer in place and flush with argon.
3. Stir at 250 rpm with a Rushton-type turbine in order to produce a vigorous agitation. Process for 10 min.
4. Transfer to a 2-L working volume separating funnel (see Note 3).

3.3. Extraction

1. Add 1 L hexane and 20 mL of 96% (v/v) ethanol (see Note 4).
2. Shake vigorously for 10 min. Let rest and allow the two phases to separate (see Note 5).
3. Recover the hexane phase. Transfer back the aqueous phase to the separating funnel and repeat the extraction with hexane. Two extractions will achieve a recovery yield over 60%. 80% yield is obtained with four extractions and over 90% with six extractions (8).
4. Filter through a 60-µm filter using a Buchner funnel and flask attached to a vacuum pump (see Note 6).

3.4. Solvent Removal

1. Place a suitable volume of the hexane phase in a rotary evaporator, approximately 25% of the evaporator flask capacity.
2. Set the bath temperature to 40°C, the rotation speed to 30 rpm (for a 2-L flask), and start the vacuum (see Note 7).

3. Add a new portion of hexane phase before the solvent is completely removed.
4. After the last addition of hexane phase, keep on eliminating the solvent until the remaining volume is 50 mL (for 100 g dry weight initial biomass).
5. Move to a suitable flask and finish the solvent removal under N_2 current at room temperature. The result is a resin extract containing up to 70% carotenoids in free form.

3.5. Purification

The resin extract obtained after step 4 of Subheading 3.4 may contain a significant part of noncarotenoid matter such as waxes, sterols or hydrocarbons, depending on the strain of microalgae used and its culture stage. At this point the carotenoid extract can be purified by the following procedure.

1. During step 4 of Subheading 3.4, keep on eliminating solvent until the extract becomes a slurry.
2. Stop the vacuum and add 20 mL of acetone.
3. Put back the flask in place and stir at 35°C for 5 min at low speed.
4. Stop and allow settling. The acetone extract should separate neatly from the slurry (see Note 7). Transfer the acetone extract to a suitable flask.
5. Repeat the extraction with acetone three more times or until the acetone extract comes clear.
6. Pool the recovered acetone fractions in a clean rotary-vacuum flask and proceed to eliminate the solvent in the same conditions as described above for hexane.
7. Remove the solvent until dry or redissolve in olive oil as described for hexane (see Note 8).

3.6. Resuspension in Oil

Obtaining the resin carotenoid extract as a final product is not advised because it is difficult to handle and the lutein present is unstable (9). Instead a resuspension in olive oil is recommended as follows.

1. After step 4 of Subheading 3.4, stop the vacuum and add 100 mL of olive oil to render a solution of lutein in olive oil of approximately 10 mg/mL (see Note 9).
2. Stir at 40°C for 10 min.
3. Start the vacuum and eliminate the rest of the solvent.
4. Place in an adequate flask. Strip with 0.2 L/L N_2 for 30 min to reduce the solvent levels to food grade requests.

This final product is easy to handle and has been proven to last for months (9) (see Note 9).

3.7. Alternative One-Step Alkaline Extraction

This alternative procedure combines the alkaline treatment and the extraction in a single step and greatly reduces the amount of solvents used. On the downside, the extracts obtained tend to contain more impurities. A scheme of this procedure is depicted in Fig. 1b. To carry out this procedure, take the mixture of microalgal biomass and alumina (100 + 100 g) as obtained after Subheading 3.1 and process as follows.

1. Prepare 1.4 L of WEH extracting solution.
2. Place the milled biomass and alumina in a 1-L working volume jacketed glass tank thermostated at 40°C.
3. Add WEH solution to complete 1 L. The solution should be at 40°C.
4. Close the tank with the stirrer in place and flush with argon.
5. Stir at 250 rpm with a Rushton-type turbine in order to produce a vigorous agitation. Process for 10 min.
6. Filter using a Buchner funnel and flask. Use 60 μm filtering medium (see Note 10).
7. Once the filtration is complete, wash carefully the cake with the remaining 400 mL WEH solution.
8. Transfer the solution to a rotary evaporator flask. Proceed to remove the solvent in the same conditions as described in step 2 of Subheading 3.4.
9. Continue with this procedure until the extract becomes completely dry.
10. Stop the vacuum and add 25 mL acetone.
11. Put back the flask in place and stir at 35°C for 10 min at moderate speed (see Note 11).
12. Stop and allow settling. The acetone extract should separate neatly from the slurry (see Note 9). Transfer the acetone extract to a suitable flask.
13. Repeat the extraction with acetone three more times or until the acetone extract comes clear.
14. Pool the recovered acetone fractions in a clean rotary-vacuum flask and proceed to eliminate the solvent in the same conditions as described for hexane.
15. Remove the solvent until dry or redissolve in olive oil as described for hexane.

4. Notes

1. It is important that the alkaline treatment of the biomass lasts exactly 10 min at 40°C, so the alkaline solution must be prepared beforehand and its temperature adjusted at 40°C. The use of an airtight container to handle the alkaline solution is advised.
2. The ball mill reduces both the microalgae and the alumina to a very fine dust bound to become airborne during manipulation that can be easily inhaled. Take the adequate precautions, including wearing the appropriate protection.
3. The separation funnel nominal volume must be over 2 L in order to hold 1 L of the KOH-treated biomass aqueous solution, and the same volume of hexane plus enough headspace to allow proper mixing. The total load can be processed using smaller separating funnels in a number of batches as long as they are processed simultaneously because lutein degrades if left in contact with the alkaline solution.
4. Special care must be taken to prevent overpressure inside the separating funnel. This can be attained by adding the hexane cooled to 4°C.
5. Cell debris can help forming very stable emulsion that will impede phase separation. This can be avoided or ameliorated by filtering in using a Buchner funnel and flask.
6. If filtering has been performed in step 2 of the extraction procedure, this filtering step can be skipped.
7. The rotation speed of the rotary evaporator is given as an indication and it should be adjusted accordingly to the size (diameter) of the flask used to promote a proper agitation.
8. Lutein is very stably suspended in olive oil. Olive oil has been proven to keep lutein unaltered for months even at room temperature (9).
9. The acetone extract should be a transparent yellow solution easily separable from the solid debris. Otherwise the procedure has failed. This may be caused by the presence of excess water that must be further eliminated.
10. Place a disk of regular filter paper over the recommended filter to retain the coarser particles. This facilitates and speeds up the procedure.

11. Water should be eliminated thoroughly in order to prevent unwanted substances from redissolving in the acetone. A color green/brown color instead of yellow in the acetone solution indicates the presence of excess water. At this point, the acetone must be eliminated in the rotary evaporator and the extract dried carefully. Successive additions and eliminations of absolute ethanol can help eliminate water.

References

1. Granado F, Olmedilla B, Blanco I (2003) Nutritional and clinical relevance of lutein in human health. Br J Nutr 90:487–502
2. Demmig-Adams B, Adams WW III (2002) Food and photosynthesis: antioxidants in photosynthesis and human nutrition. Science 298:2149–2153
3. Granado-Lorencio F, Herrero-Barbudo C, Acién-Fernández G, Molina-Grima E, Fernández-Sevilla JM, Pérez-Sacristán B, Blanco-Navarro I (2009) In vitro bioaccesibility of lutein and zeaxanthin from the microalgae *Scenedesmus almeriensis*. Food Chem 114:747–752
4. Piccaglia R, Marotti M, Grandi S (1998) Lutein and lutein ester content in different types of *Tagetes patula* and *T. erecta*. Ind Crops Prod 8:45–51
5. Fernández-Sevilla JM, Acién Fernández FG, Molina Grima E (2010) Biotechnological production of lutein and its applications. Appl Microbiol Biotechnol 86:27–40
6. Sánchez F, Fernández JM, Acién FG, Rueda A, Pérez-Parra J, Molina E (2008) Influence of culture conditions on the productivity and lutein content of the new strain *Scenedesmus almeriensis*. Process Biochem 43:398–405
7. Del Campo JA, Rodríguez H, Moreno J, Vargas MA, Rivas J, Guerrero MG (2004) Accumulation of astaxanthin and lutein in *Chlorella zofingiensis* (Chlorophyta). Appl Microbiol Biotechnol 64:848–854
8. Cerón MC, Campos I, Sánchez JF, Acién FG, Molina E, Fernández-Sevilla JM (2008) Recovery of lutein from microalgae biomass: development of a process for *Scenedesmus almeriensis* biomass. J Agric Food Chem 56:11761–11766
9. Cerón-García MC, Campos-Pérez I, Macías-Sánchez MD, Bermejo-Román R, Fernández-Sevilla JM, Molina-Grima E (2010) Stability of carotenoids in *Scenedesmus almeriensis* biomass and extracts under various storage conditions. J Agric Food Chem 58:6944–6950

Chapter 20

NMR-Based Isotopologue Profiling of Microbial Carotenoids

Eva Eylert, Adelbert Bacher, and Wolfgang Eisenreich

Abstract

C-Isotopologue profiling is a powerful tool to determine on a quantitative basis the biosynthetic origin of carotenoids in microorganisms. To this aim, the carotenoid-producing microorganism is grown in medium containing ^{13}C-labeled glucose. After growth, the ^{13}C-isotopologue distribution in a given biosynthetic carotenoid is determined by quantitative NMR spectroscopy. The labeling pattern provides a fingerprint of processes involved in the metabolism of glucose and the formation of the carotenoid. For example, the ^{13}C-profile shows whether the isoprenoid precursors, isopentenyl diphosphate and dimethylallyl diphosphate, are formed by the mevalonate or the non-mevalonate pathway. The labeling data also specify the pathways of glucose utilization, e.g., via the Entner-Doudoroff pathway or glycolysis. The method is exemplified with the analysis of zeaxanthin biosynthesis in the Alphaproteobacterium, *Paracoccus zeaxanthinifaciens*.

Key words: Zeaxanthin, Isotopologue profiling, NMR spectroscopy, Biosynthesis, Mevalonate pathway

1. Introduction

With the rapidly growing number of completely sequenced genomes, it appears safe to deduce the isoprenoid/carotenoid biosynthesis status of bacteria on the basis of sequence homologies and phylogenetic relationships. The available genome data indicate that all Archaea generate isoprenoids via the mevalonate pathway, whereas most eubacteria use either the mevalonate or the non-mevalonate pathway (1). As notable exceptions, many *Listeriae*, some *Mycobacteriae*, as well as *Nocardia farcinica* have complete sets for both pathways. Some parasitic bacteria, such as *Rickettsiae* lack the genes of either pathway.

As an alternative to the bioinformatic approach, the specific isoprenoid biosynthesis pathway used by a given microorganism can be assessed by NMR analysis of a biosynthetic isoprenoid after in vivo labeling with a ^{13}C-labeled tracer compound such as glucose.

José-Luis Barredo (ed.), *Microbial Carotenoids from Bacteria and Microalgae: Methods and Protocols*, Methods in Molecular Biology, vol. 892, DOI 10.1007/978-1-61779-879-5_20, © Springer Science+Business Media, LLC 2012

Whereas the mevalonate pathway results in the assembly of IPP and DMAPP from acetyl-CoA serving as the sole carbon source, the non-mevalonate pathway combines a two-carbon fragment (C-2 and C-3 from pyruvate) with a three-carbon compound (glyceraldehyde 3-phosphate). Therefore, a single precursor molecule can contribute three carbon atoms via the non-mevalonate pathway but only two carbon atoms via the mevalonate pathway. Hence, whenever the contribution of three carbon atoms from a single precursor molecule can be documented, a given isoprenoid molecule must have been formed via the non-mevalonate pathway. This problem can be experimentally addressed by incorporation studies using a mixture of [U-$^{13}C_6$] glucose with a relatively large excess of unlabeled glucose as carbon source (2). Fragments for the de novo biosynthesis of isoprenoids are then retrieved at random from the metabolite pools resulting from the catabolism of the proffered glucose. For statistical reasons, a precursor resulting from the breakdown of a ^{13}C-labeled glucose molecule will typically be combined with one that stems from a non-labeled precursor molecule. Isoprenoid molecules resulting from this random assembly process can therefore be described as molecular mosaics resulting from ^{13}C-labeled and unlabeled fragments. That pattern can be deconstructed by the NMR approach that is described in detail below.

As an alternative to the use of [U-$^{13}C_6$]glucose as a tracer, it is also possible to conduct experiments with glucose samples carrying a single ^{13}C label. Due to the catabolism of the proffered glucose by glycolysis, the Entner-Doudoroff pathway or the pentose phosphate cycle, specifically labeled fragments are afforded that can serve as building blocks for isoprenoid biosynthesis. The specific labeling patterns of the resulting complex isoprenoid products can again be assessed by quantitative NMR spectroscopy.

Most terminal isoprenoids (including carotenoids) consist of multiple isoprenoid building blocks (IPP and/or DMAPP). Hence, the NMR analysis affords redundant information on the basic pathway. This can serve as a quality check of the biosynthetic analysis. On the other hand, different labeling profiles of the precursor units occasionally indicate a mixed biosynthetic origin of the isoprenoid under study.

The details of the approach are illustrated below by experiments on the biosynthesis of zeaxanthin by the marine bacterium *Paracoccus zeaxanthinifaciens* (formerly described as *Paracoccus* strain PTA-3335) (3). The zeaxanthin-producing bacterium was assigned as a new species of the genus *Paracoccus* mainly on the basis of 16S RNA sequencing (4). *Paracocci* belong to the class of Alphaproteobacteria. Currently, the complete genome sequences of more than 150 Alphaproteobacteria including that of *Paracoccus denitrificans* are available in the public domain. A compilation of genomes coding for genes of the mevalonate or the non-mevalonate pathway in Alphaproteobacteria is given in Table 1. With the exception of *Rickettsiae* and *Orientiae*, all sequences including that

Table 1
Presence (+) or absence (–) of genes encoding enzymes of the mevalonate pathway and the non-mevalonate pathway in all completely sequenced Alphaproteobacteria

Species	Mevalonate pathway		Non-mevalonate pathway	
		Partial presence or absence		Partial presence or absence
Acetobacter pasteurianus	–		+	
Acidiphilium cryptum JF-5	–	+2.3.1.9	+	
Agrobacterium radiobacter K84	–	+2.3.1.9	+	
Agrobacterium tumefaciens C58	–	+2.3.1.9	+	
Alpha proteobacterium IMCC1322	–	+2.3.1.9	+	
Agrobacterium vitis S4	–	+2.3.1.9	+	
Anaplasma centrale	–		+	–2.2.1.7
Anaplasma marginale Florida	–		+	–2.2.1.7
Anaplasma marginale St. Maries	–		+	–2.2.1.7
Anaplasma phagocytophilum	–		+	–2.2.1.7
Azospirillum sp. B510	–	+2.3.1.9	+	
Azorhizobium caulinodans	–	+2.3.1.9	+	
Bartonella bacilliformis	–		+	–1.1.1.267
Bartonella grahamii	–		+	–1.1.1.267
Bartonella henselae	–		+	–1.1.1.267
Bartonella quintana	–		+	–1.1.1.267
Bartonella tribocorum	–		+	–1.1.1.267
Beijerinckia indica	–	+2.3.1.9	+	
Bradyrhizobium japonicum	–	+2.3.1.9	+	
Bradyrhizobium sp. BTAi1	–	+2.3.1.9	+	
Bradyrhizobium sp. ORS278	–	+2.3.1.9	+	
Brevundimonas subvibrioides	–	+2.3.1.9	+	+5.3.3.2
Brucella abortus 9-941	–	+2.3.1.9	+	–1.1.1.267
Brucella abortus S19	–	+2.3.1.9	+	–1.1.1.267
Brucella canis	–	+2.3.1.9	+	–1.1.1.267
Brucella melitensis 16M	–	+2.3.1.9	+	–1.1.1.267
Brucella melitensis ATCC 23457	–	+2.3.1.9	+	–1.1.1.267

(continued)

Table 1
(continued)

Species	Mevalonate pathway		Non-mevalonate pathway	
		Partial presence or absence		Partial presence or absence
Brucella melitensis biovar Abortus	–	+2.3.1.9	+	–1.1.1.267
Brucella microti	–	+2.3.1.9	+	–1.1.1.267
Brucella ovis	–	+2.3.1.9	+	–1.1.1.267
Brucella suis 1330	–	+2.3.1.9	+	–1.1.1.267
Brucella suis ATCC 23445	–	+2.3.1.9	+	–1.1.1.267
Candidatus Liberibacter asiaticus	+	–2.3.1.9; –2.7.4.2	–	
Candidatus Pelagibacter ubique	–	+2.3.1.9	+	
Caulobacter crescentus CB15	–	+2.3.1.9	+	
Caulobacter crescentus NA1000	–	+2.3.1.9	+	
Caulobacter segnis	–	+2.3.1.9	+	
Caulobacter sp. K31	–	+2.3.1.9	+	
Dinoroseobacter shibae	–	+2.3.1.9	+	+5.3.3.2
Ehrlichia canis	–		+	–2.2.1.7
Ehrlichia chaffeensis	–		+	–2.2.1.7
Ehrlichia ruminantium Gardel	–		+	–2.2.1.7
Ehrlichia ruminantium Welgevonden (France)	–		+	–2.2.1.7
Ehrlichia ruminantium Welgevonden (South Africa)	–		+	–2.2.1.7
Erythrobacter litoralis	–	+2.3.1.9	+	
Gluconacetobacter diazotrophicus PAl 5 (Brazil)	–	+2.3.1.9	+	
Gluconacetobacter diazotrophicus PAl 5 (JGI)	–	+2.3.1.9	+	
Gluconobacter oxydans	–		+	
Granulibacter bethesdensis	–	+2.3.1.9	+	
Hirschia baltica	–	+2.3.1.9	+	
Hyphomonas neptunium	–	+2.3.1.9	+	
Jannaschia sp. CCS1	–	+2.3.1.9	+	+5.3.3.2
Magnetospirillum magneticum	–	+2.3.1.9	+	

(continued)

Table 1 (continued)

Species	Mevalonate pathway		Non-mevalonate pathway	
		Partial presence or absence		Partial presence or absence
Maricaulis maris	–	+2.3.1.9	+	
Mesorhizobium loti	–	+2.3.1.9	+	–1.1.1.267; +5.3.3.2
Mesorhizobium sp. BNC1	–	+2.3.1.9	+	–1.1.1.267; +5.3.3.2
Methylobacterium chloromethanicum	–	+2.3.1.9	+	
Methylobacterium extorquens	–	+2.3.1.9	+	
Methylobacterium extorquens AM1	–	+2.3.1.9	+	
Methylobacterium extorquens DM4	–	+2.3.1.9	+	
Methylobacterium nodulans	–	+2.3.1.9; +2.3.3.10	+	
Methylobacterium populi	–	+2.3.1.9	+	
Methylobacterium radiotolerans	–	+2.3.1.9	+	
Methylobacterium sp. *4-46*	–	+2.3.1.9	+	
Methylocella silvestris	–	+2.3.1.9	+	
Neorickettsia risticii	–		+	–2.2.1.7
Neorickettsia sennetsu	–		+	–2.2.1.7
Nitrobacter hamburgensis	–	+2.3.1.9	+	
Nitrobacter winogradskyi	–	+2.3.1.9	+	
Novosphingobium aromaticivorans	–	+2.3.1.9	+	
Ochrobactrum anthropi	–	+2.3.1.9	+	–1.1.1.267
Oligotropha carboxidovorans	–	+2.3.1.9	+	
Orientia tsutsugamushi Boryong	–		–	
Orientia tsutsugamushi Ikeda	–		–	
Paracoccus denitrificans	–	+2.3.1.9	+	
Parvibaculum lavamentivorans	–	+2.3.1.9	+	
Phenylobacterium zucineum	–	+2.3.1.9	+	
Rhizobium etli CFN 42	–	+2.3.1.9	+	+5.3.3.2
Rhizobium etli CIAT 652	–	+2.3.1.9	+	+5.3.3.2
Rhizobium leguminosarum	–	+2.3.1.9	+	
Rhizobium leguminosarum bv. Trifolii WSM1325	–	+2.3.1.9	+	

(continued)

Table 1 (continued)

Species	Mevalonate pathway		Non-mevalonate pathway	
		Partial presence or absence		Partial presence or absence
Rhizobium leguminosarum bv. trifolii WSM2304	–	+2.3.1.9	+	
Rhizobium sp. *NGR234*	–	+2.3.1.9	+	
Rhodobacter capsulatus	–	+2.3.1.9	+	+5.3.3.2
Rhodobacter sphaeroides 2.4.1	–	+2.3.1.9	+	+5.3.3.2
Rhodobacter sphaeroides ATCC 17025	–	+2.3.1.9	+	+5.3.3.2
Rhodobacter sphaeroides ATCC 17029	–	+2.3.1.9	+	+5.3.3.2
Rhodobacter sphaeroides KD131	–	+2.3.1.9	+	+5.3.3.2
Rhodopseudomonas palustris BisA53	–	+2.3.1.9	+	
Rhodopseudomonas palustris BisB18	–	+2.3.1.9	+	
Rhodopseudomonas palustris BisB5	–	+2.3.1.9	+	
Rhodopseudomonas palustris CGA009	–	+2.3.1.9	+	
Rhodopseudomonas palustris HaA2	–	+2.3.1.9	+	
Rhodopseudomonas palustris TIE-1	–	+2.3.1.9	+	
Rhodospirillum centenum	–	+2.3.1.9	+	+5.3.3.2
Rhodospirillum rubrum	–	+2.3.1.9	+	
Rickettsia africae	–	+2.3.1.9	–	+5.3.3.2
Rickettsia akari	–	+2.3.1.9	–	+5.3.3.2
Rickettsia bellii OSU 85-389	–	+2.3.1.9	–	+5.3.3.2
Rickettsia bellii RML369-C	–	+2.3.1.9	–	+5.3.3.2
Rickettsia canadensis	–		–	+5.3.3.2
Rickettsia conorii	–	+2.3.1.9	–	+5.3.3.2
Rickettsia felis	–	+2.3.1.9	–	+5.3.3.2
Rickettsia massiliae	–		–	+5.3.3.2
Rickettsia peacockii	–		–	+5.3.3.2
Rickettsia prowazekii	–	+2.3.1.9	–	+5.3.3.2
Rickettsia rickettsii Iowa	–		–	+5.3.3.2
Rickettsia rickettsii Sheila Smith	–		–	+5.3.3.2
Rickettsia typhi	–	+2.3.1.9	–	+5.3.3.2

(continued)

Table 1
(continued)

Species	Mevalonate pathway		Non-mevalonate pathway	
		Partial presence or absence		Partial presence or absence
Roseobacter denitrificans	–	+2.3.1.9	+	+5.3.3.2
Silicibacter pomeroyi	–	+2.3.1.9	+	+5.3.3.2
Silicibacter sp. TM1040	–	+2.3.1.9	+	+5.3.3.2
Sinorhizobium medicae	–	+2.3.1.9	+	
Sinorhizobium meliloti	–	+2.3.1.9	+	
Sphingobium japonicum	–	+2.3.1.9	+	
Sphingomonas wittichii	–	+2.3.1.9	+	
Sphingopyxis alaskensis	–	+2.3.1.9	+	
Starkeya novella	–	+2.3.1.9	+	-1.1.1.267
Wolbachia pipientis	–		+	-2.2.1.7
Wolbachia sp. wir	–		+	-2.2.1.7
Wolbachia wBm	–		+	-2.2.1.7
Wolbachia wMel	–		+	-2.2.1.7
Xanthobacter autotrophicus	–	+2.3.1.9	+	+5.3.3.2
Zymomonas mobilis	–		+	
Zymomonas mobilis subsp. *mobilis NCIMB 11163*	–		+	

The list was assembled from the KEGG genome database (as from 4th of September 2010). The presence (+) or absence (–) of genes homologous to the known pathway genes in incomplete pathways are also indicated (2.3.1.9 = acetyl-CoA acetyltransferase; 1.1.1.88 = hydroxymethylglutaryl-CoA synthase; 2.7.4.2 = phosphomevalonate kinase; 2.2.1.7 = 1-deoxy-D-xylulose-5-phosphate synthase (Dxs); 1.1.1.267 = 1-deoxy-D-xylulose 5-phosphate reductoisomerase (IspC/Dxr); 5.3.3.2 = isopentenyl pyrophosphate isomerase (Idi))

of *P. denitrificans* indicate the presence of a functional non-mevalonate pathway. On the other hand, with the exception of *Candidatus Liberibacter asiaticus*, none of the sequences shows homologies to a complete and functional mevalonate pathway. Based on this fact, a non-mevalonate origin of zeaxanthin from *P. zeaxanthinifaciens* would have appeared highly plausible. In sharp contrast, however, the ^{13}C-labeling data of zeaxanthin clearly demonstrated that the isoprenoid building blocks are formed via the mevalonate pathway (3).

2. Materials

2.1. Strains

1. *Paracoccus zeaxanthinifaciens* (4).
2. *Paracoccus* sp. PTA-3335 (3) (see Note 1).
3. *Paracoccus* sp. R-1512 (ATCC 21588) (American Type Culture Collection, Manassas, VA, USA) (see Note 1).

2.2. Culture and Harvest of Carotenoid-Producing Bacteria

1. Growth medium for *P. zeaxanthinifaciens* (3): 30 g/L of D-glucose, 20 g/L of yeast extract, 10 g/L of peptone, 10 g/L of NaCl, 5 g/L of $MgSO_4 \cdot 7\ H_2O$, 1.5 g/L of $(NH_4)_2HPO_4$, 1.25 g/L of K_2HPO_4, 0.4 g/L of $(NH_4)_2Fe(SO_4)_2 \cdot 6H_2O$, 375 mg/L of $CaCl_2 \cdot 2\ H_2O$, 30 mg/L of $ZnSO_4 \cdot 7\ H_2O$, 25 mg/L of $FeCl_3 \cdot 6\ H_2O$, 10 mg/L of $MnSO_4 \cdot H_2O$, 1 mg/L of $NiSO_4 \cdot 6H_2O$, 30 mg/L of Na-EDTA, and 19 μL/L of 37% hydrochloric acid (see Note 2).
2. ^{13}C-Labeled glucose ([1-$^{13}C_1$]-, [2-$^{13}C_1$]-, [6-$^{13}C_1$]- or [U-$^{13}C_6$]D-glucose) (Cambridge Isotope Laboratories, Andover, MA, USA; Isotec, Miamisburg, OH, USA; Omicron Biochemicals, South Bend, IN, USA; Euriso-Top, Saint-Aubin Cedex, France).
3. Bioflo 3000 bioreactor (New Brunswick Scientific, Edison, NJ, USA).
4. Benchtop centrifuge.

2.3. Isolation of Carotenoids (e.g., Zeaxanthin)

1. Stirring equipment.
2. Saturated NaCl solution.
3. Silica gel column.
4. *n*-Hexane:ether (1:1; v/v).

2.4. NMR Analyses

1. NMR spectrometer operating at 300–600 MHz ^{1}H-frequency.
2. NMR tubes at high quality (7"; 5 mm)
3. Deuterated solvents. $CDCl_3$.

3. Methods

3.1. Labeling Experiments

Cultures of carotenoid-producing microorganisms (see Note 3), such as *P. zeaxanthinifaciens* are grown in culture flasks, chemostats, or bioreactors. In the specific example, a Bioflo 3000 bioreactor is used (3).

1. Seed culture flasks with an isolated colony of the carotenoid-producing bacterium.

2. Incubate for 24 h at 28°C under shaking.
3. Inoculate the Bioflo 3000 bioreactor (3 L growth medium) with 100 mL from the grown flasks.
4. Grow the bacteria for 24 h at 28°C. Adjust continually the pH to 7.2 by the addition of 25% H_3PO_4 or 28% NH_4OH, as required. Keep dissolved oxygen above 40% (in a cascade with agitation) via control of agitation rate (minimum 300 rpm) and aeration rate (minimum 1 vvm) (see Note 4).

3.2. Harvest of Bacteria and Isolation of Labeled Carotenoids

Typically, the ^{13}C-labeled carotenoids are isolated from cell extracts of the grown bacteria (using apolar solvents such as dichloromethane, hexane, or tetrahydrofurane) using standard protocols for isolation and purification of the carotenoid under study. In the following, an optimized protocol is given for the isolation of zeaxanthin from the *Paracococcus* strain under study (3).

1. Cool the suspension cultures to 15°C.
2. Add 500 mL of ethanol per liter and stir at 100 rpm for 20 min.
3. Centrifuge for 20 min at 5,000×*g*.
4. Discard the supernatant (see Note 5) and extract the wet pellet with five volumes of tetrahydrofuran (THF) for 20 min under stirring.
5. Centrifuge the extracted mixture and extract the pellet a second time with one volume of THF.
6. Combine the supernatants (extracts) and concentrate to 50 mL under reduced pressure.
7. Add 5 mL of hexane.
8. Turbinate the mixture until an emulsion is formed, and centrifuge.
9. Collect the aqueous phase, dilute it with an equal volume of saturated NaCl solution, and re-extract it with dichloromethane.
10. Collect the dichloromethane phase and combine it with the THF/hexane phase.
11. Concentrate the mixture of organic extracts again under reduced pressure to remove dichloromethane.
12. Apply the solution to a silica gel column which is then developed with a mixture of *n*-hexane and ether (1:1).
13. Elute the main band containing zeaxanthin (retention volume, 2 L).
14. Evaporate this fraction to dryness under reduced pressure.
15. Dissolve the residue in a small amount of dichloromethane at 40°C and allow the solution to cool slowly.

16. Add hexane to the mixture drop-wise until you observe turbidity.
17. Incubate 48 h at 4°C; the crystallization is complete.
18. Collect the crystals on a paper filter, wash them with cold methanol, and dry them under vacuum.

3.3. NMR Signal Assignments

For isotopologue profiling, an unequivocal assignment of ^{1}H and ^{13}C NMR signals is indispensable. Although NMR assignments have been reported in the literature for numerous carotenoids, it is recommended to independently assign the NMR signals by two-dimensional NMR experiments.

1. Dissolve 1–10 mg of the carotenoid under study in 0.6 mL of deuterated solvent (e.g., chloroform) (see Note 6).
2. Record one-dimensional ^{1}H NMR and ^{13}C NMR spectra of the sample (see Note 7).
3. Assign ^{1}H NMR signals by two-dimensional correlation methods (e.g., COSY, TOCSY, NOESY, and ROESY). For experimental settings and interpretation, see textbooks of NMR spectroscopy.
4. Assign ^{13}C NMR signals assigned by heteronuclear two-dimensional experiments (HMBC spectra and HSQC or HMQC). With multiply ^{13}C-labeled samples, confirm the assignments by INADEQUATE or ADEQUATE experiments. Again follow the protocols in textbooks of NMR spectroscopy. The ^{13}C NMR assignments of zeaxanthin obtained by the comprehensive analysis of the two-dimensional NMR spectra are shown in Table 2.

3.4. NMR-Based Isotopologue Profiling

1. Dissolve 2–10 mg of the ^{13}C-labeled carotenoid in 0.6 mL of the same deuterated solvent which had been used for the unlabeled reference (e.g., $CDCl_3$).
2. Perform one-dimensional ^{1}H and ^{13}C experiments using the same parameters as in the experiments with the unlabeled reference (see Note 7).
3. Carefully correct the phase and the baseline of the spectrum.
4. Determine the integrals for each signal.

3.5. ^{13}C-Abundance from Analysis of ^{1}H NMR Signals

1. Determine absolute ^{13}C abundances from the intensities of the ^{13}C-coupled satellite pairs in the ^{1}H NMR signals. To this end, choose a proton signal with a simple signature, preferably a singlet that is located in a non-crowded region of the ^{1}H NMR spectrum. Due to the size of the ^{1}H-^{13}C coupling constants, the typical distance between the components of a given pair is 140–160 Hz, i.e., the satellites are found at distances of 70–80 Hz from the central signal (see Fig. 1, for an example).

Table 2
Chemical shift and coupling constants of zeaxanthin

Position	$\delta^{13}C$ (ppm)	J_{CC} (Hz)
1, 1′	37.13	36.0 (16, 16′)
2, 2′	48.46	35.8 (3, 3′)
3, 3′	65.10	35.8 (2, 2′)
4, 4′	42.57	37.1
5, 5′	126.17	44.2 (18, 18′)
6, 6′	137.77	56.4 (7, 7′)
7, 7′	125.59	56.2 (6, 6′)
8, 8′	138.50	71.6, 55.7
9, 9′	135.69	43.1 (19, 19′)
10, 10′	131.31	59.7 (11, 11′)
11, 11′	124.93	59.7 (10, 10′)
12, 12′	137.57	70.5
13, 13′	136.48	43.1 (20, 20′)
14, 14′	132.60	60.4 (15, 15′)
15, 15′	130.08	60.4 (14, 14′)
16, 16′	30.26	36.3 (1, 1′)
17, 17′	28.73	34.9 (1, 1′)
18, 18′	21.62	44.2 (5, 5′)
19, 19′	12.82	43.1 (9, 9′)
20, 20′	12.75	42.9 (13, 13′)

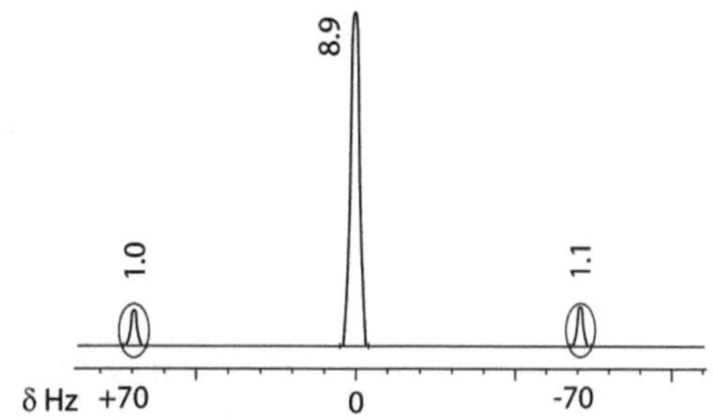

Fig. 1. Schematic view of a ^{1}H NMR signal for a ^{13}C-enriched compound. The *numbers* indicate the integral values. The ^{13}C-coupled satellite signals (coupling constant, 140 Hz in the example) are highlighted. With the values of the example, the ^{13}C-enrichment is calculated as: $\%^{13}C_{abs.} = \frac{1.0+1.1}{(1.0+1.1)+8.9} \cdot 100\% = 19.1\%.$

2. Integrate the central signal and the cognate ^{13}C satellites.
3. Calculate the absolute ^{13}C abundance in% with Eq. 1.

$$\%^{13}\mathrm{C}_{\mathrm{abs.}} = \frac{{}^{\mathrm{H}}I_{\mathrm{S}}\cdot}{{}^{\mathrm{H}}I_{\mathrm{S}} + {}^{\mathrm{H}}I_{\mathrm{C}}} \cdot 100\% \quad (1)$$

${}^{\mathrm{H}}I_{\mathrm{S}} = {}^{\mathrm{H}}I_{\mathrm{S1}} + {}^{\mathrm{H}}I_{\mathrm{S2}}$: Sum of the integrals of the two ^{13}C-coupled satellite signals (corresponding to the proton attached to ^{13}C)

${}^{\mathrm{H}}I_{\mathrm{C}}$: Integral of the central signal (corresponding to the proton attached to ^{12}C)

3.6. Relative ^{13}C Abundance from Analysis of ^{13}C NMR Signals

1. Normalize the ^{13}C NMR integrals in the labeled compound using the respective values of the same compound at natural ^{13}C-abundance using Eq. 2. This is required due to the nonlinear behavior of relaxation times for different types of carbon atoms. The calculation affords relative ^{13}C abundances for each nonequivalent carbon position in the molecule under study. The procedure is exemplified in Fig. 2.

$$\%^{13}\mathrm{Crel.} = \frac{I^{*}}{I} \times \frac{1}{f}. \quad (2)$$

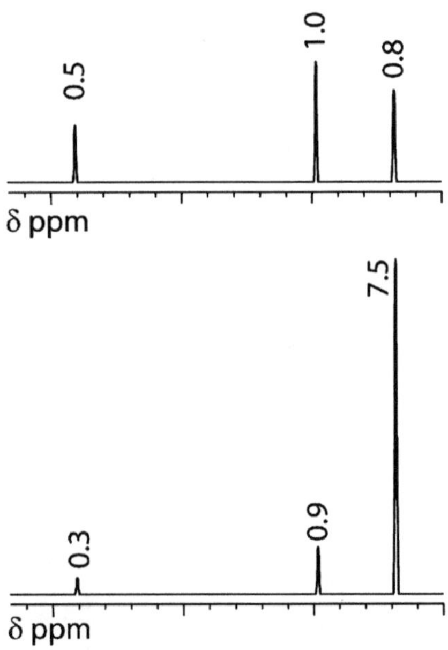

Signal	$I_{n.a}$	I^*	$I^*/I_{n.a}$	$\%^{13}C_{rel.}$	$\%^{13}C_{abs.}$
1	0.5	0.3	0.6	1.0	1.2
2	1.0	0.9	0.9	1.5	1.8
3	0.8	7.5	9.4	15.6	19.1

Fig. 2. Analysis of the relative ^{13}C abundances ($^{13}C_{rel.}$) by quantitative ^{13}C NMR spectroscopy. The example schematically shows three signals of a natural ^{13}C-abundance sample (*top*) and a positionally ^{13}C-enriched sample (*below*). The *numbers* indicate the integral intensities. The relative ^{13}C abundances are then converted into absolute ones (see Subheading 3.7).

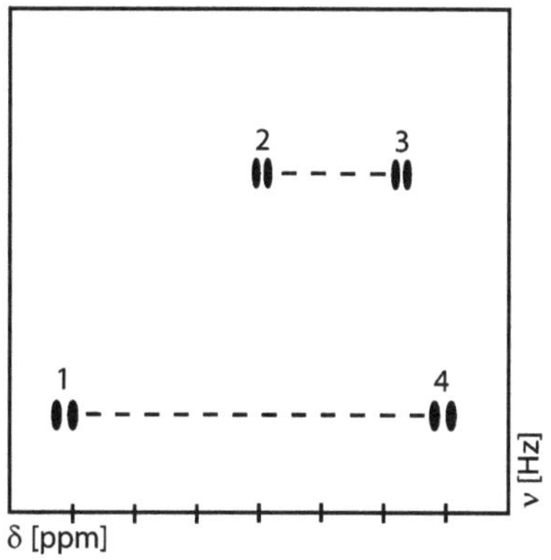

Fig. 3. Two-dimensional INADEQUATE spectrum of a multiply ^{13}C-labeled compound featuring pairs of ^{13}C-atoms between positions 1/4 and 2/3 (i.e., comprising [1,4-$^{13}C_2$]- and [2,3-$^{13}C_2$]-isotopologues).

I^*: Integral intensity of a ^{13}C NMR signal of the enriched sample

I: Integral intensity of the corresponding ^{13}C signal of the natural abundance sample

f: Calibration factor with f being the smallest ratio for I^*/I

3.7. Conversion of Relative into Absolute ^{13}C Abundance

1. Convert the relative abundances into absolute abundances (%$^{13}C_{abs.}$, see also example in Fig. 2) using the data obtained from the intensities of ^{13}C-coupled satellites in the ^{1}H NMR spectrum (see also Eq. 1 and example in Fig. 1). For this procedure, the absolute value for at least one carbon position is required (in the example, %$^{13}C_{abs.}$ = 19.1% for the carbon with the up-field shifted ^{13}C NMR signal 3 in Fig. 2).

3.8. Isotopologue Deconvolution of Multiply ^{13}C-Labeled Samples

The catabolism of multiply ^{13}C-labeled precursors (e.g., [U-$^{13}C_6$] glucose), can afford metabolites comprising blocks of two or more ^{13}C-atoms. In the ^{13}C NMR spectrum, multiply ^{13}C-labeled isotopologues give rise to satellites by ^{13}C coupling (at distances of 35–70 Hz due to the coupling constants between adjacent ^{13}C-atoms; for an example, see Fig. 3). These signals are assigned on the basis of the specific coupling constants detected for a given pair of ^{13}C-atoms (i.e., each of doublets in the ^{13}C signals of coupled carbon atoms display the same coupling constant). Alternatively, the pairs of ^{13}C-atoms are assigned by INADEQUTE spectroscopy where both signals are correlated by their double quantum frequencies. As a result, the horizontal pairs of carbon signals occur at the same position in F1 of the two-dimensional spectrum (see Fig. 3 for a schematic INADEQUATE spectrum).

In certain instances, multiply ^{13}C-labeled isotopologues are generated due to rearrangement events in the biosynthetic pathway resulting in ^{13}C at positions which are not directly attached by a carbon–carbon bond. These species can also be detected via

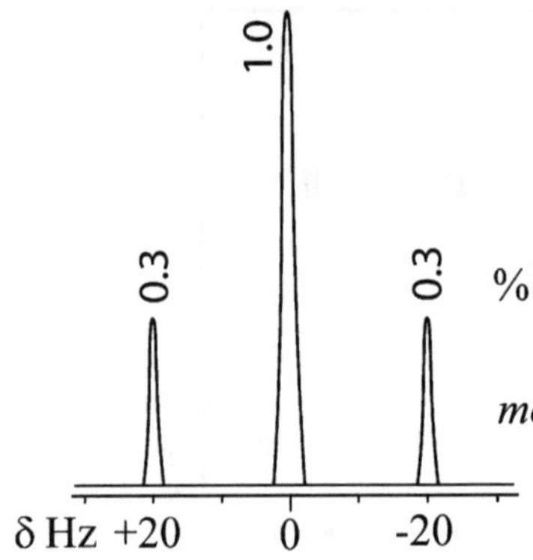

$$\%^{13}C^{13}C = \frac{0.3+0.3}{(0.3+0.3)+1.0} \cdot 100\% = 37.5\%$$

$$mol\%(^{13}C_2) = 37.5\% \cdot \%^{13}C$$

Fig. 4. ^{13}C NMR signal of a given carbon atom in a multiply ^{13}C-labeled sample. The observed carbon signals display a mixture of a ^{13}C–^{13}C species (as seen by the satellites at a distance of 40 Hz in the example) and a ^{13}C–^{12}C species (as seen by the central signal). In the example, the relative fraction of the doubly labeled species (%^{13}C^{13}C) is calculated as 37.5%. The molar amount of the corresponding isotopologue (mol%) is obtained by multiplication with the ^{13}C-enrichment of the given atom (%^{13}C from the analysis given in 3.4.4).

^{13}C–^{13}C coupling analysis, provided that the coupling constants due to long-range couplings are >2 Hz. These couplings can be typically resolved in one-dimensional ^{13}C-NMR spectra, as well as by INADEQUATE experiments optimized for small coupling constants (2).

1. For the quantitative assessment of multiply ^{13}C-labeled isotopologues, integrate separately each satellite signal due to the presence of a multiply ^{13}C-labeled species in conjunction with the central signals (i.e., due to molecules with single ^{13}C-atoms (for an example, see Fig. 4)).
2. Calculate with Eq. 3 the relative fractions of the multiply ^{13}C-labeled isotopologues (% ^{13}C^{13}C).
3. Normalize these values using Eq. 4 affording the molar contributions of multiply ^{13}C-labeled isotopologues.

$$\%^{13}\mathrm{C}^{13}\mathrm{C} = \frac{I_S}{I_S + I_Z}. \quad (3)$$

I_S: Intensity of the ^{13}C-satellites in the ^{13}C-NMR spectra
I_Z: Intensity of the corresponding central signal in the ^{13}C-NMR spectra

$$\mathrm{mol}\%^{13}\mathrm{C}_2 = \%^{13}\mathrm{C}^{13}\mathrm{C} \cdot \%^{13}\mathrm{C}_{\mathrm{abs.}}. \quad (4)$$

3.9. Isotopologue Profiling of ^{13}C-Labeled Zeaxanthin from *P. zeaxanthifaciens*

Using the methods described in Subheadings 3.4–3.8, the biosynthetic origin of zeaxanthin in *P. zeaxanthifaciens* is studied (3).

1. Grow the Alphaproteobacterium in media containing [U-^{13}C$_6$]-, [2-^{13}C]-, or [6-^{13}C]glucose for 24 h at 28°C.
2. Harvest the bacteria and isolate the carotenoids following the procedures described in Subheading 3.2.

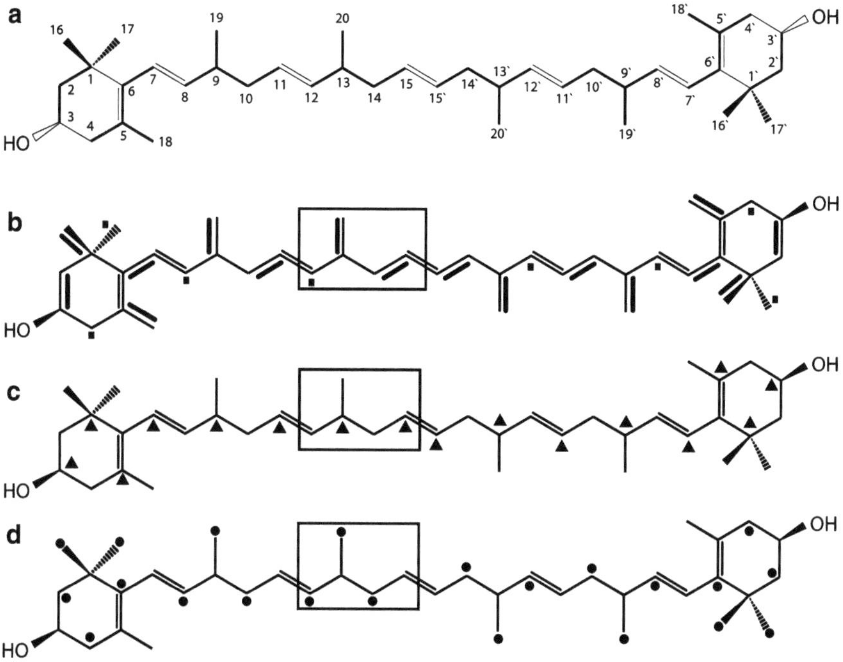

Fig. 5. Observed ^{13}C profiles of zeaxanthin in labeling experiments with ^{13}C-glucose. (**a**) structure of zeaxanthin. (**b**) Labeling pattern in the experiment with [U-$^{13}C_6$]glucose. The *black bars* symbolize pairs of adjacent ^{13}C-atoms. *Filled squares* indicate single ^{13}C-labeled isotopologues (i.e., with ^{12}C at neighbored positions). (**c**) Labeling pattern in the experiment with [2-$^{13}C_1$] glucose. The *filled triangles* indicate ^{13}C-enriched atoms. (**d**) Labeling pattern in the experiment with [6-$^{13}C_1$]glucose. The *filled dots* indicate ^{13}C-enriched atoms. For numerical values, see ref. 3.

3. Analyze by quantitative NMR spectroscopy. An example of the resulting labeling patterns is shown in Fig. 5.
4. Deduce the labeling patterns of the IPP/DMAPP precursors according to the different labeling profiles of the carotenoid at hand, following the known isoprene dissection of the carotenoid. As an example, one of the C_5-units (one of the IPP precursor units in the example) is highlighted in Fig. 5 by the box. It is immediately obvious that the labeling patterns are highly specific in the different labeling experiments. A closer inspection shows that the patterns of all C_5-units are uniform for a given experiment indicating that the pathways leading to the IPP and DMPP precursors are identical (3). A mixed biosynthetic origin of the various units is excluded on this basis.

A retrodictive approach is useful to identify the metabolic pathways leading from glucose to the C_5 precursors. Specifically, it predicts the isotopologue profiles of the precursor units on the basis of the known mechanisms of glucose utilization by the Entner-Doudoroff pathway or glycolysis (see Fig. 6), in conjunction with formation of IPP/DMAPP by the mevalonate or the non-mevalonate pathway, respectively (see Fig. 7).

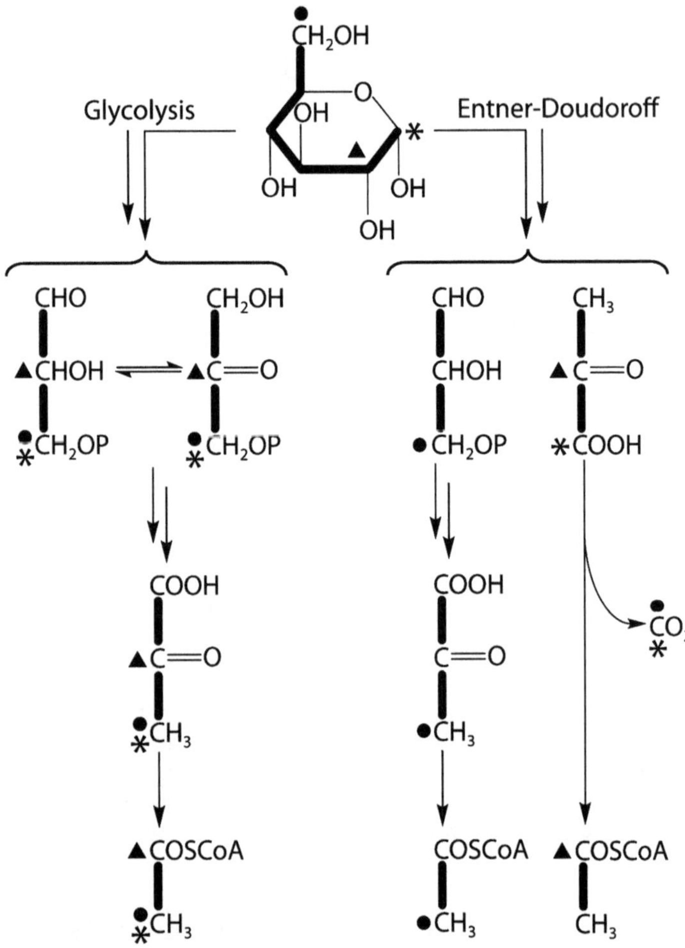

Fig. 6. Fate of glucose labeling due to glycolysis or the Entner Doudoroff pathway. The ^{13}C atom from position 1 of glucose/pyruvate gets lost by decarboxylation in the Entner Doudoroff pathway. The ^{13}C atom from position 2 ends up in position 1 of acetyl-CoA via both pathways, whereas ^{13}C from position 6 ends up in position 2 of acetyl-CoA. For more details, *see* legend of Fig. 7.

1. Characterize the pattern of the IPP/DMAPP precursor units from the experiment with [U-$^{13}C_6$]glucose by $^{13}C_2$ units (black bars in Fig. 7) via the mevalonate pathway assembling two carbon atoms from the acetate unit irrespective of glucose catabolism by glycolysis or the ED pathway. For this purpose, analyze the satellite pairs detected in the ^{13}C NMR spectrum.
2. Characterize the pattern of the IPP/DMAPP precursor units from the experiment with [U-$^{13}C_6$]glucose by $^{13}C_3$ units via the non-mevalonate pathway due to the $^{13}C_3$-labeled glyceraldehyde phosphate precursor. For this purpose, analyze the satellite pairs detected in the ^{13}C NMR spectrum with a special focus on small coupling constants (i.e., 1–10 Hz).

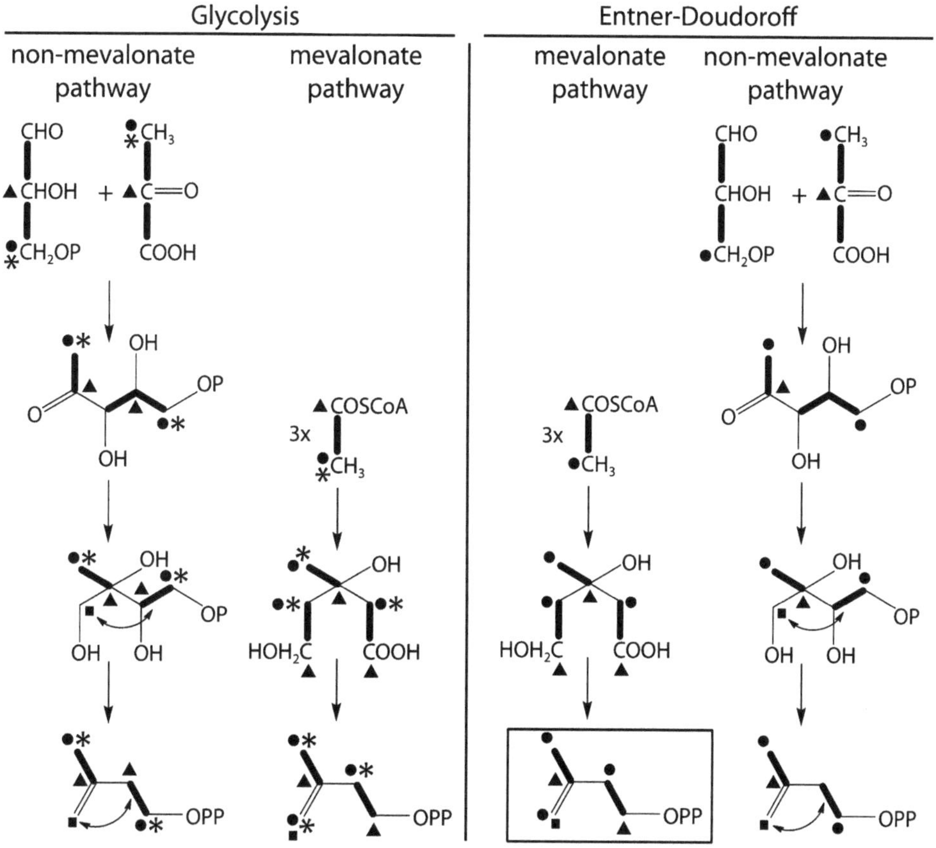

Fig. 7. Predicted labeling patterns in IPP by glycolysis or the Entner Doudoroff pathway providing the precursors for the mevalonate or the non-mevalonate pathway, respectively. The *black bars* symbolize pairs or triples of adjacent ^{13}C-atoms in the experiment with [U-$^{13}C_6$]glucose. For the same experiment, ^{13}C-triples connecting a pair of $^{13}C_2$ and an outlier ^{13}C-atom are indicated by *arrows*. *Asterisks*, *filled triangles*, and *filled circles* indicate ^{13}C-enriched atoms from [1-^{13}C]-, [2-^{13}C]-, and [6-^{13}C]glucose, respectively. The observed labeling profile in zeaxanthin from *P. zeaxanthinifaciens* is shown in the *box*.

3. Predict the profiles via the mevalonate pathway as $^{13}C_2$-units or the non-mevalonate pathway as mixtures of $^{13}C_2$- and $^{13}C_3$-units since it involves the three-carbon precursor, glyceraldehyde phosphate which acquires three ^{13}C-atoms from [U-$^{13}C_6$] glucose by glycolysis or the ED-pathway, respectively. Due to the rearrangement reaction by the IspC protein in the non-mevalonate pathway, the ^{13}C-triple becomes separated in the C_5-units with a pair of ^{13}C plus a single ^{13}C-atom in the same C_5-precursor (indicated by the arrow in Fig. 7). In other words, next to $^{13}C_2$-units, $^{13}C_3$-isotopologues are formed by the non-mevalonate pathway starting from [U-$^{13}C_6$]glucose.
4. Verify these $^{13}C_3$-triples with one ^{13}C-atom not directly attached to the other ^{13}C-atoms by long-range couplings in the one-dimensional ^{13}C-spectra at high resolution or INADEQUATE

experiments optimized for small coupling constants (for examples, see ref. 2). In the present example, ^{13}C-triples are not detected, and it is concluded that the carotenoid is made by the mevalonate pathway.

5. Confirm the result by comparison of the observed and the predicted labeling profiles in the experiments with single labeled glucose precursors (see Fig. 7). In the present example, the observed profile of IPP (shown in the box) is perfectly in line with the mevalonate prediction (see Fig. 7). Moreover, the observed patterns are specific for glucose utilization by the Entner Doudoroff pathway, but not by glycolysis.

4. Notes

1. These zeaxanthin-producing bacteria were formerly classified as *Flavobacterium* sp. (3, 4).
2. The method of isotopologue profiling is very versatile concerning the use of the basic growth medium. Typically, ^{13}C-labeled glucose is added as a sterile aqueous solution to the medium. The total amount of glucose (labeled plus unlabeled) is 30 g/L in the present application. The added amount of labeled glucose depends on the composition of the medium. In certain cases, the labeled compound can be added without dilution to a complex medium. In case of [U-$^{13}C_6$]glucose as a supplement, this holds true when the medium already contains high amounts of glucose and/or carbohydrates. When glucose is the sole carbohydrate source in the medium, [U-$^{13}C_6$] glucose should be diluted with unlabeled glucose (typically 1:10). In many studies, the dilution factor is determined experimentally in an attempt to obtain a final overall ^{13}C-enrichment of the target compound between 2 and 10% with [U-$^{13}C_6$] glucose and 5–50% with a single labeled glucose as tracer, respectively. For a full analysis of the metabolic network in carotenoid-producing bacteria, the four experimental conditions given in Table 3 are recommended.
3. The culture volume to be used for isotopologue profiling depends on the productivity of the carotenoid-producing strain. The final amount of purified carotenoid should be >1 mg for ^{13}C-NMR analysis.
4. The experimental settings should enable high metabolic activity of the organism under study and high yields of the target compound. Thus, it is preferable to use optimized bioreactors (as in the current example), although standard shake cultures may also be used. In order to afford steady-state conditions of

Table 3
Ratios of ^{13}C-labeled glucose in the overall glucose/carbohydrate amount of a given medium

	Unlabeled glucose (%)	[U-$^{13}C_6$] glucose (%)	[1-$^{13}C_1$] glucose (%)	[2-$^{13}C_1$] glucose (%)	[6-$^{13}C_1$]glucose (%)
[U-$^{13}C_6$]glucose	90	10			
[1-$^{13}C_1$]glucose	50		50		
[2-$^{13}C_1$]glucose	75			25	
[6-$^{13}C_1$]glucose	75				25

the isotopologue distribution, it is favorable to add the labeled precursor at the beginning of the bacterial growth. The bacteria should be harvested at the late logarithmic growth phase or later.

5. An equivalent of the supernatant should be checked for residual labeled glucose at the end of the experiment or for released metabolic (labeled) products.
6. The deuterated solvent is selected on the basis of the solubility of the carotenoid. Typically, deuterated chloroform is used. For carotenoids, 1–10 mg are required for NMR analysis using conventional dual ^{13}C/^{1}H probeheads (inner coil = ^{13}C). With spectrometers equipped with cryo probeheads (inner coil = ^{13}C), this amount can be lower (0.5–2 mg). In the present example, zeaxanthin was produced at high rates and the amount of the sample was not a limiting factor.
7. Typical parameter sets are:

   ^{1}H: sw = 20 ppm; td = 32 k; si = 64 k; d1 = 3 s; p1 = 30°; lb = 0.2 Hz.

   ^{13}C: sw = 230 ppm; td = 64 k; si = 128 k; d1 = 3 s; p1 = 30°; ^{1}H decoupling; lb = 0.2–1 Hz.

References

1. Eisenreich W, Bacher A, Arigoni D et al (2004) Biosynthesis of isoprenoids *via* the non-mevalonate pathway. Cell Mol Life Sci 61:1401–1426
2. Eisenreich W, Bacher A (2007) Advances of high-resolution NMR techniques in the structural and metabolic analysis of plant biochemistry. Phytochemistry 68:2799–2815
3. Eisenreich W, Bacher A, Berry A et al (2002) Biosynthesis of zeaxanthin via mevalonate in *Paracoccus* sp. strain R114. J Org Chem 67:871–875
4. Berry A, Janssens D, Hümbelin M et al (2003) *Paracoccus zeaxanthinifaciens* sp. nov., a zeaxanthin-producing bacterium. Intl J Sys Evol Microbiol 53:231–238

Chapter 21

Analysis of Diapocarotenoids Found in Pigmented *Bacillus* Species

Laura Pérez-Fons and Paul D. Fraser

Abstract

Numerous carotenoid-producing Gram positive bacteria have now been identified. Their carotenoid profiles are typically complex with C_{30} diapocarotenoid skeletons being modified by glucosidation and esterification. In the present article the protocols necessary to identify and quantify these diapocarotenoids are described. *Bacillus* species capable of forming spores will be used as generic examples and the associated procedures will cover the cultivation of the bacteria, extraction, chromatographic separations, and online detection by Photo Diode Array (PDA) and Mass Spectrometry (MS).

Key words: Diapocarotenoids, *Bacillus*, High-performance liquid chromatography, Photo diode array, Mass spectrometry

1. Introduction

In higher plants, algae, and fungi the carotenoids produced contain a C_{40} scaffold (1), while in bacteria a diverse range of carotenoids with both C_{40} and C_{30} backbones can be produced (2, 3). These carotenoid scaffolds reside from the prenyl diphosphates precursors utilized in the first committed step of carotenoid biosynthesis (4, 5) which is catalyzed by phytoene synthases. For example, if two C_{20} precursors (GGPP) are condensed together then a C_{40} backbone arises, while the conjugation of two C_{15} molecules (FPP) produces a C_{30} carotenoid backbone (see Fig. 1). The latter are generally referred to as 4,4′-diapocarotenoids and are typically found in a limited number of Gram positive bacteria such as *Methylobacterium rhodinum* (formally *Pseudomonas rhodos*) (6), *Streptococcus faecium* (7), *Heliobacteria* (5), *Staphylococcus aureus* (4), *Planococcus* (8), and *Halobacillus* (9, 10). More recently,

José-Luis Barredo (ed.), *Microbial Carotenoids from Bacteria and Microalgae: Methods and Protocols*, Methods in Molecular Biology, vol. 892, DOI 10.1007/978-1-61779-879-5_21, © Springer Science+Business Media, LLC 2012

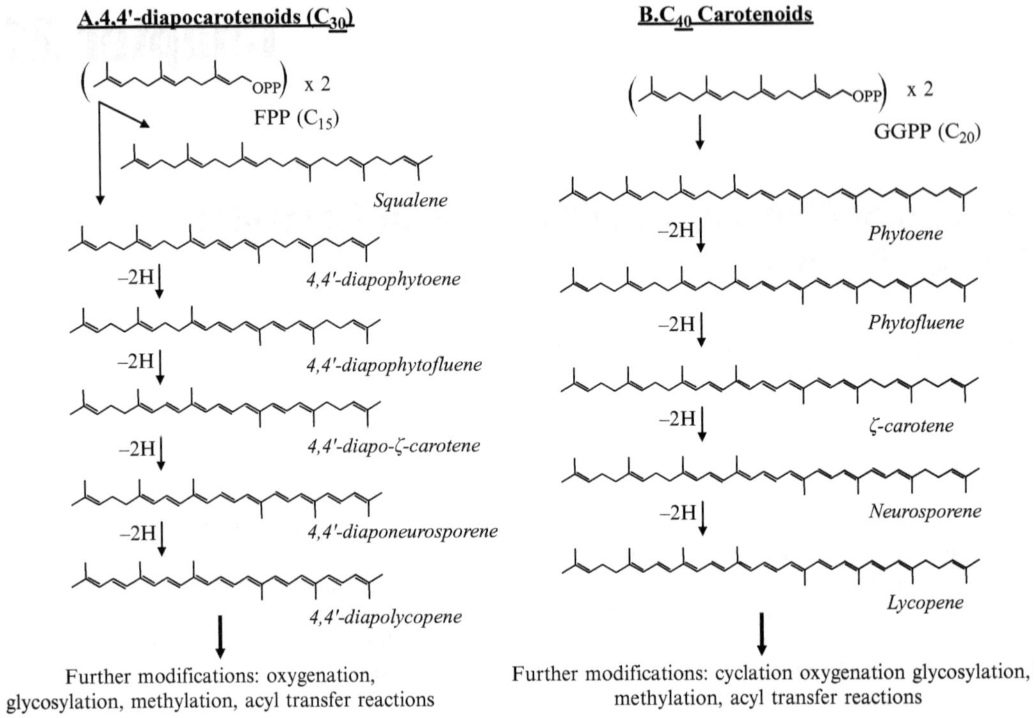

Fig. 1. The diversity of biosynthetic carotenoid pathways. Two identical prenyl diphosphate precursors, GGPP for C_{40} carotenoids (**b**) and FPP for C_{30} 4,4′-diapocarotenoid (**a**), lead to the formation of the first carotenoid of the pathway, i.e., C_{40} phytoene and C_{30} 4,4′-diapophytoene, respectively. C_{40} biosynthetic pathway is widespread in plants and fungi (1, 2) and the C_{30} biosynthesis has been described in *S. aureus* (4), *Methylobacterium* (6), *Streptococcus* (7), *Planococcus* (8), Halophilic cocci (5), *Halobacillus* (9, 10), and Bacilli (12) (figure modified from (12)).

diapocarotenoids have been identified in spore-forming *Bacillus* species (11, 12). The most abundant carotenoids present in the *Bacillus* species were oxygenated derivatives of diapolycopene, which have either undergone glycosylation and/or esterification. For example, glycosyl-4′-methyl-diapolycopenoate esters (see Fig. 2). The diapocarotenoids from these bacterial sources have in recent years gained interest due to their ability to function as virulence factors in some pathogenic bacteria (13) and as potential carotenoids bio-sources with improved solubility and stability for the food and feed industry. A diverse range of spore-forming *Bacillus* spp., which contain diapocarotenoid pigments and possess the ability to act as probiotics (14) have recently been isolated and characterized (12). In the protocol described in this chapter the *Bacillus indicus* HU36 will be used as a representative strain to demonstrate the procedures for complex diapocarotenoid analysis.

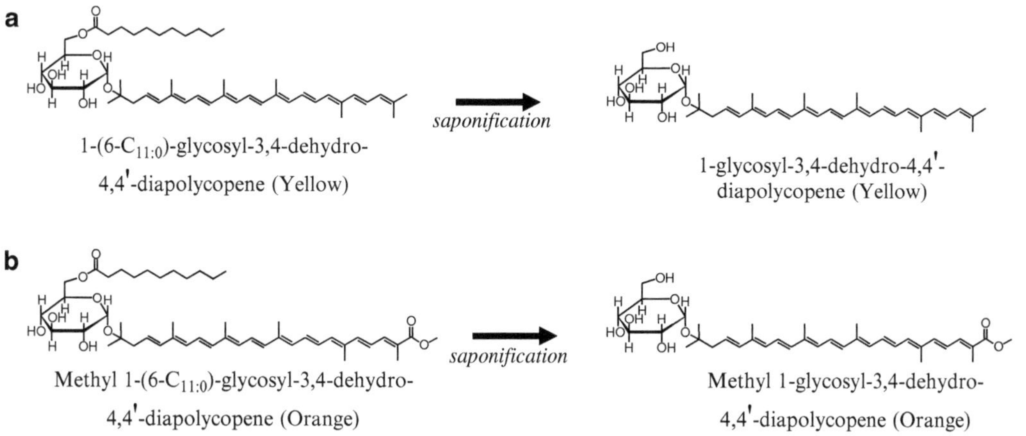

Fig. 2. Characteristic structures of the major diapocarotenoids found in saponified and non-saponified extracts prepared from *B. indicus* HU36. (**a**) Structure of the yellow pigment 1-glycosyl-3,4-dehydro-4,4′-diapolycopene before and after saponification. (**b**) Structure of the orange pigment methyl 1-glycosyl-3,4-dehydro-4,4′-diapolycopenoate previous and after hydrolysis of the esterified fatty acids (figure modified from (12)).

2. Materials

2.1. Bacterial Cultivation

1. *Bacillus indicus* HU36 (14).
2. Bacterial loops.
3. Bunsen burner.
4. Laminar flow.
5. Petri dishes.
6. Baffled flasks (250 mL).
7. Incubator capable of rotary shaking at 200 rpm and maintaining a constant temperature of 30°C.
8. Luria-Bertani (LB): 10 g/L Tryptone, 5 g/L yeast extract, and 10 g/L NaCl. Adjust to pH 7.0 with 5 N NaOH, and sterilize by autoclaving at 121°C for 15 min.
9. LB solid: LB and 15 g/L of agar.

2.2. Extraction of Carotenoids

1. Centrifuge tubes (50 mL).
2. Bench centrifuge capable of 12,000×*g*.
3. Lyophilizer.
4. Mortar and pestle.
5. Balance.
6. Screw-capped glass pyrex tubes.
7. Mixer.

8. Sonicator.
9. Centrifugal evaporator EZ-2 Plus (Genevac, Ipswich, Suffolk, UK).
10. NaOH (10% w/v) solution in water.
11. Tris-buffered saline: 50 mM Tris–HCl, and 1 M NaCl, pH 7.0.
12. Methanol (HPLC grade).
13. Chloroform or dichloromethane (HPLC/Analytical grade).
14. Nitrogen gas.

2.3. Separation, Identification, and Quantification

1. PTFE membrane filters 0.2 μm (Chromacol Ltd., Herts, UK).
2. HPLC glass vials and inserts.
3. HPLC 2600S system and PDA model 996 (Waters Alliance, Milford, MA, USA).
4. Mass Spectrometer (ion trap or Q-TOF) equipped with an APCI source.
5. Reverse phase (RP) C_{30} 5 μm column (250 × 4.6 mm i.d.) with a C_{30} guard column (20 × 4.6 mm) (YMC Inc., Wilmington, NC, USA).
6. RP C_{30} 3 μm column (150 × 2.1 mm i.d.) coupled to a 20 × 4.6 mm C_{30} guard column (YMC Inc., Wilmington, NC, USA).
7. Mobile phase A (HPLC-PDA): methanol HPLC purity.
8. Mobile phase B (HPLC-PDA): methanol:water (80:20; v:v) containing 0.2% (w/v) ammonium acetate HPLC purity.
9. Mobile phase C (HPLC-PDA): *tert*-butyl methyl ether HPLC purity.
10. Mobile phase A (LC/MS): 0.1% formic acid (v:v) in methanol HPLC purity.
11. Mobile phase B (LC/MS): 0.1% formic acid (v:v) in *tert*-butyl methyl ether HPLC purity.

3. Methods

3.1. Bacterial Cultivation

Unless otherwise specified all procedures are performed in an aseptic manner and materials are sterilized by autoclaving at 121°C for 20 min.

1. Scrape a bacterial loop across the surface of a glycerol stock of *Bacillus indicus* HU36 maintained at −80°C and use to inoculate an agar plate containing Luria-Bertani (LB) media.

2. Incubate the agar plate at 30°C for 24 h. Yellow/orange-pigmented colonies will appear.
3. Pick a single colony from the agar plate and use it to incubate a liquid culture (5 mL) of LB media previously prepared in a sterile Falcon tube (50 mL).
4. Shake the liquid culture in a rotary manner at 200 rpm, while being maintained at 30°C. After a 24-h (or overnight) incubation period a dense pigmented suspension culture results.
5. Initiate a 50 mL LB media culture in a 250-mL baffled flask with an aliquot (0.5 mL) of the liquid inoculate previously prepared.
6. Following inoculation, agitate the culture vessel in a rotary manner at 200 rpm and maintain at 30°C. After 24 h the shake culture contains a dense pigmented suspension that is ready for harvesting (see Note 1).
7. At this stage working under aseptic conditions is not necessary. Pour the suspension cultures into two precooled (20 min on ice) 50-mL centrifuge tubes, 25 mL of suspension culture per tube.
8. Harvest the cellular material from the spent media by centrifugation at 10,000 × *g* performed at 4°C for 10 min.
9. Remove the supernatant and remain frozen the pelleted cells for 3 h or overnight at −80°C. Routinely the wet cells are frozen in the centrifuge tubes.
10. After completely freezing the centrifuge tubes containing the frozen cells, lyophilize for 3 days or dry completely. The resulting material (about 1 g) is either used immediately or stored at −20°C (see Note 2).

3.2. Direct Extraction of Diapocarotenoid Pigments

1. Allow the vessel containing the lyophilized material to reach room temperature before opening.
2. Place the lyophilized material into a mortar bowl and create a fine homogeneous powder by grinding with a pestle in a circulator motion.
3. Take an aliquot (50 mg) of this ground freeze-dried material for extraction.
4. Direct extraction without saponification is carried out with methanol. Weight 50 mg of the powdered *Bacillus* material in a screw-capped glass pyrex tube.
5. Add cold methanol (2 mL; prechilled on ice, 20 min) to the material and vortex for 30 s full speed (see Note 3).
6. Sonicate then the suspension created (water bath) at room temperature (RT) for 10–15 min.
7. Carry out centrifugation (12,000 × *g*) for 5 min to remove the cell debris and create the diapocarotenoid containing supernatant.

8. Re-extract the cellular debris until no extractable color was recovered from the solid material.
9. Pool the methanolic extracts and then dry under vacuum.
10. Store the dried carotenoid and extract at −20°C under nitrogen until further analysis (see Subheading 3.4).

3.3. Extraction of Saponified Diapocarotenoid Pigments

1. Weight 50 mg of the powdered *Bacillus* material into a screw capped glass pyrex tube.
2. Saponification prior to extraction is performed with NaOH. Add a solution of NaOH (10% w/v in H_2O; 1 mL) and vortex at full power for 30 s.
3. Sonicate then the resulting suspension is for 15–20 min at room temperature.
4. Remove NaOH by centrifugation at 5,000 × *g* for 3 min from the insoluble cellular material.
5. Remove completely the residual NaOH by washing the cells with Tris-buffered saline. The washing step is optional (see Note 4).
6. Add methanol (250 μL) and chloroform (500 μL) to the pelleted material, mix by vortexing at full power for 3 min (see Note 5).
7. Create then a partition with Tris-buffered saline and centrifuging at 3,000 × *g* for 3 min.
8. Remove the diapocarotenoid yielding hypophase.
9. Re-extract the aqueous hyper-phase twice with chloroform (500 μL); at this point no color is detectable in the cellular debris.
10. Pool the organic extracts and reduce to dryness under a gentle stream of nitrogen gas.
11. Store the remaining dried residue at −20°C, until further analysis.

3.4. HPLC Separation, PDA/MS Detection and Quantification of Diapocarotenoids

1. Dissolve the dried carotenoid extract in chloroform: methanol (1:1; v:v) in a total volume of 100 μL.
2. Vortex and filter through a PTFE membrane (0.2 μm) or centrifuge at top speed for cleaning up the sample.
3. Centrifuge at 12,000 × *g* for 3 min, and transfer the resulting supernatant (50 μL) into an HPLC vial insert.
4. Perform separation of the diapocarotenoids using an appropriate HPLC unit. Typically, perform a 20-μL injection and achieve component separation using a reverse phase (RP) C_{30} 5 μm column (250 × 4.6 mm i.d.) with a C_{30} guard column (20 × 4.6 mm), which is maintained at 25°C. The mobile phase used comprises of (A) methanol, (B) methanol:water (80:20; v:v) containing 0.2% (w/v) ammonium acetate, and (C)

tert-butyl methyl ether. Elution from the column is carried out from 95% (A) and 5% (B) for 12 min, then a step to 80% (A), 5% (B), and 15% (C) followed by a linear gradient to 30% A, 5% B, and 65% C at 30 min. If multiple analyses from an autosampler are to be carried out, then return the column to the initial conditions and equilibrate over 30 min. A flow rate of 1 mL/min is employed and elution monitored continuously with the online PDA (200–600 nm). Typical HPLC traces for non-saponified and saponified extracts are shown in Fig. 3.

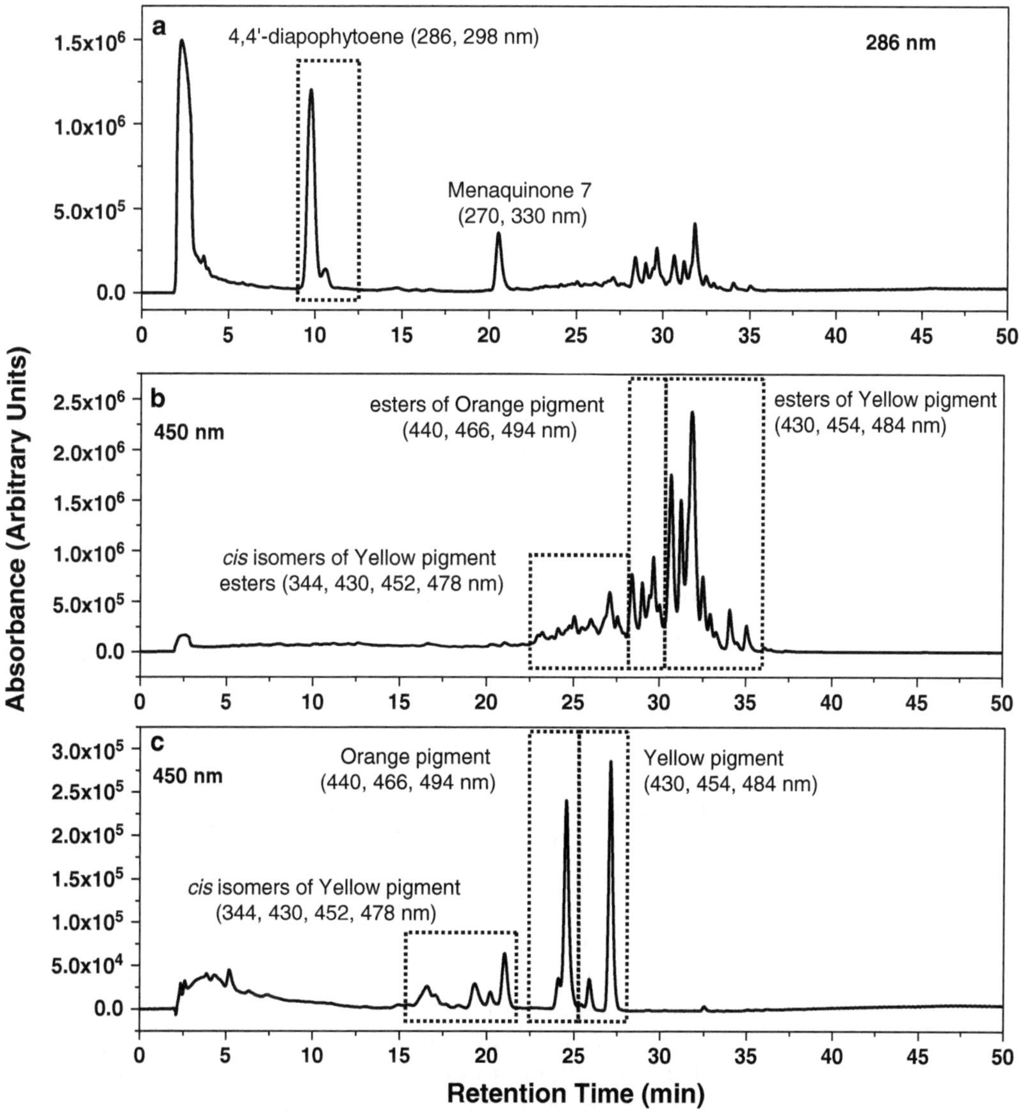

Fig. 3. HPLC analysis of extracts prepared from freeze-dried cells of *B. indicus* HU36. Chromatograms (**a**) and (**b**) represent unsaponified extracts and are recorded at 286 and 450 nm, respectively. (**a**) A profile following saponification with NaOH (10% w/v). Chromatographic peaks are labeled and grouped according to their UV/Vis spectra. Yellow pigment equals to 1-glycosyl-3, 4-dehydro-4,4′-diapolycopene and orange pigment stands for methyl 1-glycosyl-3, 4-dehydro-4,4′-diapolycopenoate (figure modified from (12)).

5. Perform identification by the comparison of spectral and chromatographic characteristics to authentic and similar carotenoids as well as reference parameters in the literature (see Note 6).
6. Carry out quantification using dose–response curves prepared from authentic standards previously purified by HPLC (see Subheading 3.4, step 4). For purification, separate by HPLC saponified extracts using identical conditions as above and pure compounds isolated using a fraction collector. The Absorption coefficients (ε) calculated for both yellow and orange pigments were 149,621 and 122,300 (M^{-1} cm^{-1}), respectively (1) (see Note 7).

4. Notes

1. Longer periods of growth or higher temperature, such as 37°C, do not affect carotenoid biosynthesis. Different pigmentation pattern/composition may occur when incubated for more than 24 h, with methyl-diapolycopenoate esters (orange pigmentation) being more abundant than diapolycopene esters (yellow pigments).
2. Wet cell mass can also be used for carotenoid extraction although the efficiency is poorer and needs to be repeated several times. Lyophilization and grinding helps to break the cells and improves the extraction of carotenoids. Other authors using not lyophilized material apply different techniques in combination with extraction, such as French-press, cell lysis with phenol or lysozyme, or glass beads for breaking the cells. In our hands freeze-drying was easier, allowed for higher recoveries, and minimizes losses of material during the process. Moreover, freeze-dried material prevents potential degradation of carotenoids during storage.
3. For extraction, mixtures of chloroform:methanol (1:1) or dichloromethane:methanol (1:1) can also be employed. Some carotenoids such as the pink/red diapolycopene diacid are best extracted using these mixtures. When the carotenoids to be extracted are unknown (C_{30} or C_{40}, glycosylated, esters, oxygenation, etc.) a recommended procedure would be: (1) Start the extraction with methanol as described in Subheading 3.2. (2) If this does not work, add 1 volume of chloroform or dichloromethane and sonicate the solution. (3) In cases where any of these methods do not work, try the methanol: water:chloroform (1:1:2). Phase separation is formed after centrifuging the mixtures and carotenoids will be extracted in the organic phase (lower layer).

4. Methanolic solutions of NaOH or KOH may cause methylation of the free acid groups after the hydrolysis of the esters and solubilize the carotenoids in the methanolic basic solution. This makes the extraction of the intact structure more difficult as the longer the carotenoids remain in the basic solution the higher is the risk for degradation and introduction of artifacts.
5. An optional step of sonication for 10–15 min can be introduced to aim the extraction.
6. When it is necessary to confirm identities Mass Spectrometry (MS) can also be employed. The authors have carried out Mass spectrometry analysis on the diapocarotenoids and a resulting reference library constructed which is provided in Table 1. Separations are performed by HPLC prior to online MS in a similar manner to that detailed above, with the exception that a RP C_{30} 3 μm column (150×2.1 mm i.d.) coupled to a 20×4.6 mm C_{30} guard column is used. The mobile phase is altered to facilitate ionization and is comprised of (A) methanol containing 0.1% formic acid (by vol.) and (B) *tert*-butyl methyl ether containing 0.1% formic acid (by vol.). These solvents are used in a gradient mode starting at 100% (A) for 5 min, then stepped to 95% (A) for 4 min, followed by a linear gradient over 30 min to 25% (A). After this gradient (A) is a step down to 10% over 10 min. Initial conditions (100% A) are restored for 10 min after the gradient to re-equilibrate the system. Use a flow rate of 0.2 mL/min and employ the ionization mode of Atmospheric Pressure Chemical Ionization (APCI) operating in positive mode. The key differential identifiers for the C_{30} and C_{40} carotenoid biosynthetic pathways are the 409 and 545 m/z corresponding to diapophytoene and phytoene $[M+H]^+$ ions respectively.
7. When determining the total carotenoid content it is best to use non-saponified extracts since saponification may cause losses.

Acknowledgments

The authors are grateful for funding through the EU FP7 Colorspore project number 207948. Prof. J. Halket is thanked for utilization of MS equipment, Prof. Bramley for helpful advice, support, and encouragement, Mr Christopher Gerrish for excellent technical assistance, and Dr Enrique Sentandreu for initial LC-MS familiarization.

Table 1
Chromatographic and spectral properties of diapocarotenoids detected in non-saponified and saponified extracts prepared from the orange/yellow *B. indicus* HU36

	Color	Rt*	l_{max} (nm)	% III/II	% A_B/A_{II}	$[M+H]^+$
Diapophytoene	NC	9.49	274, *286*, 298	<1	NA	409.20
Menaquinone-7	NC	20.65	240, 262, *270*, 330	NA	NA	649.29
cis-1-(6-$C_{11:0}$)-Glycosyl-diapolycopene	LY	24.75	344, 426, *450*, 478	53	49.8	401.2, 749.2
Methyl 1-(6-$C_{11:0}$)-glycosyl-diapolycopenoate	O	29.64	440, *466*, 492	8	NA	445.1, 793.1
1-(6-$C_{11:0}$)-Glycosyl-diapolycopene	Y	31.84	430, *454*, 484	48.3	NA	401.1, 749.2
cis-1-Glycosyl-diapolycopene	LY	20.2	344, 422, *446*, 476	62.5	49	581.12
Methyl-1-glycosyl-diapolycopenoate	O	24.53	436, *464*, 492	6.1	NA	445.1, 463.1, 625.06
1-Glycosyl-diapolycopene	Y	27.08	428, *454*, 484	64.2	NA	401.1, 419.06, 581.05

Analysis was performed by HPLC-PDA and LC/MS. *NC* not colored, Y yellow (glycosyl-diapolycopene), LY light yellow, *NA* not applicable, *Rt** retention time (min) using system II settings (Table modified from (12))

References

1. Britton G, Liaaen-Jensen S, Pfander H (eds) (2004) Carotenoids handbook Carotenoids. Birkhauser, Basel
2. Sandmann G (1994) Carotenoid biosynthesis in microorganisms and plants. Eur J Biochem 223:7–24
3. Aasen AJ, Francis GW, Liaaen-Jensen S (1969) Bacterial carotenoids XXIX. The carotenoids of two yellow halophilic cocci including a new glycosidic methyl apo-lycopenoate. Acta Chem Scand 23:2605–2615
4. Marshall JH, Wilmoth GJ (1981) Pigments of *Staphylococcus aureus,* a series of triterpenoid carotenoids. J Bacteriol 147:900–913
5. Takaichi S, Inoue K, Akaike M, Kobayashi M, Oh-oka H, Madigan MT (1997) The major carotenoid in all known species of heliobacteria is the C30 carotenoid 4,4′-diaponeurosporene, not neurosporene. Arch Microbiol 168:277–281
6. Kleinig H, Schmitt R, Meister W, Englert G, Thommen H (1979) New C-30 carotenoic acid glucosyl esters from *Pseudomonas rhodos.* Zeitschrift Fur Naturforschung C-a Journal of Biosciences 34:181–185
7. Davies BH, Taylor RF (1982) The biosynthesis of triterpenoid carotenoids in *Streptococcus faecium Unh-564p.* Canadian Journal of Biochemistry 60:684–692
8. Shindo K, Endo M, Miyake Y, Wakasugi K, Morritt D, Bramley PM, Fraser PD, Kasai H, Misawa N (2008) Methyl glucosyl-3,4-dehydro-apo-8 ′-lycopenoate, a novel antioxidative glyco-C-30-carotenoic acid produced by a marine bacterium *Planococcus maritimus.* J Antibiot 61:729–735
9. Kocher S, Breitenbach J, Muller V, Sandmann G (2009) Structure, function and biosynthesis of carotenoids in the moderately halophilic bacterium *Halobacillus halophilus.* Archives of Microbiology 191:95–104
10. Osawa A, Ishii Y, Sasamura N, Morita M, Kocher S, Muller V, Sandmann G, Shindo K (2010) Hydroxy-3,4-dehydro-apo-8′-lycopene and methyl hydroxy-3,4-dehydro-apo-8′-lycopenoate, novel C(30) carotenoids produced by a mutant of marine bacterium *Halobacillus halophilus.* J Antibiot 63: 291–295
11. Khaneja R, Perez-Fons L, Fakhry S, Baccigalupi L, Steiger S, To E, Sandmann G, Dong TC, Ricca E, Fraser PD, Cutting SM (2010) Carotenoids found in *Bacillus.* J Appl Microbiol 108:1889–1902
12. Perez-Fons L, Steiger S, Khaneja R, Bramley PM, Cutting SM, Sandmann G, Fraser PD (2011) Identification and the developmental formation of carotenoid pigments in the yellow/orange bacillus spore-formers. Biochim Biophys Acta 1811:177–185
13. Liu GY, Essex A, Buchanan JT, Datta V, Hoffman HM, Bastian JF, Fierer J, Nizet V (2005) *Staphylococcus aureus* golden pigment impairs neutrophil killing and promotes virulence through its antioxidant activity. J Exp Med 202:209–215
14. Hong HA, Huang JM, Khaneja R, Hiep LV, Urdaci MC, Cutting SM (2008) The safety of *Bacillus subtilis* and *Bacillus indicus* as food probiotics. J Appl Microbiol 105: 510–520

INDEX

A

Absorbance....86, 123, 211, 265, 266
Absorption coefficients....342
Acetoacetate-CoA ligase....140
Actinomycetes....13–20
Acyl transferase....198
Aerobic....81, 82, 103, 116, 119, 173, 180, 214
Aerobic photosynthetic bacteria....180
Affiliation....37, 103
Agar....14, 16, 19, 28–30, 37, 41, 48, 50, 65, 82, 83, 85, 86, 90, 91, 110, 116, 117, 119, 120, 122, 123, 125, 126, 136, 147, 163, 164, 176, 178, 180, 181, 187, 190, 191, 193, 209, 212, 216, 221, 227, 232, 237, 240, 248, 249, 252, 257, 260–262, 265, 276, 286, 287, 294, 299, 300, 302–304, 337–339
Agarose
  gel....17, 53, 54, 90, 124, 146, 148–149, 157, 221, 224, 228, 230–233, 238, 247, 258, 260, 272, 288, 290, 293
  gel electrophoresis....15, 17, 49, 53, 85, 89, 90, 122–124, 260, 264
Age-related macular degeneration (AMD)
  cataracts....100, 101
  glaucoma....99
  nutraceutical....99
Algae....4, 5, 9, 10, 13, 22, 24, 61, 63, 73, 102, 133, 160, 208, 271, 273, 284, 335
Alkaline treatment....299, 301, 302, 308–309
Ampicillin....140, 201, 251, 258, 261, 262, 265, 267, 286, 290, 294, 298
Anabaena sp. PCC 7120....297–305
Antheraxanthin....4
Antibiotics....210, 212, 213, 216, 221, 240, 243, 267, 271, 276, 291, 293, 294, 298, 299, 301, 302
Antioxidant....13, 14, 22, 26, 27, 47, 55, 62, 92, 100, 101, 126, 161, 256, 307
Antioxidant activity....47, 62, 64, 67, 71, 73, 88, 89, 161
Diapocarotenoid
Diapophytoene
Argon....258, 261, 308–310, 312
Astaxanthin
  derivative....61–92
  dihydroxyastaxanthin....36, 62, 66–69, 87
  diketo-dihydroxy-carotenoid....62
  dirhamnoside....61–92
  function....61–63
  importance....61–63
  microbial production
    Brevundimonas sp.,....64
    Haematococcus pluvialis....63, 64
    Halobacterium sp....64
    Paracoccus spp.....64
    Xanthophyllomyces dendrorhous
      (Phaffia rhodozyma)....63, 64
  production....63, 64
  purity....144, 145, 153, 157, 158
  structure....61–63
(3S,3'S)-Astaxanthin....63, 69
Astaxanthin-producing bacteria
  biochemical
    activity of various enzyme....81
    chemotaxonomic characterization
      cellular fatty acids....91
      cellular lipids....76
      DNA G+C content....77
      fatty acids profile....76
      quinone (Q10)....76
    identification 16S rRNA gene sequencing....85
    metabolization of carbohydrates....81
    morphological, physiological and biochemical
      characterization
      morphological
        cell shape and size....81
        gram staining....81, 85, 90
        motility....81
        spore formation....81
      physiological
        growth
          oxygen requirement....81
          pH....81
          salt concentration....81
          temperature....81
        requirement of carbon/nitrogen sources....81

José-Luis Barredo (ed.), *Microbial Carotenoids from Bacteria and Microalgae: Methods and Protocols*, Methods in Molecular Biology, vol. 892, DOI 10.1007/978-1-61779-879-5, © Springer Science+Business Media, LLC 2012

Astaxanthin-producing bacteria (*cont.*)
phylogenetic analysis 16S rRNA gene based method....76, 90
polyphasic taxonomic analysis....74, 75
radioactivity....64, 101
radio-tolerant bacteria....61–92
isolation....64–66
screening....64, 74
Athalassohaline....207
Atmospheric pressure chemical ionisation (APCI)....338, 343

B

Bacillus....41, 335–344
B. subtilis....197–204
Backbone....3, 4, 182, 194, 197, 223, 242, 243, 246, 250, 335
Bacteria....1–10, 13–17, 19, 21–55, 63–66, 69, 99–127, 133, 162, 173, 177, 178, 180, 185, 193, 197, 207–216, 220, 274, 283, 289, 294, 299, 305, 315, 322–324, 328, 332, 333, 335, 336
Bacterial sources....64, 144, 145, 151, 153, 157, 209, 336
Bacteriophytochrome....174, 178, 179, 182
Bacterioruberin....47, 208
Bergey's Manual of systematic Bacteriology....103
Beta-carotene....4–8, 10, 14, 21, 22, 24, 33, 35, 48, 51, 62, 66–68, 71, 73, 87, 123, 144, 145, 147, 151, 155, 161, 164–170, 173, 174, 208, 210–214, 219, 256, 257, 259, 269, 279, 284, 297, 303
Biochemical tests....74, 81
Biodiversity....29
Biosynthesis....1–10, 14, 133–140, 143–158, 162, 173–182, 186, 191, 198, 213, 250, 256, 257, 262, 264, 266–268, 283–295, 315, 316, 335, 336, 342
BLASTN....37, 54, 90, 106, 124
Bond labeling
Bradyrhizobium....25, 151, 161, 173–182, 317
Buffer

C

Calcium carbonate....14, 19
Canthaxanthin....8, 22, 25, 33, 35, 36, 48, 51, 62, 143–171, 173–182, 186, 189–191, 193, 208, 219, 222, 236, 239–241, 243, 257, 303
Cargo plasmid....298–301, 304
α-Carotene....14, 21
β-Carotene
hydroxylase (crtZ)....5, 6, 8, 10, 140, 144–147, 153, 154, 157, 158, 186, 187, 191–192, 194, 220, 239, 255–268, 284
hyperproduction....269
ketolase (crtW)....5, 6, 8, 10, 144–146, 151–153, 157, 174, 175, 180, 186, 187, 220, 222, 239, 256, 257, 283, 284, 293, 298, 303, 305
γ-Carotene....7, 13
Carotene oxygenase....284
Carotenogenesis....6, 7, 26, 46, 297–305
Carotenogenic....6, 9, 133, 144, 245, 246, 250, 269–280
Carotenogenic gene....5, 9, 10, 144, 145, 249
Carotenoids
analysis....10, 32, 101, 138–139, 155, 199, 202, 203, 211
antioxidant....13, 14, 22, 26, 27, 47, 71, 88, 89, 92, 126, 161, 258, 307
biosynthesis
β-carotene hydroxylase....5, 6, 10
β-carotene ketolase....5, 6, 10
canthaxanthin....8, 22, 25, 33, 35, 36, 48, 51, 62, 143–171, 173–182, 186, 189–193, 208, 219, 222, 236, 239–241, 257, 303
β-carotene....4–8, 10, 14, 21, 22, 24, 33, 35, 48, 51, 62, 66–68, 71, 73, 87, 123, 144, 145, 147, 151, 152, 155, 157, 161, 164–170, 173, 174, 208, 210–215, 219, 222, 255–269, 279, 284, 297
CD spectra....70, 87
cells disruption methods
biological....30
chemical methods....30
mechanical....30
^{13}C NMR....71, 324, 326–328, 330, 332
extract....32, 51, 52, 55, 68, 92, 126, 138, 202, 211, 261, 262, 310, 311, 340
extraction....30–32, 51, 54, 66, 103, 138, 154, 202, 215, 258, 266, 337–338, 342
fractionation (*see* Isolation and purification of novel astaxanthin derivative)
HPLC (High performance liquid chromatography)....14–18, 26, 29, 32–36, 48, 49, 51–53, 55, 64, 66, 68, 71, 83, 84, 86–89, 101, 103, 105, 121, 136–139, 155, 179, 192, 199, 202, 211, 214, 258, 262, 265, 266, 292, 303
HPLC-Diodearray-MS....64, 66, 101, 103
HPLC-Mass....66, 104
identification of microbial carotenoids
HPLC-Diodearray-MS
method 1....33, 36, 51–52, 66, 68, 103, 104
method 2....33, 34, 36, 52, 66, 68, 86, 87, 92, 103, 105, 123, 126
method 3....34–36, 52–53, 66–68, 87, 92, 104, 123, 126
retention time....35, 36
thin layer chromatography (TLC)....32
UV-vis absorption....70, 87

IR 68, 70, 71, 87
pathway 1–10, 186, 246, 336
photodamage protection 13
production 9, 19, 22–26, 45–47, 135–139, 143, 167, 197–204, 207–216, 220, 222, 245–252, 310
retention time 18, 35, 36, 66, 68, 92, 105, 123, 138, 179, 202, 262, 267
separation .... 32–34, 155, 308 (*see* also HPLC)
standards 33–35, 51–53, 68, 155, 210, 259
thin layer chromatography (TLC), ... 32, 68, 215
titer 144, 152, 156
UV-vis absorption 70, 87
zeaxanthin 4, 8, 22, 25, 33–36, 48, 51, 52, 62, 71, 99–127, 147, 155, 257, 259, 264, 265, 267, 316, 321–325, 328–333
Carotenoids-producing bacteria
identification
chemotaxonomic characterization
cellular fatty acids 109, 126
cellular lipids 110, 126
DNA G+C content 91, 110, 126
fatty acids profile 109, 111–112
quinone (Q10) 110, 123
morphological, physiological and biochemical characterization
morphological
cell shape and size 81, 114
gram staining 81, 85, 90, 122
motility 81, 114
physiological 81
spore formation 81, 125
phylogenetic analysis16S rRNA gene based method 76, 90, 110, 113–117
polyphasic taxonomic analysis 74, 75
radioactivity 64, 101
radio-tolerant bacteria 61–92
requirement of carbon or nitrogen sources
biochemical
activity of various enzyme 81
metabolization of carbohydrates 81
growth at different temperature
at different pH 116
Catalysis 3
C30 carotenoid 3, 197–204, 246, 343
C50 carotenoids 208, 210, 250
CD spectra 70, 71, 87
Cell-bound pigments 28
Cell lysis 208, 211, 342
Cell-wall disintegration
alumina 308–310
ball mill 308, 309
^{13}C-enrichment 325, 328, 332
^{13}C-glucose 328, 329, 331
Chemical extraction 31
Chemo-organotrophic 116
Chemotaxonomic characterization. *See* Chemotaxonomy
Chemotaxonomy 76
Chlamydomonas reinhardtii 5, 284–287, 290–291
Chloramphenicol 139, 147, 239, 248, 249, 258, 260, 261, 265, 267, 271, 279, 298
Chlorella vulgaris 4, 5
Chlorobium 7
Chloroform 31, 32, 70, 73, 84, 136, 138, 209, 210, 271, 272, 277, 285, 287, 293, 300, 303, 305, 324, 333, 338, 340, 342
Chlorophyll *a* 302, 303
Chlorophyte 4
Chloroplast 5, 6, 13, 208, 293, 304
Chromatographic methods 32
Chromophore 7, 66, 215, 255
Chromosomal integration 185–194, 220
13-cis-zeaxanthin 104, 105
^{13}C NMR 71, 324, 326–328, 330, 332
Colony color screening 249, 251
Color screening 246–250
Combinatorial biosynthesis 250
Commercial 22, 24–26, 35, 62–64, 68, 81, 87, 90, 91, 100, 101, 123, 125, 185, 186, 208, 209, 211, 216, 219, 252, 256, 259, 261, 262, 267
Competent bacterial cells 294
Comprehensive taxonomic characterization 64, 101
Concentration of pigment 165–168
Condensation 3–5, 8, 21, 162, 207, 236
Conjugated 4, 7, 62, 100, 245, 255
Conjugation 180, 187, 189–192, 238–239, 298, 335
CrtB. *See* Phytoene synthase
CrtE. *See* Geranylgeranyldiphosphate synthase
Crt genes 162, 174, 177–182, 190, 191, 194, 229
CrtI. *See* Phytoene desaturase
CrtM. *See* Diapophytoene synthase
Crtmnpq 198, 201
CrtO 198, 284, 298, 303, 305
CrtW 8, 140, 144–146, 151–153, 157, 174, 175, 180, 186, 187, 220, 222, 239, 283, 284, 298, 303, 305
CrtZ 8, 140, 144–147, 153, 154, 157, 158, 186, 187, 191–192, 194, 220, 239, 284
β-Cryptoxanthin 14, 25
Culture conditions 165–166
Cyanobacteria 3, 4, 7, 9, 13, 102, 133, 284, 285, 297, 299, 304, 305
Cyanobacterium 5, 297–305
Cyclization 4, 8, 162, 174
Cytoplasm 3, 5, 6, 76

D

Decarboxylation....2, 330
Degenerative diseases....22, 63
Dehydrosqualene....4, 197
Dehydrosqualene desaturase....197
Dehydrosqualene synthase....197
Deinococci....36, 37, 39, 42, 44–46
Deinococcus....26, 37, 39, 40, 44–47
Deoxyxylulose 5-phosphate (DXP)....3, 8
Desaturases....4, 246
4,4'-Diapolycopene....197, 198, 202, 204
4,4'-Diaponeurosporene....197, 198, 202, 204
4,4'-Diaponeurosporenoic acid....198
Diapophytoene synthase....246
Diatoms....4, 5, 9
Diethylpyrocarbonate (DEPC)....285, 288, 292
Digital image analysis....252
Dihydroxyastaxanthin....36, 62, 66–68, 87
Dimethylallyl diphosphate (DMAPP)....1–3, 5, 7–9, 134, 144, 316, 329, 330
Diphosphate groups....4
Directed evolution....245–252, 257, 259, 263–264
Distinct lineage....106
DNA
  amplification....15, 17
  analysis....272–273
  G+C content....74, 77, 80, 82, 83, 91, 110, 113, 115, 118–120, 126
  genomic library....177
  polymerase....15, 53, 89, 124, 146–148, 199, 200, 258–260, 263, 264, 267, 274, 278, 288, 298
  precipitation....264
  purification....49, 53, 85, 89, 121, 123, 176, 178, 271, 276–277, 293
  quantification....273, 278
DNeasy....15, 17, 199, 200
dNTP mix....15, 200, 230, 234, 247, 248, 258, 260, 263, 264, 278, 286, 288
Double bonds....4, 7, 8, 62, 100, 245, 255
Double-cross over....191–194, 298, 301–303, 305
Double-reciprocal recombination....298
DpnI....251, 259, 262, 264
Dry cell weight analysis....9, 138, 154, 156
Dunaliella
  bardawil....5, 6, 24, 102, 269, 273, 279
  culture....274–276
  environmental samples....270–271
  inter-species....270
  intra-species....270
  isolation....270–271, 274–276
  salina....5–7, 10, 24, 102, 208, 269, 270, 273, 279
  species....6, 10, 269–280
  type strains....269, 273, 279

E

Echinenone....33, 35, 48, 158, 284, 297, 303, 305
E. coli. See Escherichia coli (E. coli)
Electroporation....150, 180, 201, 227, 257, 260
Elution....17, 32–35, 67, 88, 104, 105, 155, 293, 303, 341
Entner Doudoroff....316, 329–332
Environmental water....28
Enzymatic....6, 31, 37, 85, 162, 227, 246
Error-prone PCR....248, 250, 251, 259, 263–264
Erythrobacter....46, 103, 318
Erythromicrobium....103
Escherichia coli (*E. coli*)....2, 7–9, 46, 67, 69, 76, 81, 86, 91, 102, 109, 117, 122, 126, 133–140, 143–158, 176, 180–182, 185–190, 193, 198, 202, 203, 209, 212, 213, 215, 220–222, 226–227, 232, 237–242, 245–247, 249–252, 257–259, 261–265, 267, 285, 286, 290, 294, 298–302, 304
  competent cell....135, 137, 146, 212, 259, 290, 300
  electroporation....150, 180, 227
  expression....212
  genotype....267
  TOP10....238–242, 257, 259, 261, 262, 264
  transformation....67, 69, 137, 150, 237, 238, 247, 250, 251, 263
Ethanol....31, 71, 88, 162, 165–168, 170, 212, 227, 242, 258, 260, 272, 277, 287, 288, 290, 291, 298, 309, 310, 314, 323
Ethidium bromide....15, 17, 19, 49, 53, 54, 85, 89, 92, 122, 124, 126, 224, 228, 247, 273, 277–279, 286, 288, 289, 293
Exiguobacterium....41
Expression....9, 10, 145, 151, 153, 156, 157, 173–182, 186, 189, 190, 194, 211–214, 219–243, 245, 248–250, 252, 256, 259, 262–263, 267, 268, 304, 305
Extraction
  hexane....31, 162, 166, 167, 170, 310
  KOH solution....309
  Rushton turbine....312
  separating funnel....309, 310

F

Farnesyl diphosphate synthase (FDS)....250
Farnesyl pyrophosphate (FPP)....2, 3, 8, 134, 139, 143, 144, 186, 197, 335, 336
Fatty acid....65, 76, 78–79, 82, 83, 91, 109–113, 117–120, 126, 337

Feeding
Fermentation. 15, 16, 19, 80, 82, 113, 115, 116, 120, 169, 186, 194, 214
Filtering 272, 312, 313
Fingerprinting
 exon 280
 intron 269–280
 profile 270
 sizing method 269–280
 18S rDNA 270, 273, 279, 280
Flanking DNA 193, 304
Flavobacteria 36–38, 42, 44, 101
Flavobacterium 9, 25, 45, 102, 181, 332
Flectobacillus 38, 45
French-press 342
Functional characterization 261

G

Gamma-ray irradiation 81, 116
GenBank/EMBL/DDBJ 37, 54, 76, 90, 106, 124
Gene cluster 139, 144, 145, 147, 152, 153, 156, 157, 174, 175, 177, 178, 182, 186, 189–190, 192–194, 211, 220, 222, 228–231, 236, 241
Gene disruption 298
Genetic engineering 9, 220
Genomic DNA 53, 89, 123, 177, 182, 240, 287, 291, 298, 304
Genomics 9–10
Genus 16, 29, 36, 37, 44–46, 65, 73–77, 81, 82, 101–103, 106, 108–110, 117–119, 208, 269, 316
Geranylfarnesyl diphosphate synthase 250
Geranylgeranyldiphosphate synthase 246
Geranylgeranyl PP 3, 144, 162
Geranyl PP
Glass-beads method 290
Glycerinaldehyde-3-phosphate 3
α-Glycoside 71
Glycosylation 4, 336
Glycosyl 4,4'-diaponeurosporenoate 198, 204
Glycosyltransferase 197
Gordonia jacobaea 25, 159–171
Gram-negative 26, 31, 39, 65, 81, 82, 103, 116, 119, 120, 305
Growth on serum bottles 188, 190, 193

H

Haematococcus lacustris 5
Haematococcus pluvialis 5, 6, 10, 24, 25, 63, 64, 69, 160, 170, 211, 213, 215, 284–289, 292, 293
Halobacteriaceae 207, 208
Halobacterium salinarum 208
Halomonas elongata 209, 211–214, 216
Halophiles 207, 208
Helper plasmid 212, 215, 243, 298–301
Heterologous expression 142
Heterologous host 9
High-copy expression vector 250
High performance liquid chromatography (HPLC)
 analysis 16, 51, 66, 101, 136–139, 211, 261, 266, 341
 HPLC-Diodearray-MS 64, 66, 101, 103
 HPLC-Mass 66, 104
High-throughput screening 32, 254–268
Homologous recombination 187, 189–193, 298, 304, 305
Hot spring 26, 29, 54, 98–127
HPLC. *See* High performance liquid chromatography
Hydrophobic compounds 9
Hydroxylation 4

I

Identification
 bacterial identification 36
 carotenoids identification 215
 molecular 270
 phylogenetic 45, 75
 rapid 36, 257, 269–280
 sensitive 269–280
 taxonomic 13–20
 zeaxanthin-producing bacteria
  biochemical
   activity of various enzymes 125
   metabolization of carbohydrates 125
  chemotaxonomic characterization
   cellular fatty acids 109, 126
   cellular lipids 110, 126
   DNA G+C content 91, 110, 126
   fatty acids profile 109, 111–112
   quinone (Q10) 110, 126
  morphological
   cell shape and size 114
   gram staining 122
   motility 114
   spore formation 125
  morphological, physiological and biochemical characterization 110, 113–117
  phylogenetic analysis 16S rRNA gene based method 101, 106–109
  physiological
   growth at different pH 116
   growth at different temperature 116

Incorporation experiment.....316
Incubation.....16, 28–30, 46, 50, 53, 89, 90, 110, 116, 119, 124, 125, 179, 193, 252, 260, 262, 264, 291, 302, 339
Integration vector.....187, 189–190, 192, 194
Intermediates.....3, 6, 7, 139, 144, 158, 169, 194, 204
In vivo experiment.....144, 284, 315
Ionizing radiation.....26, 47
IR.....68, 70, 71, 84, 87
Irradiation.....26, 47, 81, 86, 116, 117, 122
Irradiation resistance.....45–47
Isolation
  astaxanthin derivative.....66–67, 70
  astaxanthin producing bacteria.....61–66, 86
  carotenoids-producing bacteria.....64
Isolation and purification of a novel astaxanthin derivative
  open column chromatography (OCC).....68
  thin layer chromatography (TLC).....68, 70
Isomer.....2–4, 100, 134, 160, 194
Isomerase.....2–4, 6, 134, 139, 143, 144, 211–213, 215, 321
(3R,3'R)-Isomer astaxanthin.....63
Isopentenyl diphosphate.....1, 2, 4, 5, 134, 162
Isopentenyl diphosphate isomerase.....8
Isopentenyl pyrophosphate isomerase (idi).....2, 3, 143, 144, 321
Isoprenoid pathway.....134–138, 143, 144, 152, 186
Isoprenoids.....1–5, 7–10, 139, 222, 229, 315, 316, 321
Isotopologue profiling.....315–333

K

Ketocarotenoid.....10, 169, 283–295
4-Ketomyxol glycoside.....297

L

Labeling pattern.....316, 329, 331
Labeling profile.....316, 329, 331, 332
LB.. *See* Luria–Bertani
LC/MS.....34, 87, 123, 155, 338, 344
Levansucrase.....301, 304
Library.....28, 29, 33, 48–51, 64, 177, 178, 186, 246–252, 257, 263–265, 343
Library size.....251
Light emitting diode (LED).....176, 179
Lineage.....76, 106, 108
Lithium acetoacetate.....136–140
Luria–Bertani
  agar.....30, 110, 116, 119, 123, 126, 147, 176, 178, 187, 209, 212, 221, 240, 248, 249, 257, 260–262, 265, 286, 294, 299, 300
  broth.....176, 226, 239, 240, 257, 261, 265
  medium.....37, 151, 154, 178, 180, 190, 198, 199, 201, 202, 212, 289, 294
Lutein.....4, 22, 25, 100, 307, 308, 311, 313
Lutein-enriched extracts.....307–314
Lycopene.....4, 7, 8, 22, 24, 133, 135, 138–140, 174, 175, 179, 180, 219, 246
Lycopene cyclase *(crtY)*.....140
Lyophilization.....16, 19, 167–168, 170, 202, 210, 258, 342
Lysozyme.....17, 19, 271, 276, 342

M

Manganese chloride.....259, 263, 268
Marine.....4, 14, 17, 61, 63, 67, 69, 140, 271, 274, 316
Marine actinomycetes.....13–20
Mass spectrometry (MS).....14, 343
Maximum likehood.....44, 106
Maximum-parsimony.....106
Medium
  artificial seawater salts.....14
  172b liquid.....14, 16
  172b solid.....14
  ISP1, 14, 17
MEGAWHOMP cloning.....263, 264
Metabolic engineering.....9–10, 220, 285
Metabolic flux
Metabolite.....10, 14, 270, 316, 327
Methane.....185–194
Methanotrophs.....185, 186
Methylase.....299, 301, 304
Methylomonas.....185–194, 220
Mevalonate (Mevalonic acid)
  independent.....1, 3
  pathway.....1, 7, 134, 315–321, 329–332
D-Mevalonolactone (D-Mevalonic acid lactone).....135–137, 140
Mevinolin.....4
Microalgae.....1–10, 13, 24, 32, 61, 64, 67, 83, 100, 101, 282–295, 308, 311, 313
Microbial library.....28, 29, 48, 50
Microorganisms.....3, 10, 13, 14, 22, 24–25, 27, 28, 32, 46, 47, 50, 51, 61, 62, 64, 73, 100–102, 133, 160, 162, 181, 186, 207–209, 276, 315, 322
Misasa hot spring.....27, 29, 30, 46, 64, 83, 86, 101, 116, 119, 120, 123
Molecular weight.....32, 34, 35, 52, 66, 68, 87, 100, 104, 106, 123, 155, 178, 258, 273
Molecules.....2–4, 6–8, 13, 27, 66, 67, 162, 215, 316, 328, 355

Morphological.................36, 65, 81, 85–86, 90–91, 110, 116–117, 122, 125–126, 270
Mutant.................25, 26, 47, 64, 83, 102, 157, 179–182, 212, 216, 245, 247, 249, 250, 265, 298, 303, 305
Mutation rate.................251
Myxol 2'-fucoside.................297, 303

N

Nitrocellulose membrane.................249, 252
Nitzschia.................4
NMR spectroscopy.................316, 324, 326, 329
Non cell-bound pigments.................28
Non-mevalonate pathway.................3–5, 9, 134, 144, 315–321, 329–331
Non-photosynthetic.................8, 22, 178, 180
Nostoc sp. PCC 7120.................297
Nostoxanthin.................36, 66, 103–106, 123
Novel functions.................245
Nucleotides.................37, 44, 53, 54, 76, 77, 89, 90, 106–109, 124, 200, 203, 221, 223, 225, 235, 240, 242
Nutraceutical.................13, 63, 64, 143, 256, 283

O

Obligate bacteria for methane and methanol.................185
Oligonucleotides
  specific.................270
Olivibacter sitiensis AW-6T.................106, 110, 113, 118
Open column chromatography (OCC),.................68
Operon assembly.................153–154
Organic solvents.................30, 31, 35, 101, 162, 298
Overexpression.................9, 144, 179
Overlapping PCR.................147–151, 192
Oxidase.................80–82, 85, 90, 114, 116, 119, 120, 122, 125
Oxidative cleavage.................4

P

Pantoea agglomerans.................153, 213, 215, 260
Paracoccus zeaxanthinifaciens.................8, 25, 39, 102, 316, 321, 322, 331
Parapedobacter koreensis Jip14^T.................106, 110, 113, 118
Paromomycin.................287, 291, 293, 294
Pathway
  biosynthetic
  engineering.................134, 135
PCR. *See* Polymerase chain reaction
Pedobacter saltans.................102, 106, 113, 117
Pedobacter spp..................101, 102, 106, 110, 113, 117
pGEMT vector system.................176, 180, 199, 200
Phaeodactylum tricornutum.................5, 9
Phaffia rhodozyma.................9, 25, 31, 63, 64, 102, 160, 170
Pharmaceuticals.................22, 63
Photo diode array detector.................84, 86, 123, 137, 176, 179, 199, 202, 300
Photosynthesis.................5, 13, 174, 177
Photosynthetic bacteria.................3, 133, 178, 180
Photosynthetic carotenoid.................173, 177
Phylogenetic
  analysis.................54, 76, 90, 101, 106–109, 118, 124–125
  diversity.................46, 47
  position.................46
  tree.................37, 44, 54, 75, 77, 90, 106–108, 125
Phy300plk.................199–201
Phytochrome.................178
Phytoene
  desaturase.................4–6, 8, 10, 140, 144, 163, 181, 245, 246
  synthase.................4–6, 10, 140, 144, 162, 163, 178, 181, 246, 247, 335
Phytol.................4, 5
Pigment.................4, 11, 13, 15, 16, 21, 61, 99, 133, 159, 173, 193, 208, 239, 245, 291, 298, 308, 336
  co-expression.................267
  DNA extraction.................216
  origin of replication (Ori).................239, 267
  pAC-Beta.................215, 257, 261–264
  pAC-Zeax.................257, 260, 264, 267
  pAC-Zeax-IDI.................257, 264, 267
  pTrcHis2-TOPO.................267
  pTrc-ZEH.................262, 263
  SDM-Trc-ZEH.................262–264
  selection.................215
Plastids.................3, 5
Polar carotenoids.................33, 34, 51, 52, 297
Polar lipid.................76, 82, 110, 118, 126
Polymerase chain reaction.................177
Polyphasic taxonomic analysis.................74, 75, 103
Porphyrobacter.................45, 103
PpsR regulator.................174
Precursors
  desaturase.................246
  synthase.................246
Primary screening.................28, 48, 50
Primer.................15, 17, 20, 49, 53, 85, 89, 121, 124, 145–147, 156–158, 174, 176–178, 180, 181, 188, 189, 191, 192, 194, 199–201, 220, 223, 224, 228, 229, 231, 240–242, 247, 248, 250, 262, 264, 267, 273, 278–280, 286–288, 291
Production.................9, 14, 22, 63, 100, 133, 143, 159, 174, 185, 198, 208, 220, 246, 256, 270, 284, 310

Promoter heat shock protein 70A (Hsp70A) 293
Promoterless 220, 222, 223, 228–231, 236–239, 241, 243
Promoter probe vector 222
Promoter ribulose biphosphate carboxylase/ oxygenase 2 (RbcS2) 291, 293
Protein expression...262
Proteobacteria..8
PSY gene.... 5
Pyruvate... 3, 8, 316, 330

Q

Quantitative NMR..316, 329
Quinone (Q10),... 28, 65, 76, 82, 83, 91, 110, 118–120

R

Radioactive 27, 29, 46, 47, 86, 99–127
Radioactivity......................................26, 47, 54, 64, 101, 116
Radiotolerant..21–55, 61–92
Random mutagenesis. *See* Error prone PCR
γ-Ray..37, 45, 47
Recombinant 10, 24, 102, 134, 139, 180, 182, 189, 202, 210, 215, 298, 299, 301–305
Red-pigmented bacterium..65
Reductase...2–4, 10, 139
Reductoisomerase 3, 8, 9, 321
Retention time... 18, 35, 55, 66, 68, 71, 87, 88, 92, 104, 123, 127, 138, 166, 179, 202, 203, 262, 267, 305
Retrotranscriptase..288
Reverse-phase chromatography... .... 15, 33–35, 67, 104, 105, 202
Rhamnose residues 71
Rhodobacter... 4, 8
Rhodothermus marinus 210
Rotavapor...165, 167, 170

S

sacB.. 182, 187, 193, 194, 298, 299, 301, 304, 305
Salinibacter ruber 208–211
Salinixanthin 208, 210, 211
Saltern ponds...208
Salt saturated lagoons 269
Sandaracinobacter.. 103
Saponification.. 337, 339–341, 343
Scenedesmus... 4
Screening
- of carotenoids-producing bacteria 64
- of novel zeaxanthin-producing bacteria.......................64
- size...252

strategy 26, 28–30
Sequence.. 4, 6, 10, 14, 17, 37, 44, 49, 53–54, 64, 74–77, 82, 85, 90, 101, 106–108, 118, 119, 121–122, 124, 140, 156, 157, 177, 178, 181, 203, 228, 242, 248, 250, 251, 262, 267, 273, 279, 284, 293, 298, 315, 316, 321
Sequence similarity.................................. 37, 46, 65, 75, 106
Sequencing 10, 15, 17–18, 20, 26, 29, 36, 37, 49, 54, 74, 75, 85, 90, 122, 124, 146, 151, 152, 154, 157, 158, 228, 240–242, 249, 261, 290, 298, 316
Sesqui-terpenes 2, 5, 139
Silica gel.. 30, 68, 84, 85, 87, 88, 164, 210, 211, 322, 323
Single-crossover....................................... 189, 194, 298, 304
Site directed mutagenesis 259, 262–263
Size specificity 250
S.O.C. media 257, 260
Solvent-removal
- inert atmosphere 308
- rotary evaporator...309, 310
- vacuum evaporation 311

Soy-meal.. 162, 164–166, 168–170
Specific inhibitor 4
Spectrum.... ....32–35, 66–68, 71, 87, 88, 103–105, 123, 138, 202, 204, 208, 211, 215, 259, 265, 324, 327, 330
Sphingobacteria.. 36–38, 42, 44, 101
Sphingobacteriaceae... 44, 46, 101, 106, 107, 109, 110, 113, 116, 118, 119
Sphingobacterium spp.,..................25, 101, 102, 106, 110, 117
Sphingoglycolipid...82, 110
Sphingomonadaceae 47, 65, 66, 76, 82, 101, 103, 106, 110
Sphingomonas................................37, 39, 44–46, 65, 66, 103
- *S. astaxanthinifaciens* 80, 82
- *S. sensu stricto*.. 76, 77, 81, 82, 103, 106, 108–110, 117–119

Sphingosinicella 103
Spirilloxanthin...173–182
16S rRNA
- gene...............................29, 37, 44, 54, 90, 106–109, 124
- gene sequences.................................... 37, 46, 49, 53–54, 64, 65, 74–77, 82, 85, 90, 101, 106–108, 118, 119, 121–122, 124
- gene sequencing...36

Staphylococcus aureus.......................... 197–202, 204, 335, 336
Staphyloxanthin biosynthetic pathway 197, 198, 204
Sterols...2, 4, 5, 311
Strain
- stability 186
- TDMA-5^{T}...103, 104, 106, 107, 109, 110, 116, 117, 119, 126

TDMA-16T........................103, 105, 108, 110, 116–119
TDMA-17T........................64–68, 73, 76, 77, 81–84, 87, 90, 91
Stress........................5, 10, 26, 27, 62
SureClean........................258, 260, 262, 264
SYBR safe DNA gel stain........................258, 260
Synechocystis sp.,........................5, 25, 102
Synthesis........................1–8, 10, 22, 23, 63, 101, 133, 142, 144, 145, 147, 151–154, 157, 159–162, 173, 174, 177–180, 182, 219, 246, 267, 284–286, 288–289, 335

T

TAE buffer. *See* Tris-acetate EDTA buffer
Taxonomic diversity........................27, 46
TBE........................15, 17, 19, 49, 53, 85, 89, 121, 124, 146, 148, 221, 224, 228, 230–233, 238, 272, 277
Thalassiosira pseudonana........................5, 9
Thalassohaline........................207
Thin layer chromatography (TLC),........................32, 68, 70, 77, 84, 86, 87, 91, 122, 126, 167, 168, 170, 210, 211, 215
Tolerance. *See* Radioactive tolerance
Topology........................106
Transformation........................4, 9, 10, 137, 150, 216, 226–227, 232, 237, 247, 250, 251, 257–261, 264, 285–287, 290–291, 294, 298, 300–302
Transformation efficiency........................251
Transgenic........................292
Transgenic microalgae........................10, 283–295
Transposon........................220, 222–226, 228–234, 236–238, 240–243
Transposon ends........................223
Triparental mating........................189, 220–221, 238–240, 298–301, 304
Tris-acetate EDTA (TAE) buffer........................247, 258, 260, 286, 288
Triterpenes........................2, 3, 5
Tryptone soya broth........................162, 198
Two plasmid system........................250

U

Ubiquinone Q-10........................65, 76, 83, 110, 120
Unique diversity........................47, 101
Unnatural carotenoids........................245–252
UV-transilluminator........................260, 264, 286, 289
UV-visible
  absorption........................66, 103, 104, 211, 215
  photodiode array........................32
  spectrophotometer........................259, 264, 265

V

Violaxanthin........................4, 292
Volvox carteri........................5

W

Wavelength........................17, 18, 86, 123, 166, 176, 179, 181, 203, 211, 255, 259, 264, 265, 279
Wave scan........................264–265
WEH solution........................312

X

Xanthophyllomyces dendrorhous........................9, 24, 63, 64, 69, 160, 170
Xanthophylls........................21, 99, 138, 161, 283

Y

Yield........................2, 9, 24–25, 31, 54, 63, 64, 88, 101, 102, 134, 140, 162, 170, 198, 245, 248, 251, 256, 267, 279, 310, 332
YM-modified medium........................176, 178, 180, 181

Z

Z-Competent cell........................249, 251
Zeaxanthin
  analysis and identification
    HPLC-Diodearray-MS........................101
    HPLC-Mass........................104
    retention time........................104, 123, 127
    UV-vis absorption........................103, 104
  derivative........................106
  function........................99–100
  importance........................99–100
  polyphasic taxonomic analysis........................103
  producing bacteria........................99–127
  production
    extraction from plant tissues
      alfalfa........................100
      marigolds........................100
    microbial production
      bacteria........................101
      Brevundimonas
      Flavobacterium multivorum........................101
      microalgae........................101
      yeast........................101
  radioactivity........................101, 116
  radio-tolerant
  screening........................101–103
  structure........................99–100
Zymomonas........................103

MIX
Papier aus verantwortungsvollen Quellen
Paper from responsible sources
FSC® C105338

If you have any concerns about our products,
you can contact us on
ProductSafety@springernature.com

In case Publisher is established outside the EU,
the EU authorized representative is:
Springer Nature Customer Service Center GmbH
Europaplatz 3, 69115 Heidelberg, Germany

Printed by Libri Plureos GmbH
in Hamburg, Germany